Observational Astrophysics provides a comprehensive and accessible introduction to the whole of modern astrophysics beyond the solar system. It
combines a critical account of observational methods (telescopes and instrumentation) with a lucid description of the Universe, including stars, galaxies
and cosmology.

The first half of *Observational Astrophysics* describes the techniques used
by astronomers to observe the Universe: optical telescopes and instruments
are discussed in detail, but observations at all wavelengths are covered, from
radio to gamma-rays, and there are sections on cosmic rays, neutrinos and
gravitational waves. After a short interlude describing the appearance of the
sky at all wavelengths, the role of positional astronomy is highlighted. In the
second half of the book, a clear description is given of the contents of the
Universe, including accounts of stellar evolution and cosmological models.
An extensive list of further reading is also provided, with valuable comments
on the level and contents of the books mentioned.

Fully illustrated throughout, with exercises given in each chapter, this
textbook provides a thorough introduction to astrophysics for all physics
undergraduates, and a valuable background for physics graduates turning to
research in astronomy.

D0997781

OBSERVATIONAL ASTROPHYSICS

OBSERVATIONAL ASTROPHYSICS

Robert C. Smith
Astronomy Centre, University of Sussex

CAMBRIDGE
UNIVERSITY PRESS

Published by the Press Syndicate of the University of Cambridge
The Pitt Building, Trumpington Street, Cambridge CB2 1RP
40 West 20th Street, New York, NY 10011-4211, USA
10 Stamford Road, Oakleigh, Melbourne 3166, Australia

© Cambridge University Press 1995

First published 1995

Printed in Great Britain at the University Press, Cambridge

A catalogue record of this book is available from the British Library

Library of Congress cataloguing in publication data

Smith, Robert C. (Robert Connon), 1941–
Observational astrophysics / Robert C. Smith.
p. cm.
ISBN 0 521 26091 4 – ISBN 0 521 27834 1 (pbk.)
1. Astrophysics – Technique.   I. Title.
QB461.S56   1995
523.01–dc20   94-21723   CIP

ISBN 0 521 26091 4 hardback
ISBN 0 521 27834 1 paperback

To the memory of
my father-in-law

Dr T.S. Graham

who would have written
a better book than this
if he had not been
so much of a perfectionist

Contents

List of illustrations

List of tables

Preface

This book was first conceived as a written version of a lecture course with the same name given to second-year undergraduates at the University of Sussex. The aim of the course was to provide a general introduction to astrophysics for students who were studying physics but who had not as yet been introduced to astronomy. Within the confines of 20 lectures I tried both to give them some idea of the way that astronomical observations are made, with emphasis on the physics involved, and to outline the main properties of a representative selection of the Universe. To save space, I jumped straight from the Earth's atmosphere to the stars, and included nothing on the solar system.

While I have been writing the book, in the gaps between many other commitments, the course on which it was based has been replaced by a longer but more elementary course for first-year undergraduates. The aim of this course is similar, and I have not attempted to restructure this book to follow the new course. However, I have included one or two topics from this course, such as the discussion of observing sites (Section 3.6). I have also drawn on other undergraduate and graduate courses that I have given at Sussex over the years.

In a book, it is possible to be more detailed than in a lecture course, and I hope that I have included enough detail to make the book of interest to first-year graduate students as well as to undergraduates. However, I have tried to avoid being too technical: this is neither a monograph on observational techniques, nor a detailed textbook on any particular branch of astrophysics. Rather, it is a broad-brush coverage of astrophysics, with an emphasis throughout on the physical principles and on the observations. Experts in most of the fields covered will no doubt wish I had included their own pet topic; to them I apologise, but the book is not aimed at them. I have tried to be succinct rather than comprehensive, and this has inevitably meant the omission of many interesting facts.

My only indulgence has been to expand on the topic of stellar structure and evolution (Chapter 10). This is partly because I know far more about that than about (say) inflationary cosmology! However, I justify the length of Chapter 10 on different grounds. It is intended to introduce the student in detail to one topic, which is sufficiently well-understood that the details I have given should stand the test of time. Further, a good background in stellar structure and evolution is essential for the understanding of other topics, such as the origin of the chemical elements and the evolution of galaxies, which I have had space to touch on only briefly.

I have, as in the original lecture course, deliberately omitted the solar system. There are plenty of excellent books on the planets, and several on the Sun, and there is so much new material that it would be impossible to do justice to it in the space I had available. I have given a list of some suitable books in Appendix 3.

I am grateful to Simon Mitton of Cambridge University Press for encouraging me to write this book, for comments on an early draft and for his patience as I missed yet another deadline. Several of my Sussex colleagues have read parts of the book, and I am grateful to David Betts, Andrew Cameron, Martin Hendry, Leon Mestel and in particular Peter Thomas, who has read well over half the book at several stages. The following have also read various chapters, and have saved me from at least some errors: John Baldwin, Malcolm Longair, Andrew Murray and Chris Tout; I greatly appreciated their comments. Chris Tout suggested the argument in the caption of Fig. 10.3. Mark Garlick helped with the drawing of Fig. 4.7. Most of the exercises have been attempted by several generations of Sussex students, but any remaining errors are my responsibility, not theirs, as are any blunders in the text. I am greatly indebted to all those who gave permission for their diagrams and photographs to be reproduced, and in several cases provided better originals; a full list of sources is given in Appendix 5.

Finally, my thanks must go to my family, who have seen me writing this book for too many years and probably thought I would never finish it, and my especial thanks to my wife, Eleanor, without whose patient encouragement I probably wouldn't have done.

Robert Smith
Lewes, January 1994

1

Seeing through the fog

Prelude

What is the farthest you can see with the naked eye? This is a familiar trick question and the answer, of course, is an astronomical one. If your eyesight is good, and the sky is dark, you can see as far as the Andromeda galaxy, two million light years away. Huge though this distance seems, it is a tiny fraction of the size of the known Universe, which we now believe to be some ten to twenty thousand million light years in extent. One of the purposes of this book is to show, in outline, how it is possible to arrive at such results and to give some idea as to how reliable they are.

Because the Universe is so vast, it is impossible to grasp its scale expressed in kilometres, or even in light years, so it is useful to think of the Universe as a hierarchy, building up gradually from human scales to astronomical ones, always relating the next distance to the previous one. One way of doing this is shown in Table 1.1, which starts with an hour's train journey. A hierarchy is also a useful way of thinking about both the structure and the contents of the Universe (Appendix 1). On the smallest scale there is our solar system and many double and multiple star systems. On larger scales, stars are often found in clusters of from a few hundred up to a million stars. These clusters, together with individual stars, form the building blocks of galaxies, which themselves are grouped into clusters and superclusters on scales up to the size of the observable Universe. As we move up in scale, the units of distance used on Earth become inconveniently small, and astronomers introduce new ones: the astronomical unit, the light year and the parsec.

The *astronomical unit* (AU) is the semi-major axis of the Earth's orbit around the Sun, and is approximately 150 million kilometres. It is commonly used as the unit for describing distances within the solar system, and also for those in binary star systems. The solar system has a diameter of

1

Table 1.1. *Some astronomical distances and sizes. Note that* $R_*/R_\odot \gg$ $AU/R_E \gg R_A/R_G$. *That is, stellar separations are relatively larger than any other astronomical distance. In the entry for* R_V, $h = H_0/100\,km\,s^{-1}\,Mpc^{-1}$ *where* $H_0 = Hubble's\ constant$ (*Chapter* 15)

Distance or size	Symbol	Value	Relative value
London to Brighton	R_L		80 km
Earth radius	R_E	6371 km	80 R_L
Earth to Moon (nearest neighbour)	R_M	384,000 km	60 R_E
Solar radius	$R_\odot$	696,000 km	100 R_E
Earth to Sun (astronomical unit)	AU	149.6×10^6 km	400 R_M ($2 \times 10^4\ R_E$)
Solar system radius	R_S	5.9×10^9 km	40 AU
One parsec	pc	3.0856×10^{16} m	206265 AU
Nearest star	R_*	1.275 pc	7000 R_S ($7 \times 10^7\ R_\odot$)
Sun to Galactic Centre	R_G	10 kpc (10^4 pc)	8000 R_*
Andromeda galaxy (radius of Local Group)	R_A	670 kpc	70 R_G
Nearest cluster of galaxies (Virgo)	R_V	$11\,h^{-1}$ Mpc	30 R_A
Radius of observable Universe	R	3000 Mpc	300 R_V

about 100 AU and typical binary star separations are tens or hundreds of AU.

Interstellar distances are usually expressed in *light years* or in *parsecs*. Although they differ numerically only by a factor of about 3, they are conceptually quite different. A light year is the distance travelled by light in one year, and is about 10 million million kilometres. A parsec is defined geometrically (see Chapter 7) as the distance at which 1 AU subtends an angle of 1 arcsecond, and is about 30 million million kilometres. Because 1 radian = 206 265 arcseconds, a parsec can also be expressed as 206 265 AU (Chapter 7). Stellar separations are typically a few parsecs. A galaxy has a typical size of about 100 kiloparsecs (kpc) and galactic separations are typically a few megaparsecs (Mpc). The size of the observable Universe is about 3000 Mpc. More precise values of these new units are given in Table 1.2.

We will look at the different scales in the Universe in different detail, jumping straight from the top of the Earth's atmosphere to the nearest stars without pausing in the solar system. We will spend most of our time among

Table 1.2. *Units of distance in astronomy (see text)*

1 astronomical unit (AU) = 149.6 × 10⁶ km
1 light year (ly) = 9.460 × 10¹² km ≃ 6 × 10⁴ AU
1 parsec (pc) = 206 265 AU (≃ 2 × 10⁵ AU) = 3.086×10¹³ km ≃ 3¼ ly

the stars, but will also discuss the general properties of galaxies, those great collections of stars, dust and gas of which Andromeda and our own Milky Way are local examples. At the end of the book we will extend our grasp to the whole Universe and reach back in time to the origins of the universal microwave background – the last whisper of the Big Bang which astronomers believe began the Universe.

Throughout, the emphasis will be on what can be, and is, observed, so I will spend the first half of the book discussing how astronomers observe the Universe. This is not, however, a practical handbook for observers, either amateur or professional – it is a theoretician's view of observational astronomy, in which I will concentrate on the principles involved rather than on all the practical details.

1.1 The naked-eye sky

If you are lucky enough to live far from city lights, and look up into a clear, dark sky, your first impression will be of a myriad points of light: the stars. As your eyes become accustomed to the dark, you will notice that some of them are concentrated into a band across the sky, faint in the northern hemisphere but quite marked in the southern hemisphere. This is the Milky Way, our own local galaxy, whose bright central regions are in the southern sky. Some brighter 'stars' will gradually reveal themselves by their motion against the star background as planets (from the Greek word *planetes*, meaning 'wanderers'). Very occasionally one of the points will display a fuzzy tail that shows it to be a comet. Much more commonly, fuzzy patches will turn out to be star clusters or gas clouds (such as the nebula in the sword of Orion).†

On moonlit nights, the Moon itself is the dominant object; because scattered moonlight makes the background sky much brighter, we can then

---

† The eighteenth-century French comet-hunter Messier found so many fuzzy objects that weren't comets that he compiled a list of about 100 of them, as objects for comet-seekers to avoid! The Messier catalogue contains many of the brightest gas clouds, star clusters and galaxies, and the names are still used; the Orion nebula is M42.

only see the brightest stars and planets, especially in regions of the sky close
to the Moon. In the daytime sky, all the stars disappear: the Sun is so bright
that only the Sun (and the Moon when up) are normally visible, although
Venus can be seen if you know where to look. This is not only because the
Sun is bright but (as for the Moon) because the sky itself is bright: scattered
sunlight causes it to appear blue (or grey on a cloudy day); I will explain
this in Section 1.3.

If you look at the sky for long enough, you will find that it is constantly changing, on many timescales, and the changes reveal new objects
and phenomena. 'Shooting stars' or *meteors* streak across the sky in a few
seconds; these are the debris of dead comets, and other interplanetary dust
particles, burning up as they enter the atmosphere. Typically, you might
see about six per hour; occasional rich showers may display one a minute
or more. Frequently you will see a more slowly moving point of light: an
artificial satellite, crossing the sky in 15–20 minutes.

After a few hours, you will become aware of the rotation of the Earth:
at night, the whole pattern of stars will move westward across the sky at a
rate of about 15°/ hour (15°corresponds to about the width of two fists held
together at arm's length); the Sun shows the same motion during the day,
as does the Moon when visible. But the Moon has its own motion against
the star background, which accounts for the fact that it sometimes appears
in daytime and sometimes at night: it moves eastward at a rate of about
13°/ day, rising nearly an hour later each day.

The planets, comets and the Sun itself also have their own motions relative
to the stars, but they are much slower and become apparent only after weeks
or months. The planets and comets move at comparable rates against the
stars, from about 30°/ week for Mercury to 1°/ month for Saturn. The Sun,
because of its annual motion around the Earth, has a motion of about
1°/ day, or 30°/ month. I will discuss the Sun's motion in more detail in
Section 7.3.

1.2 Absorption in the Earth's atmosphere

All these qualitative features can be discovered by simple naked-eye observations, and have been known for thousands of years. To learn more about
the Universe, we must begin to make more quantitative measurements. A
fundamental constraint on the observational astronomer is then imposed by
the Earth's atmosphere, which absorbs all radiation from space except in two
narrow windows in the visual and radio regions (Fig. 1.1). Of course, rockets
and satellites now allow observation from above the atmosphere over the

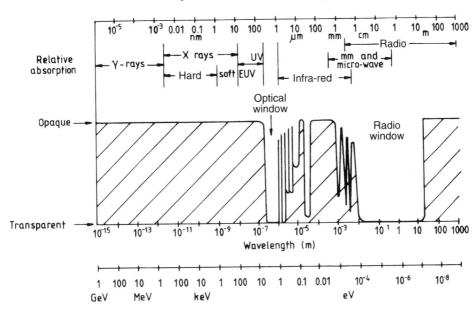

Fig. 1.1. The electromagnetic spectrum, showing the absorption by the Earth's atmosphere. At the Earth's surface only radiation in the radio, infrared and optical windows can be detected. The divisions between the various named wavebands are somewhat arbitrary, and there is some overlap in usage. The wavelength scale is shown at the top of the diagram in the units normally employed in the wavebands shown below them. At the very shortest wavelengths it is more common to describe photons by their energies, shown on the lowermost scale. Even above the atmosphere, the interstellar medium in our Galaxy effectively cuts out radiation in the EUV and soft X-ray region, and is also opaque to the very lowest energy radio waves and the very highest energy gamma-rays.

whole range of the electromagnetic spectrum, and many of the most exciting developments in astronomy in the last 20 years have come from gamma-ray, X-ray, ultraviolet and infrared observations (see Chapter 6). However, space research remains very expensive and the bulk of astronomical observations are still made using ground-based telescopes. I will therefore concentrate in this book on visual and radio techniques, and will start by discussing the effects of the Earth's atmosphere.

Radiation is absorbed in the Earth's atmosphere by a number of atomic and molecular processes. Absorption at particular frequencies – line or band absorption – arises from excitation of molecules and atoms, while the more drastic ionization or molecular dissociation processes give rise to continuous absorption at all frequencies above the threshold energy needed to knock the atom or molecule apart. The extinction is total, except in the visual and radio windows (Fig. 1.1).

The visual window extends from about 300 nm (3000 Å) to about 1.4 μm. The cut-off in the ultraviolet is caused by a thin layer of ozone molecules (O_3) at a height of about 25 km – the now infamous ozone layer. In the infrared, the cut-off is more gradual, with a series of narrow windows extending up to about 24 μm between bands of absorption caused mainly by water vapour (H_2O) and carbon dioxide (CO_2). Infrared astronomers take full advantage of these windows (see Chapters 5 and 8).

The radio window extends from about 8 mm to about 15 m, although the screening by water vapour and oxygen (O_2) molecules starts to decrease at wavelengths longer than about 300 μm. The long wavelength cut-off is caused by critical reflection in the ionosphere, a region in the atmosphere above about 100 km which has a high density of free electrons and ions. Long-wavelength radio waves cannot penetrate this 'plasma' because their wave frequency is too low to excite oscillations at the natural vibrational frequency (the 'plasma frequency') of the ionosphere. It is this phenomenon, frustrating to radio astronomers, that allows communication round the Earth by bouncing long-wavelength radio waves off the ionosphere; however, long-wave radio is seriously affected by variations in solar activity, which alter the level of ionization in the upper atmosphere, and satellite communications are now more reliable.

Not only photons are absorbed in the atmosphere. The Earth is also being bombarded by a steady flux of high-energy charged particles ('cosmic rays'), mainly protons and electrons. Despite their high energies (10^{10}–10^{20} eV), most primary cosmic ray particles are completely stopped by collisions with air molecules. These collisions often produce many secondary particles ('air showers') that can be detected directly, but both primary and secondary cosmic rays can also be detected by radiation from fluorescing air molecules along the tracks of the particles. Particles with enough energy may be travelling through the atmosphere at speeds in excess of the local speed of light ($= c/n$, where c is the speed of light in a vacuum and n is the refractive index of the atmosphere; $n > 1$, so this does not violate the relativistic upper limit of c for the propagation of information) and can be detected by the resulting blueish Čerenkov radiation. Some fraction of the primary and secondary particles reach the ground, though with much reduced energies (the atmosphere has an absorbing power equivalent to about one metre of lead), and can even be detected in deep mines.

Although our own atmosphere is the major source of absorption for as-tronomers, it is not the only one. Even satellite observations are handicapped by absorption, either by dust and charged particles within the solar system (a fairly small effect) or by dust and gas in the interstellar medium in our own and other galaxies. As we will see later, absorption by interstellar dust is important enough in the visual and ultraviolet to prevent our seeing the centre of our Galaxy. In the extreme ultraviolet (EUV) and soft X-ray regions, the photons have just the right energies to ionize the neutral hydrogen which pervades almost all the space between the stars, and the extinction is almost total in directions where the hydrogen is dense; the ROSAT X-ray satellite launched in 1990 was the first one to make observations in the EUV and to reveal that there were more gaps in the hydrogen distribution than had been expected.

1.3 Bouguer's method of allowing for absorption

Even in the 'windows' in the spectrum, the atmosphere is not completely transparent, and it is important to understand the effects of absorption on what we observe and to be able to allow for them.

There are two different effects. *Absorption* by atoms or molecules reduces the observed brightness of a source, because photons are actually destroyed. *Scattering* of photons by molecules or dust degrades the image of an extended source, and may change the wavelength distribution of the radiation, but it has little effect on the net flux of photons reaching the ground, although of course it gives an effective absorption *in the direction of the source.*

The properties of scattering depend on the relative sizes of the scattering particles and the wavelength λ of the radiation:

1. *Particle size $\gg$ wavelength.* The scattering is independent of wavelength, e.g. (in the visual) clouds, mist or fog. This explains why the sky is grey on a cloudy day: the scattered sunlight is scattered equally at all (visual) wavelengths.

2. *Particle size $\simeq$ wavelength.* The scattering depends strongly both on wavelength and on the actual particle size. In the visual, the relative change in brightness caused by dust varies roughly as $1/\lambda$, so blue light is scattered more than red. Interstellar dust (Chapters 11 and 12) produces a similar reddening.

3. *Particle size $\ll$ wavelength.* The scattering depends very strongly on wavelength, with the scattered intensity varying as $1/\lambda^4$ (Rayleigh scattering). In the visual, the blueness of the sky, and of distant horizons (an effect well known to landscape painters), is caused by the Rayleigh scattering of sunlight by air molecules: blue light reaches us from all directions in which there are air molecules, and the blue is more intense in directions in which there are many molecules. Because

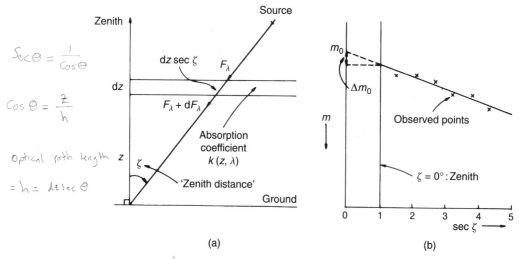

$\sec\theta = \dfrac{1}{\cos\theta}$

$\cos\theta = \dfrac{z}{h}$

Optical path length z

$= h = dz \sec\theta$

Fig. 1.2. (a) The Earth's atmosphere, represented as a series of plane parallel layers of thickness dz at height z. Neglecting refraction, light from an astronomical source passes in a straight line through these layers, being progressively attenuated by the amount given in equation (1.1). (b) The apparent magnitude of a star plotted as a function of zenith distance (equation (1.6)) and extrapolated to the value above the atmosphere.

the blue light has been scattered out of the direct beam, the Sun itself appears reddened, especially when seen near sunset or sunrise through a long path-length in the atmosphere.

The third case is the only one of interest at radio wavelengths.

It is clear that the amount of absorption by the atmosphere depends on whether we are looking directly overhead, through a minimum thickness of atmosphere, or along a long path-length to the horizon. The variation of absorption with the altitude of the source (its height above the horizon – see Chapter 7) can be used to estimate the total absorption, using a method due to Bouguer.

If we neglect refraction and the curvature of the atmosphere, we may picture the atmosphere as made up of a series of plane parallel layers whose properties depend on their height z above the ground (Fig. 1.2(a)). Suppose radiation of wavelength λ passes through a thin layer, thickness dz, at an angle ζ to the vertical (ζ is known as the 'zenith distance'). If the incident flux is F_λ then the decrease in flux dF_λ is clearly proportional to F_λ and to the path-length through the layer:

$$dF_\lambda = -kF_\lambda \sec\zeta \,|\, dz\,|, \qquad (1.1)$$

where k, which is a function $k(z, \lambda)$ of the zenith distance and wavelength, is the *absorption coefficient* in the layer (this equation is really a definition of k). If F_0 is the flux outside the atmosphere, then by integrating equation (1.1) with respect to z we find at ground level ($z = 0$):

$$F_\lambda(\zeta) = F_0 \exp(-\sec\zeta \int_0^\infty k\,dz). \tag{1.2}$$

The unknown integral can be eliminated in terms of the flux at the zenith ($\zeta = 0$):

$$F_\lambda(0) = F_0 \exp(-\int_0^\infty k\,dz), \tag{1.3}$$

(handwritten: $\ln\left\{\dfrac{F_\lambda(0)}{F_0}\right\} = -\int_0^\infty k\,dz$)

giving

(handwritten: $(1.2) \Rightarrow F_\lambda(\theta) = F_0 \exp(+\sec\theta \ln\{F_\lambda(0)/F_0\})$)

$$\log_{10} F_\lambda(\zeta) = \log_{10} F_0 + \sec\zeta \log_{10} \frac{F_\lambda(0)}{F_0}. \tag{1.4}$$

(handwritten: ? See below)

As we will see in Chapters 2 and 8, astronomers conventionally express visual fluxes in magnitudes, a logarithmic scale defined by:

(handwritten: $m_1 - m_2 = -2.5 \log\left(\dfrac{F_1}{F_2}\right)$)

$$m_\lambda = -2.5 \log_{10} F_\lambda + \text{constant}. \tag{1.5}$$

(handwritten: $\Rightarrow m_1 = -2.5 \log F_1 + (2.5 \log F_2 + m_2)$ — constant)

(The same convention is sometimes used for radio magnitudes as well.) Then equation (1.4) can be written as:

$$m_\lambda = m_0 + \Delta m_0 \sec\zeta \tag{1.6}$$

where m_λ is the magnitude that is observed from the ground, m_0 is the magnitude that would be observed from outside the atmosphere and Δm_0 ($= -2.5 \log_{10}(F_\lambda(0)/F_0)$) is the total absorption in magnitudes at the zenith ($\zeta = 0$, $\sec\zeta = 1$). Since this expression is linear in $\sec\zeta$, m_0 and Δm_0 can easily be estimated by observing m_λ at various zenith distances, plotting these values against $\sec\zeta$, fitting a straight line to the observations and extrapolating to $\sec\zeta = 0$ (Fig. 1.2(b)). Since Δm_0 depends quite strongly on λ, observations need to be made in carefully defined wavebands.

For zenith distances exceeding about 60°, we cannot neglect refraction or the Earth's curvature. However, we can use an exactly analogous method, simply replacing equation (1.6) by

$$m_\lambda = m_0 + \Delta m_0 M(\zeta) \tag{1.7}$$

where $M(\zeta)$ is the *air mass* – a measure of the absorption along the curved light-path. The air mass is determined empirically from balloon observations of pressure and temperature, and tables are available of air mass as a function of zenith distance (for small zenith distances, the air mass reduces to $\sec\zeta$). Then $M(\zeta)$ can be treated as a known function, and extrapolation

(handwritten at bottom:)
make $\int \cdot \cdot \cdot$ and obtain $\quad F_\lambda = F_0 e^{\sec\zeta \ln \frac{F_\lambda(0)}{F_0}}\quad$ and take $\log_{10}$ $\quad \left[e^{\ln x} = x \text{ and all that} \right.$

$\Rightarrow \log F_\lambda = \log F_0 + \log e^{\sec\zeta \ln \frac{F_\lambda(0)}{F_0}} = \log F_0 + \sec\zeta \log e^{\ln \frac{F_\lambda(0)}{F_0}} = \log F_0 + \sec\zeta \log \frac{F_\lambda(0)}{F_0}$

to zero air mass again gives m_0 and Δm_0. Because the atmospheric structure is not static, the tables give a mean value, so estimates of absorption are less satisfactory for objects near the horizon. For this reason, and because the total absorption is less, astronomers try to observe objects when they are as close to the zenith as possible, that is, when they are on the observer's meridian (Chapter 7).

1.4 Scintillation and 'seeing'

In the last section I tacitly assumed the atmosphere to be static and horizontally stratified. Neither assumption is true – the real atmosphere is in constant motion on many spatial and time scales, from the slow, large-scale air motions involved in weather fronts to the rapid, small-scale turbulence present in winds. These restless changes cause *scintillation*. There are two effects: variations in air mass along the line of sight cause fluctuations in intensity, while variations in the refractive index along the line of sight cause variations in the position of the image. Small-scale turbulence in the atmosphere causes a stellar image, for example, to 'dance' rapidly and randomly with time on a scale of a few arcseconds – a point source is smeared out into a 'seeing disc', unless observed with very high time resolution. Extended sources, with images much larger than an arcsecond in diameter, are less affected; thus stars 'twinkle', but the major planets, with diameters of 10–30 arcseconds, do not unless the air is extremely turbulent.

Rapid scintillation is a major limitation for ground-based optical telescopes, affecting both the maximum resolving power and the faintest detectable sources (see Chapter 2). Optical telescopes are therefore often said to be 'seeing-limited'. The effects are largest near the horizon, which is another reason why astronomers try to observe as near to the zenith as possible.

At radio wavelengths, very-small-scale turbulence has less effect and scintillation arising from the Earth's atmosphere is less important. In particular, the maximum resolving power is set by diffraction, not seeing (Chapter 4) and radio telescopes are said to be 'diffraction-limited'. However, radio telescopes are significantly affected by scintillation arising in the interplanetary and interstellar medium. High-time-resolution observations, planned by Hewish in the late 1960s to use scintillation to study the properties of the interstellar medium, led incidentally to the discovery of pulsars (see Chapter 9).

1.5 Summary

In this first chapter, I have tried to set the scene and outline some of the difficulties that face the observational astronomer. The vastness of the Universe forces astronomers to think about it on many different scales, and to use units of distance (the astronomical unit, the light year and the parsec) hugely larger than any terrestrial scales.

Observing from the ground is a bit like looking at the world from the bottom of a swimming pool: the view is partial and distorted. Fortunately, the Earth's atmosphere is more transparent than water. The existence of the Sun and Moon, of stars, planets, comets and meteors has been known for many centuries, and optical telescopes have revealed a multitude of faint galaxies. Radio telescopes opened up a whole new window on the sky, and some high-energy particles (cosmic rays) reach the ground.

Observations at other wavelengths are impossible from the ground and absorption and scattering in the atmosphere affect even optical observations. However, the increase of absorption towards the horizon can be used to estimate how much light is lost in the atmosphere by extrapolating a star's magnitude to zero air mass. Finally, we saw how turbulence in the atmosphere also affects observations, by distorting images and smearing out point sources into a seeing disc.

Exercises

1.1   What is the distance of the Earth from the Sun in solar radii?

1.2   What is the distance to the nearest star in astronomical units?

1.3   Use the following data to draw a graph of air mass against zenith distance; for zenith distances z less than $40°$, use the relation airmass = $\sec z$ to complete the table.

Apparent zenith distance (degrees)	Air mass	Apparent zenith distance (degrees)	Air mass
88	19.79	75	3.82
87	15.36	70	2.90
86	12.44	65	2.36
85	10.40	60	2.00
84	8.90	50	1.55
82	6.88	40	1.30
80	5.60		

1.4   A star's apparent visual magnitude m_{vis}, measured photoelectrically
       with high precision, depends on zenith distance z according to the
       following table:

Zenith distance (degrees)	m_{vis}
10	9.105
20	9.119
30	9.146
40	9.192
50	9.267

Use Bouguer's method to find m_{vis} above the atmosphere.

2

Optical telescopes

2.1 Introduction

Before 1600, all astronomical observations were made with the human eye, aided only by sighting devices, such as the large quadrants used by Tycho Brahe, to enable directions to be determined. Precisely who the genius was who invented the telescope seems to be lost in controversy, but there is no doubt that Galileo was the first person to use it systematically for astronomy, starting in about 1610. Despite the development of radio astronomy since the 1930s, and of space astronomy since the 1950s, optical telescopes remain major workhorses in modern research and we will start our discussion of observational methods by reviewing how they work.

Originally, telescopes were used visually, and the most important component was the *objective* – the lens, or 'object glass', in a refracting telescope, or the main mirror in a reflecting telescope (Figs. 2.2, 2.4, Section 2.4). The main purpose of the objective is to collect as much light as possible. Nowadays the human eye has been replaced by a great range of ways of analysing and detecting this light, and the design of the analyser and detector is now just as important as the design of the light-collector. For a particular application, astronomers try to optimize the total *system*, which consists of a combination of:

telescope	reflector *or* refractor
+	
	image analyser *or*
analyser	flux measurer *or*
	spectrograph *or* polarimeter *or* ...
+	
	eye *or*
detector	photographic plate *or*
	photoelectric device *or* ....

13

Despite these many combinations, a telescope has really only two basic modes of operation:

1. *Imaging*. This mode produces direct pictures of star fields and extended objects (such as gas clouds or galaxies). The telescope's optics need to be carefully designed if sharp images are to be obtained over a wide field of view (several degrees).

2. *Photometry*. This mode is used for measuring the total brightness, spectrum, etc., of a single object. Poorer images are therefore acceptable, though the image of a star, for example, still needs to be small enough that all (or most) of the light from the star can be fed through the entrance aperture of the detector or analyser (for example, the slit of a spectrograph).

The use of telescopes revolutionized astronomy for two main reasons: compared to the naked eye, telescopes collect more light and can therefore detect fainter objects, and they have better angular resolution. I will discuss these points in the next two sections, before going on to describe telescope optics and mountings.

A further revolution occurred in the nineteenth century, when photography was introduced, enabling for the first time both an objective permanent record of an observation and the possibility of observing much fainter objects by integrating over time. I will mention the effect of this on the limiting magnitude but postpone a full discussion of detectors until the next chapter.

2.2 Limiting magnitude

The system used today to measure stellar brightnesses is based on one which was probably originally devised by Hipparchus in the first century BC. No manuscript by Hipparchus survives, but his observations formed the basis of the famous catalogue of visible stars (those we still call the 'naked-eye' stars) published by Ptolemy in the second century AD. Ptolemy divided all the stars in his catalogue into six 'magnitude classes', those of the first magnitude being the brightest and those of the sixth magnitude being the faintest visible to the eye. Because the eye has a logarithmic response, the equal magnitude differences defined by Ptolemy actually correspond roughly to equal ratios of brightness: the magnitude scale is a logarithmic scale of brightness. Ptolemy's approximate scale was made quantitative in 1854 by Norman Pogson at the Radcliffe Observatory in Oxford and the visual magnitude scale is now defined by

$$m_{\text{vis}} = -2.5 \log_{10} F_{\text{vis}} + \text{constant} \qquad (2.1)$$

where F_{vis} is the *flux* (energy per unit time per unit area) falling on the eye and the factor 2.5 and the additive constant are both chosen to make the modern scale approximate to that introduced by Ptolemy and in use ever since. The constant is fixed by arbitrarily assigning a magnitude of $+6.55$ to the star λ Ursae Minoris, which is near the north celestial pole and so is always at nearly the same zenith distance (see Section 1.3 for why that is important). Note that a difference of five magnitudes corresponds to a brightness ratio of exactly 100; one magnitude corresponds approximately to a brightness ratio of 2.5. Because the constant is negative, faint objects have larger (positive) magnitudes than bright ones. Extremely bright stars have negative magnitudes; for example, Sirius, the brightest star in the sky apart from the Sun, has a magnitude of -1.4.

With the advent of linear detectors that measure flux directly (Section 3.2), the use of a logarithmic scale is becoming less important. However, it is interesting that there is another logarithmic scale in common use in everyday life, for measuring sound levels. The often-quoted decibel scale is defined in a similar way to the magnitude scale, but with a positive constant, so that large numbers correspond to loud noises. In fact, the scale is also more finely divided, with 20 db corresponding to a factor of 100 (the factor -2.5 is replaced by $+10$: 1 mag $\equiv$ 4 db).

With Pogson's definition of magnitude, the limiting magnitude for the naked eye is about $+6$, just as it was for Ptolemy. This corresponds to the faintest object visible under normal observing conditions, with a dark-adapted eye. It normally takes about 25 minutes for the eye to become completely dark-adapted, and an unadapted eye will typically see only down to third or fourth magnitude; a city-dweller, contending with street lighting, is unlikely ever to do much better than fourth or fifth magnitude. However, if the field of view is restricted, the eye is capable of doing much better; experiments by Heber Curtis and Henry Norris Russell in the early 1900s showed that a dark-adapted eye, looking at a patch of dark sky through an aperture 5 arcminutes across, could see stars as faint as $m_{\text{vis}} = +8.5$. This corresponds to a flux of about 200 photons s^{-1}, and represents the practical limit of detection for the unaided eye.

Because a telescope has a larger collecting area than the eye, it can collect more energy from an object with a given flux, so the object will appear brighter if viewed through a telescope. If a star seen by the naked eye has *brightness* B_{e} (total energy collected per unit time), then if viewed through a telescope it will appear to have brightness B_t given by:

$$\frac{B_t}{B_e} = \frac{D^2}{d^2} \qquad (2.2)$$

where D and d are, respectively, the diameters of the telescope objective and the pupil of the eye; all the energy from area πD^2 is now being fed into the smaller area of the eye (see Section 2.4). The eye therefore sees an object of given flux F_e as well through the telescope as if it had flux F_t, where

$$\frac{F_t}{F_e} = \frac{D^2}{d^2}, \qquad (2.3)$$

and were seen by the unaided eye. If m_t and m_e are the corresponding magnitudes, then

$$m_t - m_e = -2.5\log_{10}(F_t/F_e) = -5\log_{10}(D/d). \qquad (2.4)$$

The telescope thus makes a very faint star, of true magnitude m_e, appear to have a magnitude of m_t. Although the telescope restricts the field of view somewhat, the restriction is not enough to affect the limiting magnitude significantly and the faintest star detectable by the telescope–eye combination may be assumed to have $m_t = +6$. The corresponding value of the 'true' visual magnitude m_e is then called the *limiting (visual) magnitude* of the telescope:

$$m_{\text{lim}} = 6 + 5\log_{10}(D/d). \qquad (2.5)$$

The diameter of the pupil of a dark-adapted eye is typically 8 mm, so (measuring D in metres) we find $m_{\text{lim}} \simeq 16.5 + 5\log_{10} D$. This is rather optimistic, because of light losses in the telescope optics, and the practical limit is about half a magnitude brighter:

$$m_{\text{lim}} \simeq 16 + 5\log_{10}(D/\text{m}). \qquad (2.6)$$

A large telescope for amateur use might have $D = \frac{1}{3}$ m, giving $m_{\text{lim}} \simeq 13.6$, or a little fainter in ideal conditions.

For larger telescopes, the detection is much more likely to be photographic or photoelectric (Chapter 3) and much fainter objects can be detected by time integration of the signal. Even under ideal conditions, the eye needs to receive about 200 photons s^{-1} before it can detect an image. This corresponds to the extreme limiting visual magnitude of $+8.5$ found by Curtis and Russell. An integrating device can detect a much weaker flux if it accumulates photons for long enough.

Modern photoelectric equipment can detect individual photons, and the limiting magnitude is set by background noise; I will return to this in

Section 3.5. For the moment, consider a photographic plate. Without special treatment, only about 0.1% of the incident photons is recorded by blackening a grain (see Section 3.1), and about 50 grains need to be blackened to produce a detectable image. So an image will be detectable only after about 5×10^4 photons have been received. At 200 s^{-1}, this takes about four minutes, so the photographic plate can match the eye's limit with an exposure of some four minutes and can detect much fainter objects by integrating for longer. The limit to the flux which is detectable is clearly given by

$$F_{\text{lim}} \propto \frac{1}{D^2 t},\qquad(2.7)$$

where t is the exposure time and D is again the telescope aperture. Thus

$$
\begin{aligned}
m_{\text{lim}} &= -2.5 \log_{10} F_{\text{lim}} + \text{constant}\\
&= 12.5 + 5 \log_{10}(D/\text{m}) + 2.5 \log_{10}(t/\text{s}),\qquad(2.8)
\end{aligned}
$$

where I have fixed the constant by requiring that $m_{\text{lim}} = +8.5$ (corresponding to 200 photons s^{-1}: see above) for $t = 4$ minutes and $D = 8$ mm, and I have again allowed for light losses of 0.5 magnitudes in the telescope's optics. Thus a three-hour exposure ($\sim 10^4$s) with a 1 m telescope should detect stars with visual magnitudes down to 22.5. In practice the sky background and plate fog (see Section 3.1, Fig. 3.2) make the limit rather brighter; the 1.2 m UK Schmidt telescope (Section 2.4) reaches a limit at a visual magnitude of about 21 after about 60 minutes' exposure, while the above formula gives $m_{\text{lim}} = 21.9$. With modern photoelectric detectors, much fainter limits are possible; for example, the 3.5 m New Technology Telescope at the European Southern Observatory in Chile can detect galaxies with magnitudes as faint as $+28$.

2.3 Angular resolution

A star is such a distant source that light from it can be treated as a parallel beam, so we have a plane wave incident on the circular aperture of the telescope objective. The plane wave is diffracted at the edges of the aperture and a point source does not produce a point image at the focus of the objective but a circular diffraction pattern (Fig. 2.1(a)). This pattern was studied in detail by the great nineteenth-century Astronomer Royal, Sir George Airy, and the central bright spot, which contains 84% of the light, is named after him. The surrounding bright rings fall off rapidly in brightness, even the first ring having a maximum brightness less than 2% of that in the centre, so only a few rings are seen even in ideal conditions. The radii of the

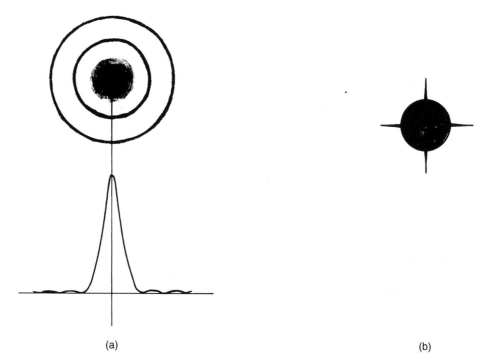

(a)   (b)

Fig. 2.1. (a) Theoretical diffraction pattern of a point source. Most of the light is concentrated into the central bright spot (the *Airy disc*). (b) The image of a bright star on a photograph taken by a large reflector. The four spikes are the diffraction pattern of the supports of the secondary mirror; the central blurred image is much larger than the Airy disc and is a combination of the seeing disc and of a halo arising from the scattering of photons in the photographic emulsion.

minima are given by the zeros of Bessel's function of order unity, and the first minimum occurs at a radius α given by

$$\alpha = \frac{1.22\lambda}{D} \text{ radians} \tag{2.9}$$

where λ is the wavelength of the light and D is the diameter of the telescope aperture (so the diffraction rings for white light would show a red outer edge and a blue inner one).

The size of the Airy disc puts a limit on the resolution of the telescope: two point sources can only be resolved if their Airy discs are sufficiently separated to be seen as distinct. Rayleigh's criterion for resolution is that the central maximum of one should lie on the first minimum of the other; this corresponds to about a 20% drop in intensity between the maxima and yields a *theoretical resolving power* given by equation (2.9). This is usually referred to as 'diffraction-limited' resolution. For a typical visual wavelength of 550 nm, the angle becomes

$$\alpha \simeq 0.14 \frac{1\,\text{m}}{D} \text{ arcseconds.} \tag{2.10}$$

For small telescopes ($D \leq 0.1$–0.2 m) this is a realistic, or even slightly pessimistic formula. For the large telescopes used by professional astronomers the resolvable angle is smaller than the seeing disc (Section 1.4). The effective resolution is then set by the size of the seeing disc and is independent of the size of the telescope; the resolution is said to be 'seeing-limited'. Thus photographs of star fields do not usually show the Airy diffraction pattern, although if taken by reflectors they may show diffraction 'spikes' produced by diffraction of light by the supports of the secondary mirror (Fig. 2.1(b)).

Telescopes in high-altitude balloons or in satellites are not seeing-limited and can therefore achieve much better resolution. For example, the balloon-borne Stratoscope telescopes, first launched in 1957, took the first really sharp pictures of convective 'granulation' on the solar surface. More recently, superb pictures of the granulation were obtained during the 1985 Spacelab 2 mission on the Shuttle.

The other way of minimizing seeing is to observe at longer wavelengths, since the size of the seeing disc decreases slowly towards the infrared while the Airy disc grows larger. Large telescopes become diffraction-limited at wavelengths longer than a few μm (exercise 2.3). This does not improve the angular resolution (the diffraction-limited resolution is worse than in the visual waveband) but it does help to remove the time-dependent distorting effects of the atmosphere.

In the absence of seeing, the resolution is ultimately limited only by the size of the telescope and by how well the mirror is constructed; the various aberrations discussed in the next section usually conspire to prevent the full diffraction limit being reached. It was one of the major hopes for the Hubble Space Telescope, launched in 1990, that it would be able to detect fainter objects than a ground-based telescope, partly because the light would be concentrated into a smaller, and therefore brighter, image. In the event this hope was initially only partially realized, because of the serious spherical aberration in the primary mirror (Section 2.4).

2.4 Principles of telescope optics

Galileo's telescope was a refractor, focusing the light entirely by lenses. Modern refracting telescopes use essentially the same principles, which are

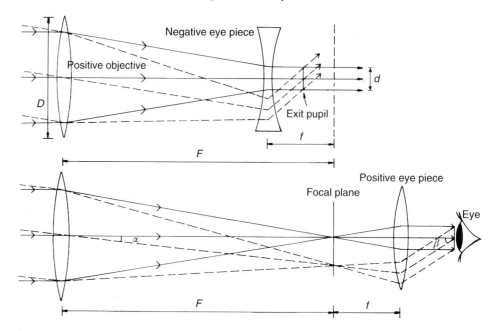

Fig. 2.2. Principles of refracting telescopes. *Upper*: Galilean design. *Lower*: Keplerian design. The objective lens, of diameter D, focuses the parallel light from a distant star in the focal plane, at distance F (the focal length) from the objective; the eyepiece magnifies the image and produces a parallel beam at the eye. The image of the objective formed by the eyepiece is known as the exit pupil and is marked in both diagrams by a vertical bar. The magnification is $m = \beta/\alpha = F/f$. If d is the diameter of the exit pupil, then $d = D/m$. The Galilean design produces an upright image while the more common Keplerian one produces an inverted image. A typical aperture ratio, or focal ratio, is $F/D \simeq 15$. (The diagram is not drawn to scale.)

illustrated in Fig. 2.2; the main difference is that Galileo's telescope used a diverging eyepiece and was therefore able to be rather shorter than the later design used by Kepler. For visual use, Kepler's design has the advantage that cross-wires can be placed in the focal plane. The image is inverted in an astronomical telescope, since the extra lens required to avoid this would introduce light losses that are more important than having an upright image. For most professional purposes, a photographic plate or other recording device is placed in the focal plane, converting the telescope into a large camera. As in a conventional hand-camera, the speed is determined by the focal ratio, F/D, where F is the focal length of the objective lens. The larger is F/D, the slower is the camera, because the photon flux on a given area of plate is smaller (for larger F at fixed D, the photons are spread out over a larger area because the plate scale is larger; for smaller D at fixed F, fewer photons per second reach the same area on the plate).

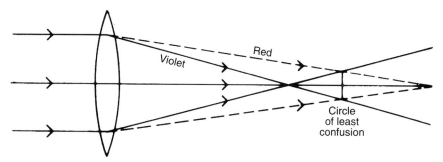

Fig. 2.3. Chromatic aberration (lenses only). The refractive index of glass varies with wavelength and so violet rays are brought to a focus nearer the lens than are red rays. The distance between the two foci is greatly exaggerated in the diagram, but it is several per cent of the mean focal length. The best image of a point source is a filled circle (the circle of least confusion).

Note that how much light enters the eye is determined by the size of the image of the objective that is formed by the eyepiece: all the light that enters the objective is concentrated into this circular image, known as the *exit pupil*. If light is not to be wasted, the size of the exit pupil must be matched to the size of the pupil of the eye, that is, it must be less than about 8 mm. In deriving equation (2.2), I tacitly assumed that the exit pupil of the telescope was small enough for all the light which came through the objective to enter the eye. This is not only important for visual observations: when instruments are attached to a telescope, their entrance apertures should also match the exit pupil of the telescope if light is not to be wasted. This puts a lower limit on the magnification; if all the light received through an aperture of diameter D is to emerge through an exit pupil of diameter $d < 8$ mm, the magnification must be $D/d > D/8$ mm.

Refracting telescopes generally have fairly long focal lengths and so, with F/D typically about 15, are fairly slow cameras, making them unsuitable for searching for faint objects. However, the large plate scale makes them ideal for measuring star positions accurately and most large refractors nowadays are used for that purpose, known as *astrometry*. Two other properties of refractors are particularly useful for this application: they are thermally stable, that is, their optical properties are not sensitive to temperature changes, and they need very little maintenance, allowing the optical alignments to remain undisturbed for many years at a time. This enables star positions recorded on photographic plates taken years apart and under different temperature conditions to be compared and the resulting small changes in positions to be attributed with confidence to motions of the stars.

The major snag of refracting telescopes is *chromatic aberration* (Fig. 2.3): light of different wavelengths comes to a focus at different distances from the objective. This aberration, by far the largest one to affect astronomical telescopes, is usually corrected by using an achromatic doublet consisting of two thin lenses made of glass of different refractive index and placed in contact. The lens of smaller refractive index, usually crown glass, is strongly convergent, while the other lens, usually flint glass, is made divergent, but less strongly, so as to make the dispersion of the two lenses the same. The aberration then cancels out while the compound lens remains convergent.

Even an achromatic doublet, or triplet, only makes the focal length exactly the same for two, or three, wavelengths and there is a residual aberration at intermediate wavelengths. Refractors also have mechanical drawbacks which prevent very large refracting telescopes from being built. Because the objective can be supported only round its edge, and not at the centre where it is thickest, large lenses tend to bend under their own weight and distort the image. The largest refractor in use is the 40 inch (1 m) telescope at Yerkes Observatory in Wisconsin. The other drawback is a financial one: the length of the tube is necessarily longer than the focal length, so a very large, and expensive, dome is required. For all these reasons, most modern professional telescopes are reflectors.

The main element in a reflecting telescope is a parabolic mirror, which brings all the light from a point source to a focus at a single point. Because secondary mirrors can easily be used to deflect a light beam, a variety of possible foci can be arranged, with different focal ratios, so that a single telescope can be used equally efficiently for several different purposes. The various foci in common use are shown in Fig. 2.4.

The *prime* focus involves no secondary mirror and has the least light losses. This, combined with the small focal ratio (see above), makes it the best focal arrangement for observing very faint objects. In large telescopes built in the 1970s or earlier, there is an observer's cage at the prime focus. Originally, the observer sat cramped in this cage throughout the observing session; nowadays his or her place is taken by instruments controlled remotely from a comfortable observing room, and more recent telescopes have smaller and lighter secondary mirror assemblies with no space for an observer.

The *Newtonian* focus is effectively a 'bent prime', with the light beam deflected out of the side of the tube, without change of focal ratio, by using a flat secondary mirror; a focal ratio of 5 is common. Although this is still a popular arrangement for small telescopes, used visually, access is not convenient on large telescopes and attaching instruments there would unbalance the telescope. It is therefore not now used on professional telescopes,

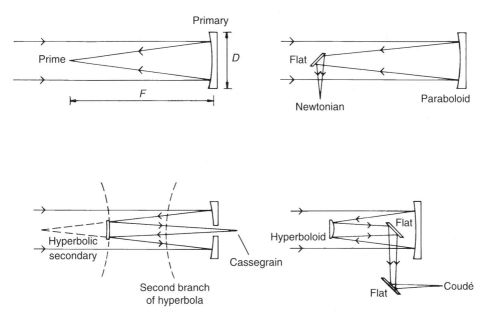

Fig. 2.4. The main foci used in reflecting telescopes. The main mirror in each case is a paraboloid, which brings the parallel light from a distant star to a focus at the *prime* focus, which usually has a focal ratio of $F/D \simeq 3$. The converging beam can instead be diverted to a variety of other foci, described in the text.

although a variant was successfully used by Sir William Herschel in the late eighteenth and early nineteenth centuries.

The commonest focus on a large modern telescope is the *Cassegrain*, since the focal plane is in a convenient place for attaching instruments. In this arrangement, the focal length of the main mirror is effectively increased by introducing a slightly diverging hyperboloidal mirror into the beam, which is then reflected back down the tube and through a hole in the centre of the main mirror to a focus just behind the mirror support. Because the light beam is 'folded', it is possible to have a long focal length without a long tube and the secondary mirror is chosen to make the focal ratio considerably larger than at the prime focus, allowing a larger plate scale. The reason for choosing the secondary mirror to be a hyperboloid is that a hyberbola has two foci, and light emitted from one focus would be brought to a focus again at the other focus. The hyperbolic secondary forms part of one branch of a hyperbola, and is placed so that the focus of that branch coincides with the focus of the primary mirror. The converging beam from the primary is then reflected from the secondary mirror, becoming equivalent to a beam emitted from the focus of the hyperbola. The reflection properties of conics then ensure that the light is brought to focus again at the focus of the other branch of the hyperbola; the two branches are shown as dashed lines in Fig. 2.4. The second focus is arranged to be just behind the main mirror. The focal length is automatically longer in this arrangement than for the prime focus, because the hyperboloid is convex, and the focal ratio at a Cassegrain focus is typically about 15.

Very-high-dispersion spectroscopy requires a large, heavy spectrograph which it would be inconvenient to mount on a telescope. The *coudé* focus is designed to produce a stationary image from a moving telescope, so that the detecting equipment can be mounted off the telescope (usually in a separate room). This is achieved by deflecting the light beam along the support axes of the telescope, using two flat mirrors; in an equatorial mounting (Section 2.5) the beam travels first along the declination axis and then down the (fixed) polar axis. Because the telescope rotates around these axes as it tracks a source, the field of view rotates about its centre. This is unimportant for single-object spectroscopy, so long as the object of interest is at the centre of the field, but if the whole field of view is to be available (for multi-object spectroscopy or imaging) a rotating detector or a field-rotating lens is necessary. For an alt–azimuth mounting, there are two such foci, located on the horizontal support axis. A single flat mirror can deflect the light horizontally onto a platform which rotates with the telescope around the vertical support axis. These *Nasmyth* platforms (one at each side of the telescope (Fig. 2.11)) are often large enough for complete laboratories to be constructed on them (for example, the GHRIL laboratory attached to the 4.2 m William Herschel Telescope on the island of La Palma in the Canaries: *Ground-based High-Resolution Imaging Laboratory*). In the same way as for the Cassegrain focus, a hyperboloidal secondary mirror is used to increase the focal length; focal ratios of 30 or more can be used.

The major advantage of reflectors, and the reason that they were introduced by Newton, is that they have no chromatic aberration. They can also operate in the near ultraviolet, which is totally absorbed in the lenses of refractors, and can be made with much larger apertures since large mirrors can be back-supported to prevent distortion. A disadvantage is that the shape of the mirror is sensitive to temperature changes, so much effort has had to go into controlling the temperature in the dome and finding low-expansion materials such as fused quartz and glass/ceramic compounds. The mirror surfaces also need regular re-aluminizing, which requires removing the mirror cells from the telescope mounting and so regularly disturbing the optical alignment.

Although mirrors do not suffer from chromatic aberration, there are

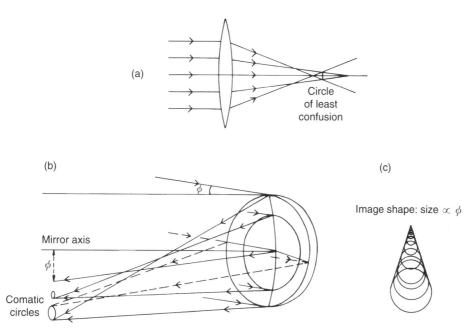

Fig. 2.5. (a) Spherical aberration, illustrated for a lens; the difference in the focal lengths for rays from different parts of the lens is greatly exaggerated, but is typically greater than 1%. (b) Coma for a paraboloidal mirror; each annular zone of the mirror forms a circular image whose size decreases with the size of the zone; for the positive coma illustrated, the smaller images are nearer the axis; within each image, each point is formed by rays from diametrically opposite sides of the mirror. (c) The overall image shape formed by comatic aberration; the overlapping circles form a 'tear-drop' shape whose size is proportional to the off-axis angle (*contd over*).

several other aberrations that affect both lenses and mirrors. The most important ones are shown in Fig. 2.5.

Because lenses nearly always have optical surfaces that are sections of spheres, they suffer from *spherical aberration* (Fig. 2.5(a)): rays from the edge of the lens come to a focus nearer the lens than do rays through the centre of the lens, causing a blurring of a point source into a circular image. The effect is quite large, usually a 1%, or larger, difference in focal length, although, as Newton discovered, the minimum image size (the 'circle of least confusion') is much smaller than for chromatic aberration. Spherical mirrors also suffer from spherical aberration, but for mirrors the effect can in principle be entirely removed by making the cross-section parabolic: all rays parallel to the axis of a paraboloidal mirror will be reflected to meet at the focus of the parabola. In practice, a parabolic shape is more difficult to grind precisely than a spherical one, and some residual distortion is likely;

(d)

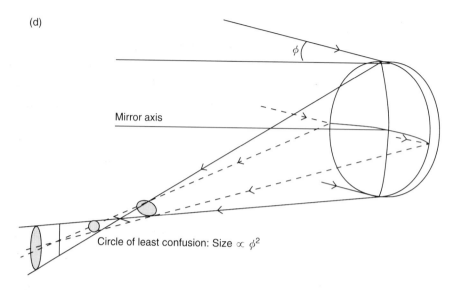

Mirror axis

Circle of least confusion: Size $\propto \phi^2$

Fig. 2.5 (*contd*). (d) Astigmatism for a paraboloidal mirror; rays in the plane defined by the incoming beam form a line-image in an orthogonal plane; the rays in that plane form a second line-image at a different distance from the mirror; the distance between the line-images, and so the size of the circle of least confusion, is proportional to the square of the off-axis angle.

the most infamous example of recent years has been the Hubble Space Telescope, where exhaustive pre-launch tests of the mirror failed to reveal a serious departure from parabolic shape. The effect was that the image of a point source was spread out over a circle of radius about two arcseconds, instead of being concentrated into the Airy disc of about one-hundredth of an arcsecond. Most of the light was in a central core about a tenth of an arcsecond in radius, so good images were still possible, but the full angular resolution of the telescope could only be recovered by extensive computer simulations of the aberration and the use of sophisticated algorithms to 'clean' the images. At the end of 1993, a complex and well-publicized repair operation, which added corrective optics, was carried out during a Shuttle mission. This was completely successful, and the telescope now produces images which are truly diffraction-limited.

For spectroscopic or flux measurements on a single object, it may be possible to use only rays parallel to the axis of the telescope by having the object at the centre of the field of view. But for many purposes, and especially for survey work, we want to use a significant field of view, so many of the rays will be from objects off the axis of the telescope. Even for parabolic reflectors, such rays do not all come to the same focus and other aberrations appear as a result.

The largest 'off-axis' aberration is *coma*; the size of the aberration increases linearly with the angle ϕ between the incoming ray and the axis. If we imagine the mirror (or lens) divided up into annular zones centred on the axis, then each zone forms a separate circular image called the comatic circle (Fig. 2.5(b)); each point on the circle corresponds to the meeting of two rays from diametrically opposite points of the zone. Zones nearer the axis form smaller comatic circles centred at positions nearer the axis (positive coma) or farther from it (negative coma); in the limit, rays through the centre of the lens or mirror form a point image at angle ϕ from the axis. The comatic circles of decreasing size overlap to form a comet-like image (Fig. 2.5(c)) – hence the name of the aberration. Because it is difficult to locate the centre of such an image, this aberration is particularly troublesome for accurate positional work.

The next most important off-axis aberration is *astigmatism* (Fig. 2.5 (d)), which varies quadratically with the off-axis angle ϕ and is only really serious for wide-field imaging. The incoming beam defines a plane through the axis, which I will for convenience describe as the vertical plane; rays in this plane come to a focus, not in a point but in a horizontal line; rays in a horizontal plane come to a focus in a vertical line. Elsewhere the image is elliptical, except that between the two line images there is a circle of least confusion. (This aberration, which effectively limits the field of view of large telescopes, should not be confused with the eye defect of the same name, which arises from the lens of the eye being slightly cylindrical instead of spherical and also produces two line images, which cannot be simultaneously focused on the retina.)

All these aberrations can be minimized, at the expense of some light loss, by careful optical design and/or the addition of correcting lenses. Modern telescope design is greatly aided by computers, which enable ray-tracing diagrams to be drawn rapidly for many possible designs; in this way the system parameters can be optimized to give the smallest possible image over the required field of view. Even so, aberrations can never be completely eliminated and the useful field of view of most telescopes is less than $1°$.

A remarkable exception is the Schmidt telescope, or Schmidt camera, in wide use for photographic survey work. Surprisingly, this uses a *spherical* mirror, but has a stop at the centre of curvature, C (Fig. 2.6(a)). The advantage of this combination is that all rays are now effectively 'on-axis', because they pass through C, or are parallel to a line through C, so there

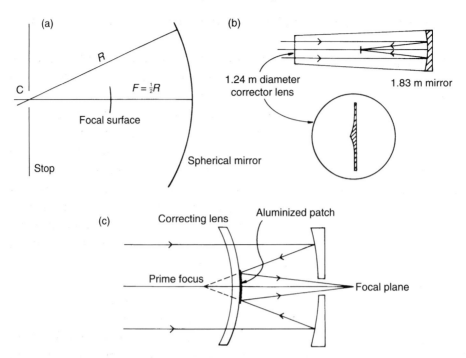

Fig. 2.6. (a) The principle of the Schmidt camera; all rays pass through the centre of curvature, C, because other rays are cut out by a stop at C, and so there are no off-axis aberrations; if the spherical mirror has radius R then the focal surface, which is also spherical, is at $R/2$ from the mirror. (b) A schematic scale drawing of the UK Schmidt Telescope on Siding Spring Mountain, Australia; the inset shows the approximate shape of the corrector lens; note that in practice the clear aperture at C is a fair fraction of the mirror diameter (cf. (a)). (c) A schematic drawing of a Maksutov–Cassegrain camera in which the tube is made considerably shorter because the light is reflected off an aluminized patch on the back of the meniscus correcting lens and is brought to a focus behind the mirror.

are no off-axis aberrations. If the mirror has radius of curvature R, then the focal length is $\frac{1}{2}R$ and the image is focused on a spherical surface.

Of course, we now need to correct for spherical aberration; this is done by placing a thin correcting lens at C (Fig. 2.6(b)). Because its surfaces are not spherical, the correcting lens introduces some off-axis aberrations, but only very small amounts. The curved focal surface is mildly inconvenient but the curvature is small and can be dealt with by bending the photographic plate slightly or (at the expense of light losses and further aberrations) adding a field-flattening lens. Further small aberrations are introduced by the fact that in practice the stop must have a comparatively large hole in it, or the telescope would be able to gather very little light; thus in practice not all rays actually pass through C.

This design gives excellent images over a wide field of view at small focal ratios, making it ideal for sky surveys to faint limits. Two well-known examples are the Palomar Observatory Schmidt in California and the UK Schmidt on Siding Spring Mountain in Australia. Both have apertures of 1.2 m (48 inches) and can photograph fields 6° or 7° square at a focal ratio of 2.5 or less. Near the edge of such a wide field some of the outer rays in the incident beam are obstructed by the telescope structure and do not reach the main mirror. This effect, known as *vignetting*, causes the edges of the field (at $\geq 2.7°$ from the axis) to be darker than the middle and makes the calibration of stellar magnitudes more complicated. At the outer edges of the field there is also some *distortion*, caused by the magnification varying with distance from the axis; this effect varies as ϕ^3, and is only important for these very-wide-field cameras.

The Palomar Observatory telescope completed a survey of the northern sky in 1956 (the Palomar Observatory Sky Survey – POSS) and the UK Schmidt, in collaboration with a similar instrument at the European Southern Observatory in Chile, has recently completed a similar survey of the southern sky, to fainter limits (see Chapter 6). A second-epoch series of POSS plates is now being taken, also to fainter limits than the original. Some 1200 plates are needed to cover the entire sky.

There are two disadvantages of the Schmidt design. Firstly, the stop sets the effective aperture, but the actual mirror must be considerably larger: the 1.2 m Schmidts have 1.8 m mirrors. Secondly, the tube necessarily has a length $2F$ (cf. F for a parabolic mirror used at prime focus); this is compensated for by a short focal length, which is in any case desirable to make the system a fast camera. A distinctly shorter tube is available in the Maksutov design shown in Fig. 2.6(c), in which the correcting lens is inside the prime focus and the light is reflected off an aluminized patch on the rear of the lens back through the main mirror, as in the Cassegrain design. The meniscus correcting lens has entirely spherical surfaces which are easier to make and also eliminate coma and astigmatism. Although this is quite a common design for compact amateur telescopes, no major observatory has used such a telescope for a systematic sky survey.

2.5 Telescope mountings

For all but the smallest telescopes a rigid mounting is essential for serious observations. Most large telescopes use some form of *equatorial* mounting

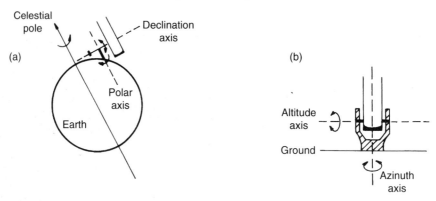

Fig. 2.7. (a) Schematic drawing of an equatorial telescope mounting. The polar axis points to the north (or south) celestial pole and so is parallel to the Earth's rotation axis. The declination axis is perpendicular to the polar axis and to the optical axis of the telescope. The telescope is moved around both axes until it is pointing at the object of interest. It is then clamped in position and kept pointing in the same direction by driving the entire mounting about the polar axis at the same rate as, but in the opposite direction to, the Earth's rotation. (b) A schematic alt–azimuth mounting. The telescope is mounted between two vertical supports and moves in a vertical plane about a horizontal axis (the altitude axis). The supports themselves are mounted on a platform, which can rotate about a vertical axis (the azimuth axis). To keep the telescope pointing at a given object, it is necessary to drive simultaneously about both axes to compensate for the Earth's rotation.

(Fig. 2.7(a)) in which there is a polar axis, parallel to the Earth's rotation axis, and a declination axis at right angles to the polar axis. The reason for this design is that it is possible to compensate for the Earth's rotation, and keep the telescope pointing at a particular object, by driving the telescope round the polar axis without any adjustment of the setting of the other axis. In practice, fine adjustments in the settings of both axes are often necessary during a long exposure to keep the object of interest precisely at the centre of the field. This guiding always used to be done manually by the observer looking through a small finder telescope attached to the main telescope and accurately aligned with its optical axis. On small telescopes this is still a common method, but for large instruments the observer is replaced either by a television camera, so that manual guiding can be done from a chair in front of a television screen, or by an auto-guider, which uses a servo-mechanism to keep a particular guide star at a fixed position in the field of view.

Although it is convenient to drive round just one of the axes, the simple offset mounting shown in Fig. 2.7(a) is not suitable for large telescopes because of large torques which cause bending of the telescope and axes and stresses on the bearings. A popular modern solution is to support the

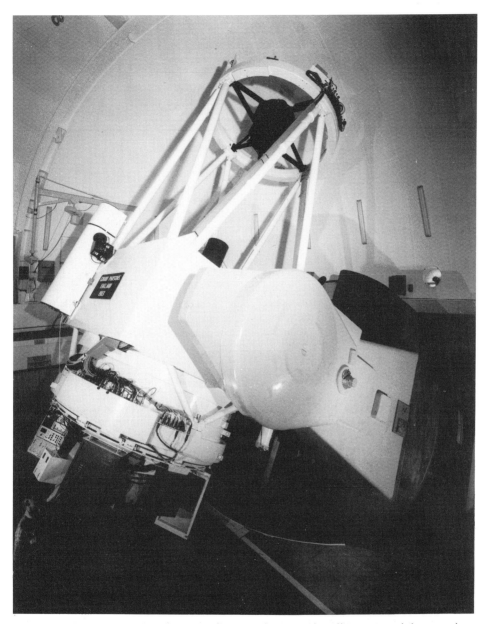

Fig. 2.8. The 2.5 m Isaac Newton Telescope has a polar disc equatorial mounting: the declination axis supports the telescope between the arms of a fork, mounted on a disc which rotates about the polar axis. Originally designed for the latitude of Herstmonceux (about 51° N) it has been adapted to its present latitude of about 29° N in the Canary Islands. (Royal Greenwich Observatory.)

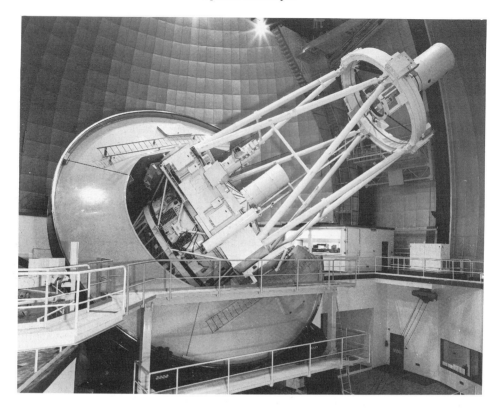

Fig. 2.9. The 3.9 m Anglo-Australian Telescope is an example of a horseshoe mounting: the horseshoe rotates about the polar axis and the declination axis is supported between the sides of the horseshoe. The Cassegrain focus (Fig. 2.4) is enclosed in an observing cage, accessible when the telescope points to the zenith. (Anglo-Australian Observatory.)

declination axis at either end (a fork mounting) and to mount this fork either on a disc or as a horseshoe, both at the upper end of the polar axis. The 2.5 m Isaac Newton Telescope at the Roque de los Muchachos Observatory on La Palma in the Canary Islands (Fig. 2.8) is an example of the polar disc mounting, while the 3.9 m Anglo-Australian Telescope at Siding Spring Observatory (Fig. 2.9) employs the horseshoe design, used also for the 4 m telescopes at the Kitt Peak National Observatory in Arizona and the Cerro Tololo Interamerican Observatory in Chile.

Even these designs have residual flexure problems, because the stresses on the axes vary as the telescope moves. With the advent of fast computers, there is now no difficulty in driving about two axes simultaneously, and for very large telescopes it is better to adopt the mechanically simpler *alt–azimuth* design shown in Fig. 2.7(b). There the main support axes are always

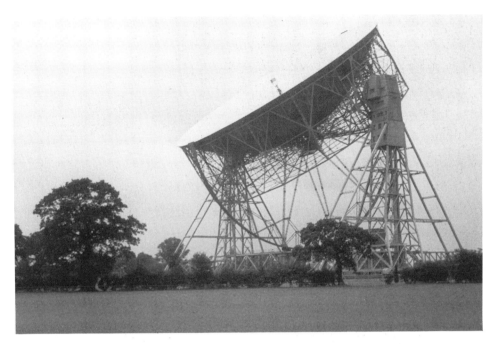

Fig. 2.10. The 76 m Lovell radio telescope at Jodrell Bank was the first very large, fully steerable telescope. It has an alt–azimuth mounting; the whole telescope is mounted on wheels on a circular track, and the dish moves about a horizontal axis between the two towers. (Photograph by the author.)

vertical and horizontal and the stresses on the axes hardly vary with the direction in which the telescope is pointing. This design has long been used for large radio telescopes, such as the 76 m (250-foot) dish at Jodrell Bank (Fig. 2.10), and has been used for the Russian 6 m optical telescope in the Caucasus mountains (which replaced the 5 m (200 inch) Hale telescope on Palomar Mountain as the world's largest single-mirror telescope), for the 4.2 m William Herschel Telescope on La Palma (Figs 2.11 and 2.12) and for most of the new large telescopes currently under construction. The main disadvantage of an alt–azimuth mounting is that it is impossible to follow a star exactly through the zenith (the point vertically overhead – see Chapter 7). The driving rate needed to compensate for the Earth's rotation becomes formally infinite at the zenith and there is an inaccessible sky area several degrees square centred on the zenith.

Any motion round an axis causes slight uncertainties in the direction in which the telescope is pointing, for example arising from misalignment or flexure of the support or optical axes. For very precise positional work special telescopes are used which have as little motion as possible and are

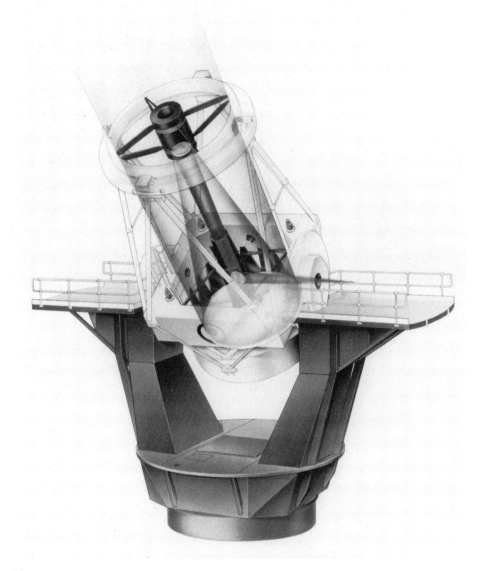

Fig. 2.11. An artist's impression of the 4.2 m William Herschel Telescope on La Palma, which has an alt–azimuth mounting. The two Nasmyth platforms are used for mounting heavy equipment. The light path to one of the Nasmyth foci is shown. (Royal Greenwich Observatory.)

not driven to follow the star being observed. Instead, the telescope is pointed in a well-defined direction and then the time at which a star drifts across the optical axis (due to the Earth's rotation) is measured precisely. The classic instrument of this kind is the *transit circle* (Fig. 2.13), which has

Fig. 2.12. The 4.2 m William Herschel Telescope soon after its installation. *Left:* test instruments are shown at the Cassegrain focus, and a person is standing on the Nasmyth platform which is now occupied by the GHRIL (see text). The other platform is used for a high-resolution echelle spectrograph. *Right:* the telescope seen from outside the dome, which has been rotated during the exposure to allow the whole interior to be seen through the open slit. (Royal Greenwich Observatory.)

a single, horizontal, east–west axis allowing the telescope to move only in elevation in the north–south plane. The mounting piers are rigidly fixed and very accurately positioned to ensure that the telescope's optical axis remains as exactly as possible in a vertical, north–south plane. Cross-wires are positioned in the focal plane (Fig. 2.13, inset) and the star is timed as it passes across the wires. The time of passage across the optical axis is known as the time of *transit*. The position of the star is then deduced from this time and the angle of elevation of the telescope. By observing stars of known position, the position of the telescope (latitude and longitude) can be found. The zero of longitude is arbitrary, and was defined by international agreement in 1884 to be the longitude of the Airy transit circle at the Royal Observatory in Greenwich Park, which still defines the Greenwich Meridian (Fig. 2.14). Modern transit circles (e.g. Fig. 2.15) use automatic recording techniques to avoid systematic timing differences between observers.

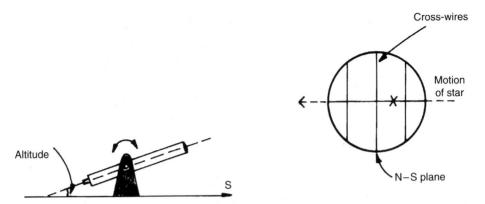

Fig. 2.13. View along the east–west horizontal axis of a transit telescope in the northern hemisphere. The telescope can move only in a vertical plane which is aligned north–south. The inset shows the view through the telescope. The star of interest is placed on the horizontal cross-wire by adjusting the altitude of the telescope and then it is timed as the Earth's rotation causes it to drift from east to west across the vertical cross-wires. The average of the three timings for the case shown gives the time of transit across the optical axis of the telescope. The apparent motion is from right to left (in the northern hemisphere) because of the inversion of the image in an astronomical telescope (see Fig. 2.2).

2.6 New telescope designs

Attempts to improve the design of telescopes all share one main goal: to maximize the rate of gathering information. There are two routes to this goal:

(a) larger apertures, which gather more light;
(b) better detectors, which make more efficient use of the light.

I will discuss the second route in the next chapter.

There are two distinct, but related, types of problem in building very large telescopes: engineering problems and financial problems. The engineering problems arise from the difficulties of producing large homogeneous mirrors and of constructing large but precise structures that can be pointed and driven with an accuracy of one arcsecond or better. Existing large mirrors have already had to overcome the difficulties of casting large blanks from a single melt; the usual solution is to cast a number of smaller hexagonal and triangular pieces which are later re-heated to a sufficient temperature that they can be fused into a single disc. Both stages of cooling must be prolonged for many months to minimize internal stresses. The sheer weight of these huge mirrors creates structural engineering problems, and is minimized by

Fig. 2.14. The Greenwich Meridian is defined by the optical axis of the nineteenth-century Airy transit circle; it is still on display behind the panel which carries the line that now marks the meridian for tourists. (Photograph by the author.)

casting the mirrors with a honeycomb structure (Fig. 2.16) which reduces weight without sacrificing rigidity.

The financial problem with large telescopes is that the cost increases very rapidly with the aperture, D. To a first approximation,

$$cost \propto D^3, \tag{2.11}$$

simply because of the increased volume of material needed to construct, support and house the telescope. If new engineering techniques are needed, or special manufacturing facilities (for example, a purpose-built casting

Fig. 2.15. The Carlsberg Automatic Meridian Circle on La Palma is a fully automatic transit telescope which follows a pre-set programme of observation on every clear night. It swings only in the north–south plane, moving from star to star about an accurately aligned horizontal east–west axis. (Royal Greenwich Observatory.)

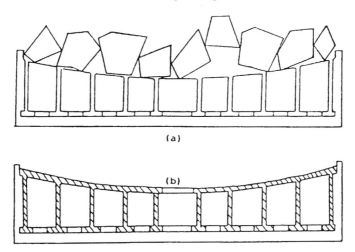

Fig. 2.16. A sketch of the rotating casting vessel used in the process of 'spin-casting'. The vessel containing the molten glass is rotated around a vertical axis so that the surface takes up the required paraboloidal shape, and is then allowed to cool slowly, while the rotation rate is maintained constant. The solidified glass then has roughly the correct shape for a primary mirror and needs minimal further shaping. To make a honeycomb mirror, a mould is first made with blocks of ceramic fibre attached to the base; (a) shows the glass pieces before melting and (b) shows the glass filling the mould after melting and spinning. After cooling, the mould is removed, using a high-pressure water spray to break up the blocks. (*Q.J.R.Astr.Soc.*, 1990.)

works), the associated research and development and capital costs may cause the power of D to be greater than 3, or produce a large, discontinuous jump in cost at some critical aperture.

Since the light-gathering power increases only with the area, that is, $\propto D^2$, we can say roughly that:

$$cost\ per\ photon\ \propto D, \qquad (2.12)$$

making the construction of very large, single-mirror telescopes appear to be uneconomic. The solution to this problem must be an engineering one, which effectively reduces the power of D to less than 3. Research in this area is very active at present and developments can be divided into three approaches.

1. Construction of a single, large mirror (a 'monolithic' mirror) which is also *thin*, so that it uses less glass. Since the glass plays no intrinsic role in the image-forming process, but merely acts as a rigid support for the extremely thin reflecting layer of aluminium, all that is needed

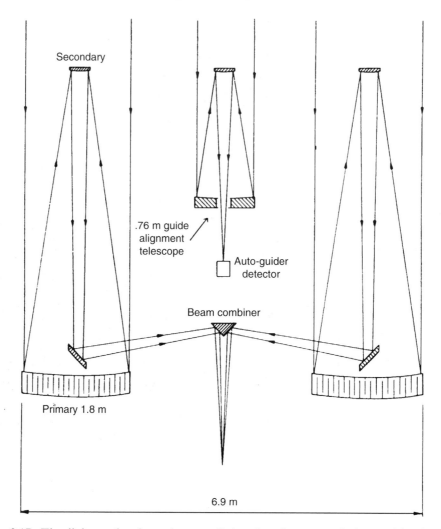

Fig. 2.17. The light paths through two of the six telescopes of the Multi-Mirror Telescope, showing how the light is brought to a common focus at the centre of the hexagonal array (Fig. 2.18). The guide telescope is mounted centrally above the beam combiner. (Reproduced, with permission, from *Telescopes for the 1980s*, ©1981, by Annual Reviews Inc.)

is for the glass support to retain its parabolic shape. The design involves three elements:

(a) 'Spin-casting'. This process, devised by Roger Angel in Arizona, uses the fact that a free surface rotating about a vertical axis in a gravitational field has a naturally *parabolic* shape. The glass is therefore melted in a rotating casting vessel (Fig. 2.16),

Fig. 2.18. An aerial view of the Multi-Mirror Telescope in Arizona, showing the six separate mirrors arranged in a hexagonal array. The entire building rotates to move the telescope in azimuth. (Reproduced, with permission, from *Telescopes for the 1980s*, ©1981, by Annual Reviews Inc.)

and the vessel is kept rotating throughout the cooling phase, so that the mirror blank starts with an approximately parabolic shape. This both saves glass and shortens the grinding time.

(b) A thin honeycomb mirror (see above) is used, which again saves glass and also allows a much shorter cooling time.

(c) An active multi-point support system is used for the mirror. The shape of the mirror is continually monitored as the telescope moves about the axes and the back-supports are adjusted to keep the mirror a constant shape. This active support allows the mirror to be thinner and less rigid than was formerly necessary, and again saves material and money.

This is the favoured design for the Gemini project, in which the UK, the USA, Canada and other countries will construct two 8 m telescopes, one in each hemisphere.

2. Combining the light from several smaller telescopes, to simulate a large telescope. Two examples of this are:

(a) The Multi-Mirror Telescope (MMT). This instrument, on Mt Hopkins in Arizona, was the first successful attempt to create a new design of telescope (Figs. 2.17 and 2.18). It consisted of six 1.8 m bent-Cassegrain-type telescopes arranged in a hexagonal frame in such a way that the light was brought to a common focus. A servo-mechanism was used to maintain accurate superposition of the images. The mounting is alt–azimuth, and the entire building rotates about the vertical axis.

This is a case where the cost advantage can be neatly illustrated. The MMT has a collecting area equivalent to that of a single-mirror 4.4 m telescope, but achieves it at about 40% of the cost (exercise 2.6). For an n-mirror telescope, the diameter of the equivalent single mirror is D_t, where $D_t^2 = nD^2$, while the cost (equation (2.11)) is $£nAD^3 = £ADD_t^2$, where A is a constant. Thus, for a fixed component-mirror size, D, the cost of providing a total effective aperture D_t increases only as the square of the aperture rather than as the cube.

Unfortunately, the problems with keeping the images aligned were so great that it was really only successful as a spectroscopic instrument. It is currently being rebuilt with a single, spin-cast mirror.

(b) The Very Large Telescope, under construction for the European Southern Observatory. This is an extension of the MMT concept, but also involves very large, single mirrors. There will be four 8.2 m Cassegrain telescopes, which will be able to be used either as four separate telescopes or as an optical interferometer with a maximum baseline of 130 m. If all four mirrors were to feed their light to a common focus, the total area would be equivalent to a 16.4 m single-mirror telescope. An even larger interferometric array is planned, with the addition of at least three 1.8 m mobile telescopes and a maximum baseline of 200 m.

3. Construction of a *mosaic* mirror. This design has been adopted for the 10 m Keck telescope (CalTech and the University of California) on Mauna Kea on the island of Hawaii. The segmented primary mirror is constructed of 36 1.8 m hexagonal mirrors, fitted together in a mosaic around the central (hexagonal) hole that gives access to the Cassegrain focus. Each mirror is supported independently and the light is brought to a common prime focus. The mirror shape is preserved by using 'active optics' – as described above, this involves a support system where the shape of the mirror is constantly monitored and the supports adjusted to maintain both the alignment of the components and the overall shape of the reflecting surface.

The figuring of the individual mirrors presented a new problem, since each mirror is effectively a section of a large mirror and is in general not symmetrical about its own axis; there are in fact six distinct shapes. The adopted solution has been to use a technique called 'stressed grinding', where the mirror is mounted on a support and precisely calculated weights are hung from the edges of the mirror blank so that it is held under an asymmetric stress while being ground symmetrically about its axis. After grinding, the stress is released and the mirror takes up its correct asymmetric shape.

Note, finally, that it is also possible to make better use of the light by two forms of telescope *control*:

1. Active optics – as described above, this involves a continual adjustment of the primary mirror shape, on a timescale of a few minutes, to maintain the sharpest image. On this timescale, it is possible to correct for small errors in following the source (the *tracking*) and for changes in the mechanical support as the telescope drives to follow the source and bends slightly under gravity.

2. Adaptive optics – in this case the aim is to correct the image for the effects of atmospheric turbulence, which involves timescales of a few times 10–100 *milli*seconds. The concept was suggested in 1953 by H.W. Babcock, but it was only some 30 years later that serious work began to implement the idea. This technique deliberately distorts the reflecting surface away from its theoretically best shape to minimize the effects of atmospheric seeing – the surface is distorted to match the distorted wavefronts produced by the turbulent atmosphere. The timescales are too short for it to be possible to adjust the shape of the primary mirror. The overall motion of the image can be removed by tilting the secondary mirror, but detailed wavefront correction is

achieved by re-imaging the telescope's exit pupil onto a much smaller active mirror, which can be distorted on the very short timescales involved. The wavefront distortions are monitored by looking at a guide star close to the object being observed.

2.7 Summary

Optical telescopes have two major advantages over the naked eye: they enable the detection of much fainter objects and they have much better angular resolution.

Astronomers often use a logarithmic scale (the magnitude scale) to measure the brightness of a star, with faint stars having larger magnitudes (very bright stars can have negative magnitudes). The typical limit for the naked eye is a magnitude of +6, but a telescope can extend this to +16 for a 1 m diameter mirror. If the signal can be added up over time, as with a long-exposure photograph, even fainter objects can be recorded: the 1.2 m UK Schmidt telescope can record stars down to magnitude 21 in an hour's exposure. Larger telescopes, with sensitive photoelectric detectors, can detect objects as faint as magnitude 25 or 26.

The theoretical limit to the angular resolution of a telescope is set by the size of the diffraction pattern produced by the circular aperture of the telescope, and is about 0.1 arcseconds for a 1 m telescope at visual wavelengths. However, this is only relevant for small telescopes. For large telescopes, the turbulence in the atmosphere smears out the image of a point source into a seeing disc, which is usually much larger than the theoretical resolution: the resolution is seeing-limited.

Refracting telescopes focus the light entirely by lenses. They are still used for astrometry, but suffer from chromatic aberration and from various mechanical snags. Most large, modern telescopes are reflectors.

Reflecting telescopes use a parabolic primary mirror and a variety of different secondary mirrors. The prime focus (no secondary) is used for observing very faint objects. The commonest focus is the Cassegrain, where a hyperboloidal secondary reflects the light back down through a hole in the primary mirror to a focus just behind the primary. For high-dispersion spectroscopy, this beam is sometimes diverted by one or more flat mirrors along the telescope support axes into the coudé (equatorial mount) or Nasmyth (alt–azimuth mount) focus.

Mirrors do not suffer at all from chromatic aberration, and perfectly parabolic mirrors are free of spherical aberration. However, there are several off-axis aberrations that affect mirrors just as much as lenses. The most serious is coma, which produces comet-shaped images of a point source, but astigmatism and distortion also play a role and the useful field of view of most telescopes is restricted to about 1°. The major exception is the Schmidt camera, which uses a spherical mirror to eliminate off-axis aberrations and a corrector plate to remove spherical aberration, and has a field of view of 6° or 7°.

Medium-sized telescopes use equatorial mountings, in which one axis is parallel to the Earth's rotation axis. This enables a star to be followed, as the Earth rotates, by driving about only one axis, but it is mechanically complex. For many large, modern telescopes an alt–azimuth mounting, with vertical and horizontal axes, is preferred, since it is mechanically simpler and computer control now makes driving about two axes quite straightforward. For very precise positional measurements, transit telescopes are used, with movement about a single horizontal axis.

New telescope designs aim to gather more light while not increasing the cost too much. Single, large mirrors are being made very thin by using a honeycomb structure and forming the mirror blanks in a parabolic shape by spin casting. Some telescopes simulate a large aperture by combining the light from several smaller telescopes, while another design uses a mosaic of small segments with an active support system to keep them aligned in the shape of a single, large mirror. Active support is also used in other telescopes to keep the image as sharp as possible. Adjustment of the effective mirror shape on very short timescales (adaptive optics) can be used to try to compensate in real time for the effects of atmospheric turbulence on the image.

Exercises

2.1 Taking the aperture of the human eye to have diameter 8 mm and the faintest magnitude visible to the naked eye to be +6.0, calculate the limiting magnitude of a 0.5 m telescope if used visually. (Allow 0.5 magnitudes for light losses in the telescope.)

2.2 Taking the aperture of the human eye to have diameter 8 mm, calculate the magnification needed if the 5 m Hale telescope were to be used visually, with all the incident light entering the eye. Now calculate the corresponding angular size of the image of a point source, assuming a 1 arcsec seeing disc, and compare it with the angular resolution limit of the human eye (40–60 arcsec). (This example explains why such large telescopes are never used visually!)

2.3   The diffraction limit on the angular resolution of a telescope of mirror diameter D may be written as

$$\Delta\theta_d = 0.14 \frac{\lambda}{0.55\mu m} \frac{1m}{D} \text{ arcsec,}$$

while the size of the seeing disc at a good observing site can be approximated as

$$\Delta\theta_s = 0.7 \left(\frac{0.55\mu m}{\lambda}\right)^{1/5},$$

where λ is the wavelength of observation. Find an expression in terms of D for the wavelength at which these two expressions are equal, and verify that an 8 m telescope is diffraction-limited at wavelengths longer than about 12 μm. At what wavelength would a 1 m telescope become diffraction-limited?

2.4   Calculate the theoretical diffraction-limited resolution of a 10 m telescope at wavelengths of 300 nm (UV) and 1 μm (IR). Under what circumstances might each be realized in practice?

2.5   Suppose that it is planned to construct a multi-mirror telescope whose total collecting area is equivalent to that of a single 6 m aperture mirror. How many 2 m mirrors would be needed?

2.6   Suppose that the cost of building a single telescope of mirror diameter D is £AD^3, where A is a constant. The Multi-Mirror Telescope in Arizona (MMT) consists of six such telescopes, arranged to bring light to a common focus. Find the mirror diameter of the single telescope that would have the same collecting area as the MMT, for which $D = 1.8$ m. Find the ratio of the cost of that single telescope to the cost of the MMT.

2.7   A Schmidt telescope with focal ratio $F/D = 2.5$ has an effective aperture of 1.2 m and a $7° \times 7°$ field. What size of photographic plate is needed? (Round up to the nearest 10 mm to allow an edge for handling.)

3

Optical detectors and instruments

So far I have been discussing telescopes in isolation. On their own, however, telescopes merely collect photons: the photons must then be recorded in some way. Originally, of course, this was done visually, and visual use continued to be important for more than a century after the first use of photography in the 1840s, especially for observing planetary surfaces, where the eye could take advantage of rare moments of good seeing. The eye, however, is a poor detector at the low light levels typical in astronomy and is now rarely, if ever, used directly by professional astronomers.

None the less, all other radiation detectors operate in essentially the same way as the eye: energy is transferred from an electromagnetic wave to an electron, or to some other atomic particle (a nucleus in the case of X-rays and harder radiation). In the case of the eye, the radiation excites molecules in the retina. These changes create electrical impulses in the optic nerves, which transmit signals to the brain, where they are interpreted as a picture.

In the optical window (near ultraviolet to near infrared), the detectors now in use essentially all fall into one of two main categories, photographic and photoelectric, with photoelectric and related techniques becoming increasingly dominant.

3.1 Photography

The formation of a photographic image is a complicated chemical process whose details are still not fully understood. Here I will just sketch the process in broad outline. The active constituents of a photographic emulsion are micron-sized grains of some silver halide crystal, such as silver bromide (AgBr). The grains are suspended in a thin layer of gelatin (Fig. 3.1(a)), which acts to keep the grains well separated and fixed in position. It also protects

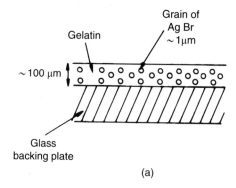

(a)

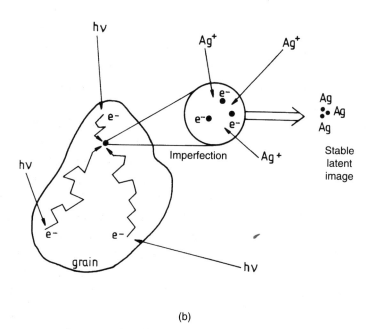

(b)

Fig. 3.1. (a) Schematic cross-section of a photographic plate, showing micron-sized silver halide crystals suspended in a thin layer of gelatin on the surface of a glass plate. The gelatin and the glass plate play (literally) only supporting roles. Glass is used in astronomical photography because it gives greater rigidity than the plastic used in conventional roll film. (b) Schematic representation of the formation of a latent photographic image. Incident photons detach electrons from the atoms in the crystal lattice of a grain. Some of these free electrons may become trapped at an imperfection in the lattice and, while there, recombine with migrating silver ions to form a stable clump of silver atoms – a latent image, which can subsequently be developed.

the developed image from damage. For astronomical use, it is common to put the emulsion on a glass base, for further stability (for example, preventing shrinkage of the film during development), and astronomers speak of a photographic *plate*.

There are two main stages in the formation of a photographic image. When a photon strikes a grain in the emulsion, its energy is absorbed by the crystal lattice and an electron is excited out of the filled *valence band* into the *conduction band* (Section 3.2, Fig. 3.4), which is otherwise generally empty for a silver halide crystal. The electron is then able to move freely through the crystal. Also present in the crystal is a small number of silver ions which have been displaced from the lattice by thermal excitation and are also migrating through the lattice. If the electron meets one of these ions a neutral silver atom is formed. A single atom is easily broken apart again (within about a second) by thermal effects, but if two or more can be formed together they are stable and form a *latent image*. It appears that defects or dislocations in the crystal act as electron traps, so latent images are preferentially formed there (Fig. 3.1(b)).

These groups of silver atoms are much smaller than the halide crystal and are not directly visible. If a chemical reducing agent is added (in general, one which adds hydrogen or removes oxygen or some similar element: in this case, one which removes bromine), then the small groups of atoms at imperfections act as catalysts for the conversion of the entire grain to pure silver. Thus the process of development amplifies the original signal by a very large factor, with a few photons finally producing more than 10^{10} silver atoms. The pattern of developed silver grains forms the visible image, whose spatial resolution is limited by the size and spacing of the grains.

The process is complicated by the fact that all the grains are in fact converted to silver by development. The conversion occurs rapidly solely for the catalysed grains, but it also occurs gradually for the other grains. This means that it is necessary to stop developing once a clear image appears, but that even so there is always a background of partially-developed grains, which produce a 'fog' on the film.

There are two main disadvantages of the photographic process. First, most of the photoelectrons do not survive long enough to meet a silver ion, being captured by positive holes or halogen atoms, so the process is very inefficient. Until recently, a typical photographic emulsion required about 1000 photons for each developed grain in the image. Modern astronomical emulsions are treated by a variety of processes to increase their sensitivity, from adding chemical sensitizers to the emulsion to baking or soaking in nitrogen or hydrogen. Even with these hypersensitizing techniques, an

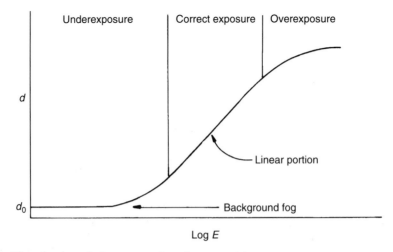

Fig. 3.2. The characteristic curve of a photographic emulsion. At very low values of the exposure E the density d tends to a constant low background value – a 'fog' which is a property of the plate rather than a measure of the exposure. Over a short range of exposure, the density varies linearly with exposure, but overexposure causes saturation and the density again ceases to give any information about the exposure.

efficiency of about 4% is generally the best that can be achieved, although a new emulsion with an efficiency of nearly 10% was introduced in 1991.

The second disadvantage of a photographic plate is its non-linear response: the strength of the image is not linearly related to the number of incident photons. The silver grains absorb light and appear dark, so the image on the film is a negative one. It is usual to measure the strength, or density, of the image by passing a beam of light through the developed plate. If a flux F_{in} is incident on the plate and a flux F_{out} is transmitted, then the density d is defined by

$$d = \log_{10}(F_{in}/F_{out}). \tag{3.1}$$

Thus, for example, $d = 0.3$ corresponds to 50% transmission. Clearly, the image density must be related to the exposure of the plate. The exposure value E is defined as the product of the exposure time t and the incident energy per unit area I : $E = I \times t$. The relation between d and $\log_{10} E$ is known as the characteristic curve of the photographic plate (Fig. 3.2). This curve does have a linear portion, but only over a restricted range of exposure, and the quantity which varies linearly with the exposure is a complicated function of the incident flux of photons. At the faint end of the characteristic curve, faint stars disappear into the background fog, while at the very bright

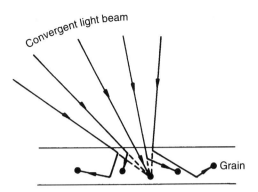

Fig. 3.3. Ideally, a convergent light beam would strike a single grain and produce essentially a point image on the developed plate. In practice, different photons in the beam will be scattered by different atoms along their path through the emulsion and so will diverge and strike different grains. Thus a point source produces many developed grains and hence an image of finite size: the higher the photon flux, the more scattering occurs and the larger is the final image.

end the image saturates and its brightness cannot be estimated accurately. A plate exposed for a given time will then be correctly exposed (i.e. on the linear part of the curve) for some stars, but will be overexposed for brighter stars and underexposed for fainter ones. It is therefore difficult to measure accurately the relative brightness of two stars of very different brightness – they must be related through a series of exposures of different lengths which enable them to be compared indirectly via a set of stars of intermediate brightness.

Even for a correctly exposed star, the density of the image is not simply related to the brightness of the star, because of internal scattering (Fig. 3.3); incident photons are scattered within the emulsion before being absorbed on a grain surface, so a point source gives rise to a roughly circular image whose size increases (non-linearly) with the number of scattered photons. The brightness of the star can be determined by measuring the size and total density of the image, but a complicated calibration procedure is needed. Internal scattering also puts a practical limit on the angular resolution of the system, especially near bright stars.

Despite all these disadvantages, photography still possesses one major advantage which cannot yet be matched by electronic devices. This is the ability to take photographs of large areas of sky for survey purposes. Schmidt telescopes (Section 2.4) use photographic plates that are more than 300 mm square, corresponding to an area of sky of about $6° \times 6°$; plates of up to 500 mm square are used on some telescopes. By contrast, electronic detectors cannot yet easily be made more than 50 mm square, giving a sky area which is only 1° square, 36 times smaller. As you will see later, photographic plates also have a better spatial resolving power and so a large Schmidt plate gathers vastly more information at a single exposure than even the best electronic device (exercise 3.1). However, electronic devices win in almost every other point of comparison.

It is only recently that it has been possible to take full advantage of the huge quantity of information on a typical Schmidt plate. Originally, the information had to be extracted visually, by inspection or by laborious hand-measurement of individual images. Since 1970 various fast, automatic plate-measuring machines have been developed which can scan the plate in two dimensions and convert positions and image densities and density profiles into digital form for computer processing. In a matter of hours, more information can be extracted from a plate by such a machine than could be extracted by a human being in a lifetime, thus making routine all sorts of statistical analyses that were once quite impossible.

3.2 Photoelectric detection

Electronic recording in astronomy began as early as the 1920s, but did not develop seriously until the 1950s. For many years the main recording device was a photoelectric cell with a light-sensitive surface which could respond to light in one of two ways.

Photoemissive devices make use of the well-known photoelectric effect, in which an incident photon causes an electron to be emitted from the detector surface, or *photocathode*. The energy of the emitted electron is the difference between the energy of the incident photon and the energy needed to overcome the forces binding the electron to the surface. The cathode is placed in an evacuated cell together with an anode and when a potential difference is applied the emitted electrons flow to the anode and are recorded as a current, whose strength increases linearly with the number of incident photons.

The photoelectric effect was first studied for metals, which are good conductors because the most energetic outer, or valence, electrons of the atoms in the metal lattice are free to move through the lattice. It is useful to express this in terms of energy levels (Fig. 3.4). Most of the valence electrons lie in a range of energies known as the *valence band*, while the free electrons, which are responsible for the conductivity of the metal, lie in a higher range of energies known as the *conduction band*. Because of the Pauli exclusion principle, only two electrons can occupy a particular energy level

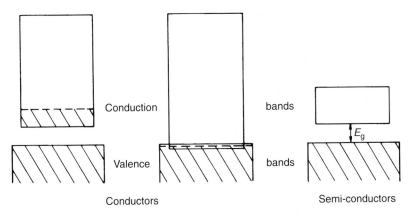

Fig. 3.4. Schematic energy diagrams for (a) a metal (with and without overlapping energy bands – see text) and (b) a semi-conductor. In a semi-conductor there is always an energy gap, E_g, between the valence and conduction bands, but the conduction band is narrower than in a metal. Electrons fill the energy bands up to the top of the Fermi sea (shaded regions). The top of the Fermi sea is always in the valence band for a semi-conductor. The top of the conduction band represents the energy of the free surface.

in the lattice. The myriads of electrons in the lattice fill up the energy levels like water filling a bath, and the most energetic electrons are at the surface of this 'sea' of electrons (known as the *Fermi sea*, after the physicist who first studied the behaviour of particles that obey the exclusion principle). In some metals, the valence band is filled to the top, and the top of the Fermi sea is in the conduction band. In other metals, the valence band is not quite filled to the top, but the conduction and valence bands overlap. In either case it is easy for electrons to escape from their parent atoms, for example by thermal excitation. It is much more difficult for them to escape from the metal altogether, because the energy difference between the top of the Fermi sea (Fig. 3.4) and the free surface is large. This energy difference is known as the *work function*, or *electron affinity*, and is the threshold energy for the photoelectric effect in metals.

Most photoelectric cells use cathodes made of a semi-conductor material, in which there is always an energy gap E_g between the conduction and valence bands (Fig. 3.4), and the top of the Fermi sea is always in the valence band; E_g is then the minimum energy needed to detach an outer electron from a lattice atom and enable it to travel freely through the lattice. There are several reasons for using semi-conductors as photocathodes. Firstly, they absorb a large fraction of the incident photons, whereas metals are good reflectors. Secondly, there are few free (thermally excited) electrons in the conduction band, so photoelectrons are likely to reach the surface and escape without collision with free electrons. Thirdly, although the electrons have to be given enough energy to cross the energy gap, the work function of a semi-conductor surface is generally much less than the several electron volts typical of a metal: the range of energies in the conduction band is much smaller than in a metal. For many years the commonest photocathode material was caesium antimonide (Cs_3Sb).

Semi-conductor detectors that use the photoelectric effect have a low-energy threshold equal to the sum of the energy gap and the conduction band energy width. For typical semi-conductors this long-wavelength cut-off is at about 600 nm, in the red part of the optical spectrum. It is possible to extend this limit somewhat farther into the red by coating the semi-conductor with a thin layer of an electro-positive metal, which reduces still further the width of the conduction band. The currently most popular photocathode, caesiated sodium potassium antimonide, $(Cs)Na_2KSb$, uses this effect and is sensitive out to about 900 nm.

At lower photon energies only the most loosely bound electrons can be detached from the surface, so for infrared work photoemissive devices cannot be used. However, infrared photons still have enough energy to free an electron from the valence band and raise it to the conduction band, so that it can travel freely through the material. This is the basis of photoconductive devices, in which incident photons cause an increased current to flow in the material itself. I will discuss these devices again in Section 3.3. Formerly, the commonest photoconductive material was lead sulphide (PbS), but it is now indium antimonide (InSb).

In both photoemissive and photoconductive devices a photon affects exactly one electron, so there is an exactly linear relation between the incident photon flux and the resulting photocurrent. This linearity is one of the most important advantages of photoelectric devices.

There are two other important advantages of photoelectric techniques:

1. *Dynamic range.* Because there is a far greater number of available electrons in a photocell than there are grains on a photographic plate, a photoelectric cell can record accurately and linearly over a far greater range of photon flux than can a photographic plate: a photoelectric cell does not saturate nearly so easily as a photographic plate. This large 'dynamic range' makes it easy to measure accurately the relative brightness of stars over a large range of magnitude.

2. *Quantum efficiency.* You have seen that, at best, photographic plates record only about 4% of the photons that fall on them. By contrast,

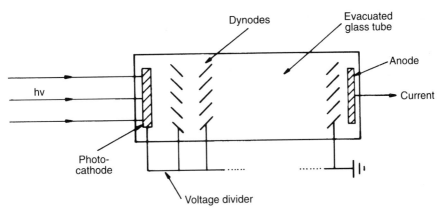

Fig. 3.5. An idealized photomultiplier. Incident photons strike a photocathode, which is held at a negative potential of about one kilovolt with respect to the anode at the other end of an evacuated tube. A series of dynodes separates the cathode and anode and are at successively more positive potentials, the final dynode being connected to earth. Electrons emitted from the cathode by the photoelectric effect are accelerated to the first dynode, where they each have enough energy to release several electrons. This is repeated at all the dynodes so that, finally, each photon results in a cascade of electrons reaching the anode, where they are registered, either as a direct current or as a pulse.

photoemissive devices routinely record 20% of the incident photons, while photoconductive devices may have 'quantum efficiencies' of up to 80% (a more precise definition of quantum efficiency is given in Section 3.5.2). This 20-fold advantage puts a 1 m telescope using a photoconductive cell on equal terms with a 4 m telescope using a photographic plate, at least for looking at single objects.

Although photoelectric cells make efficient use of the incident photons, the typical photon flux from an astronomical source is often so low that the resultant current is far too small to be measured directly: a flux of a few tens of electrons per second corresponds to a current of only about 10^{-17} A! The current is therefore amplified using a *photomultiplier tube* (Fig. 3.5), a photoemissive device in which the cathode and anode are separated by a series of dynodes at successively higher potential with respect to the cathode. Each electron emitted from the photocathode is thus accelerated by a strong electric field, which gives it enough energy to eject several electrons from the first dynode. Each of these electrons ejects several electrons from the next dynode, and so on in a cascade, so that one incident photon results in a large burst of electrons reaching the anode. The voltage difference between each dynode is about 100 V, to ensure that each electron is accelerated to an energy of about 100 eV,† well above the threshold for the photoelectric effect, which is a few eV.

If the yield of secondary electrons at each dynode is y per incident electron, and there are n dynodes, the overall gain is

$$G = y^n. \qquad (3.2)$$

In the direct current mode, the average number of electron bursts per second is detected as a current in an external circuit. The gain is typically 10^7, giving a final current of about 0.1 nA: still small, but large enough to be recorded by conventional circuitry and, if necessary, amplified further.

Each incident photon gives rise to a single burst of electrons at the anode, so the linear advantage is retained. Bursts of electrons may also be produced by events other than the arrival of a photon, for example thermal emission of an electron from the photocathode. This thermal emission, which occurs even in the absence of a light source, is indistinguishable from a true photon event and is known as the *dark current*. It can be minimized by cooling the photomultiplier tube.

Other sources of noise, for example thermal emission of an electron from a dynode or the occasional arrival of a cosmic ray, also contribute to the average current. However, they normally produce electron bursts, or pulses, of a different strength from those produced by a photoelectron, and can be distinguished. For this reason, photomultipliers are often operated in a pulse-counting mode, with a pulse-height discriminator to cut out spurious pulses. This reduces the noise considerably but causes saturation at high count rates because there is inevitably a 'dead time' between the recording of successive pulses, during which the system recovers from the previous pulse.

The introduction of photoelectric cells transformed the measurement of accurate stellar magnitudes. However, a conventional, single photomultiplier tube or single photoconductive device has the disadvantage that only one object can be observed at a time. In normal use, all the light coming through a small aperture in the focal plane is directed onto the photocathode in a 'defocused' spot‡ covering most of the cathode area so as to average over any non-uniformities on the cathode surface. Thus each star requires a separate exposure and no imaging is possible; for extended objects, such as galaxies and gas clouds, contour maps must be laboriously constructed by taking

† 1 electron volt (eV) is the energy given to an electron when it is accelerated through a potential of 1 volt. $1 \, \text{eV} = 1.6 \times 10^{-19}$ joule and has the energy of a photon with a wavelength of about 1240 nm.

‡ The 'defocused spot' has to be the telescope exit pupil, so that small movements of the star image in the focal-plane aperture don't cause the spot to move. A small lens is used to re-image the pupil onto the photocathode; since this is the image of the uniformly illuminated primary mirror, it is actually a sharply focused, circular spot of light of uniform surface brightness, which doesn't move.

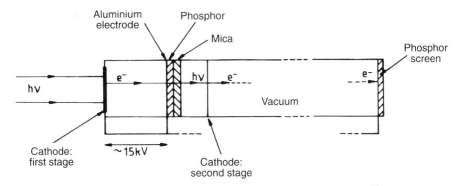

Fig. 3.6. Principles of a multi-stage image intensifier (not to scale). Electrons emitted from the first cathode are accelerated down an evacuated tube by a voltage difference of about 15 kV and focused magnetically onto a phosphor screen. The magnetic focusing coil, which surrounds the tube (see Fig. 3.7), is omitted from the diagram for clarity. The photons emitted from the phosphor strike a second cathode, and the process repeats, with the intensity of the image on the phosphor increasing at each stage. The mica layer acts to absorb any electrons not stopped in the phosphor, but has no significant effect on the emitted photons.

many observations with the aperture centred at successive points of a grid of positions across the object. So for survey work, where astronomers want to look at many hundreds or thousand of stars, and for studying extended objects, photography retains a considerable advantage.

3.3 Two-dimensional electronic recording

Ideally, we would like to combine the areal advantage of photography with the linear response, large dynamic range and high quantum efficiency of photoelectric detection. Since the early 1970s, this has become increasingly possible with the development of a variety of electronic imaging devices which either use a photocathode to produce an intermediate electron image or else record the incident photon image directly. I do not have the space here to discuss them all in detail, so I will just describe some examples that illustrate the range of techniques and how they have developed.

3.3.1 Image intensifiers

In an image intensifier, or image tube, photons are again detected by a photo-cathode, but the emitted electrons are then recorded directly, for example on a phosphor screen. Because there are no intervening dynodes, the image is preserved, as in photography. The electrons are accelerated down an evacuated tube by a voltage of several tens of kilovolts, and the image is produced by electrostatic or magnetic focusing of the electrons onto a recording device at the far end. Electrostatic focusing is simpler and more compact, and is commonly used in commercial and military applications, but magnetic focusing produces a higher-quality image, with much less spherical aberration, and so it is often preferred for astronomical use. If the accelerated electrons are made to strike a phosphor screen, then their high energy ensures that many secondary photons are emitted from the screen for each primary photon incident on the cathode, thereby strongly enhancing the image. The gain can be increased further by placing several intensifiers in series, as shown in Fig. 3.6, which shows the main features of a practical image tube. The quantum efficiency is determined by the properties of the photocathode and is typically 20–30%.

Image tubes can also be used as image converters, in which the cathode and the phosphor target are chosen to have different wavelength responses. This is the principle of 'night-vision' glasses, long used militarily but now also available commercially, in which infrared radiation is detected by the cathode and a visible image is produced on a phosphor screen. Astronomically, this can be used to provide a visible image for guiding an infrared telescope.

Each incident photon initiates a single train of electron and photon inter-actions along the image tube, so the final image on the last phosphor screen has one 'electron event', consisting of the arrival of a bunch of electrons, for each primary photon. This not only guarantees that the response is accurately linear, but means that in some circumstances it is possible to count the arrival of individual photons. I will describe a photon-counting system in the next section.

3.3.2 Television systems

A television picture is made up of lines, produced by an electron beam rapidly scanning across and down the phosphor screen in what is known as a *raster scan*: a two-dimensional image is built up from a one-dimensional time-sequence of electrical signals which determine the intensity of the electron beam.

The converse happens in a television camera, where a two-dimensional image is produced on a target which is then scanned, or 'read', by an electron beam and converted into the one-dimensional time-sequence of signals that is transmitted and used to form the final television picture. This procedure is easily adapted to astronomical purposes, where the signal can be used either to produce a television picture of the sky, for example for guiding

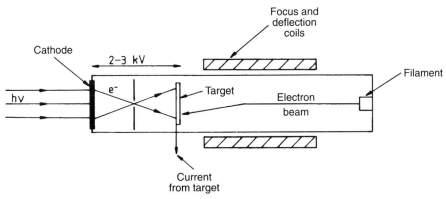

Fig. 3.7. Principles of a simple television camera. The electrons emitted from the photocathode are accelerated through a few kilovolts and then focused onto a target, creating a two-dimensional charge distribution. The target is scanned continuously by an electron beam emitted from a filament at the other end of the tube and controlled by magnetic coils. The output current from the target at any instant depends on the charge at that part of the target being swept by the electron beam and so the time variation of the current forms a one-dimensional record of the charge distribution.

the telescope, or to provide a permanent record. Because the signals form a one-dimensional record, they are easily stored magnetically. An advantage of two-dimensional electronic detectors of this type is that the data are immediately available in digital form; this facilitates both accurate copying and computer processing.

The essential parts of a television camera tube are shown schematically in Fig. 3.7. Photons strike a photocathode, where electrons are released and accelerated through a few kilovolts. When these high-energy electrons strike the target, which is a semi-conductor, they produce electron–hole pairs deep within the target, building up a two-dimensional charge pattern that is an analogue of the photon image on the cathode: the charge density production rate in the pattern is proportional to the photon arrival rate at the corresponding point of the image. When the electron beam scans the target, there is an output current from the target which is modulated by this charge density pattern; the modulation provides a record of the pattern.

Targets are of several kinds. In secondary electron conduction (SEC) tubes, the energetic photoelectrons produce many secondary electrons, which gives a gain of several hundred in the usual target material of potassium chloride. In vidicon tubes, photons strike the photoconductive target directly, so that it acts both to form the image and to store it, without any intermediate photoelectrons. This makes the camera more compact, but has the disadvantage that there is no gain. Common materials for vidicon targets are lead oxide and silicon. The silicon target has a better red response but a much higher dark current, so silicon target vidicons are usually cooled, for example using dry ice at $-78\,°C$.

The lack of target gain in vidicon tubes puts severe demands on the amplifier needed to detect the very small output currents. For this reason an electrostatically focused, single-stage image intensifier is often placed in front of the target, as in the silicon intensified target vidicon (SIT). This is similar to the potassium chloride SEC tube, but has a gain some ten times larger. The SIT tube is therefore better at detecting a wide range of brightnesses in a short time. The potassium chloride SEC tube has an advantage for long integrations on very faint objects, because it has a very low dark current at room temperature and can be run uncooled for several hours.

In all these tubes the spectral response and quantum efficiency are set by the initial photocathode. Unintensified vidicons usually have higher quantum efficiencies and broader spectral responses. The lead oxide vidicon has a quantum efficiency of 50% at 500 nm, but the response falls off sharply at other wavelengths; the silicon vidicon retains a quantum efficiency of 50% over the range 400–800 nm and the fall-off at other wavelengths is more gradual, with a significant response out to 1.1 μm.

A significant limitation of all television-type systems is the finite spatial resolution imposed by the finite number of scan lines and in some cases by the structure of the target. The smallest resolvable element in the final image is known as a picture element or *pixel* and is typically 20–40 μm, some ten times worse than the resolution of a good photographic plate. The major advantage over photography is the accurate linear response over a wide dynamic range (typically a ratio of 1000 to 1 in brightness).

The sensitivity of television camera tubes is often enhanced by adding an image intensifier stage in front of the tube. With the advent of fibre optics (see also Section 3.4), the intensifier is often fully integrated into the camera tube, with fibre-optic coupling between the output phosphor and the input photocathode of the camera. These compact cameras are often used for finding the target for observation with a large telescope and for guiding the telescope during a long exposure.

The addition of an intensifier allows the detection of individual photons. I will complete this section by describing the Image Photon Counting System (IPCS) developed at University College London. This highly successful instrument (Fig. 3.8), first introduced in the early 1970s and still in service 20 years later, has a four-stage magnetically focused image tube, with a gain of about 10^7, coupled by a lens system to a lead oxide vidicon camera.

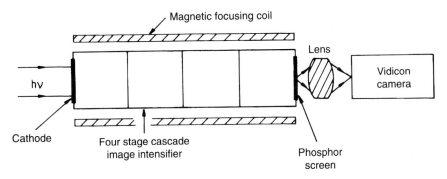

Fig. 3.8. A block diagram of the image photon-counting system (IPCS) developed at University College, London. The incoming signal is amplified by a four-stage cascade image intensifier, similar to the one shown in Fig. 3.6. The enhanced image on the final phosphor is focused by a lens system onto a lead oxide vidicon camera, which has special logic designed to record only photon-induced pulses (see text).

Each incident photon produces a single electron event (a 'splash' of incident electrons) on the output phosphor, of a size that covers several camera scan lines. The camera target is scanned in the usual way, but the special feature of this instrument is the treatment of the target output signal. Special electronic hardware compares consecutive scan lines so that the output from a single photon is not counted more than once. Event-centring logic simultaneously determines the position of the pulse. The size of the pulse is also determined and a pulse-height discriminator is used to cut out both the large pulses from thermal ions and the smaller pulses from thermal electrons. This reduces the background noise very considerably and makes the system very suitable for spectroscopy of faint objects. In fact, because several scan lines must be read before a photon event is properly registered, there is a limit to the photon-arrival rate that can be handled and the system is actually *less* suitable for bright objects.

All photocathodes show inhomogeneities and variations in sensitivity over their surfaces. These variations can be removed by taking an exposure of a uniform light source and dividing all the data frames by this frame, a process known as 'flat-fielding'. Once this has been done, two-dimensional detectors have the advantage for spectroscopy that a spectrum of the night sky can be recorded at the same time as the spectrum of the object being studied, which also of course includes a contribution from the night sky. After flat-fielding, the night-sky spectrum can simply be subtracted from the observed spectrum, leaving the intrinsic spectrum of the source. The possibility of accurate night-sky subtraction has enabled the study of spectra of very faint objects, such as distant quasars (Section 11.5), which may be considerably fainter than the sky (cf. Section 3.5, especially Fig. 3.27).

3.3.3 Silicon arrays

A possible way of extending photoelectric detection to two dimensions is to have an array of photomultipliers. Small clusters of photomultipliers have been used for multi-colour photometry (Section 8.1) or for simultaneous observation of a variable star and its comparison stars.

Silicon arrays are a way of miniaturizing this approach so as to obtain a reasonable spatial resolution. Essentially, an array of photomultipliers is replaced by an array of solid-state detectors embedded in a single silicon chip. The entire array, containing some 10^5–10^6 pixels, occupies an area of only a few square centimetres. The detectors convert the incident photons into a charge distribution over the array, which is then recorded electronically by a variety of techniques. By far the most common array for astronomical purposes is the *charge-coupled device* (CCD), in which the accumulated charge distribution is read out by transferring it along the lines of detectors and recording it as it reaches the ends of the lines.

To understand in a little more detail how a CCD works, I must first return to semi-conductors, introduced in Section 3.2, and describe how photoconductive devices work. When a photon of sufficient energy strikes a semi-conductor crystal lattice an electron is released from the valence band and given enough energy to raise it to the conduction band. In this process a lattice bond is broken and a vacancy, or 'hole', is left behind in the valence band. Because a negative charge has been removed, this hole is effectively positively charged and is free to move around the lattice if an electric field is applied. Thus in reality an electron–hole pair is created. If the pair is to be detected before they recombine, an external electric field must be applied, which sweeps holes and electrons to opposite ends of the semi-conductor. The local increase in conductivity represented by the electron–hole pairs is then detected as an increase in the voltage across a load resistor connected in series with the external voltage and the crystal.

The properties of a semi-conductor can be altered by deliberately adding small amounts of impurities, usually by replacing a lattice atom by an atom of a different valency. This process, known as 'doping', will create either an excess or a deficit of valence electrons. Doped semi-conductors with an excess of electrons (*n*egatively charged) are known as n-type. Those with a deficit can be thought of as having an excess of holes (*p*ositively-charged) and are known as p-type. If a p-type and an n-type semi-conductor are

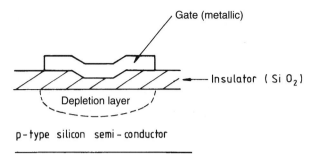

Fig. 3.9. The structure of a metal oxide semi-conductor (MOS) storage capacitor. When the gate is at a positive potential, any electrons freed by incident photons collect in the depletion layer, which behaves like a potential well.

placed in contact, an intrinsic electric field is produced which causes the combination to act as a diode, called a p–n junction diode. Incident photons still release electron–hole pairs, but these are now swept apart without any need for an external voltage, although of course an external circuit is still needed to detect the increased potential across the diode. The diode can be made more sensitive by including a layer of undoped (*intrinsic*) semiconductor between the p and n doped layers, creating what is called a p–i–n diode. The quantum efficiency of a typical silicon p–i–n diode is about 70% over the $0.5\,\mu\mathrm{m}$–$1.1\,\mu\mathrm{m}$ spectral range.

The targets of television cameras are often made of doped photoconductive material; the lead oxide vidicon target is in the form of a p–i–n diode and the silicon target is made into a p–n diode array by diffusing small islands of p-type silicon at regular intervals onto one face of a slice of n-type silicon. The electron read beam closes the external circuit, and the scanning provides a timebase that enables the signals from the individual diodes to be related to their position and so enables the image to be reconstructed.

A CCD detector works in a similar way, the main difference being that the readout does not use an electron beam. However, there are other detailed differences. Instead of the array element being a diode it is a charge storage capacitor, whose basic construction is shown in Fig. 3.9. A metallic electrode, or gate, is separated from a p-type silicon semi-conductor by an insulating layer of silicon dioxide. If the gate is raised to a positive potential, the excess holes in the p-type layer are repelled from the Si–SiO$_2$ junction, forming a depletion layer. Any free electrons present are attracted towards the positive voltage of the junction and collect along the Si–SiO$_2$ interface. Within the depletion layer, the electrons are surrounded by a positive potential and can be thought of as being in a potential well, or reservoir, which fills downwards into the p-type layer from the junction. In a cooled device the only electrons available will be those released by incident photons and so the electron charge stored in the potential well is a direct measure of the number of incident photons. The optical image falling on a CCD is therefore stored as a set of charges in an array of potential wells. To read this stored image, the voltages on the electrodes are pulsed in sequence between high and low levels. This causes the potential well depth to oscillate and each well empties its charge in turn into the next well. In this way an entire charge image can be transferred across the array, with essentially no loss, and registered in external circuitry at the edge of the silicon chip. The small errors introduced by stray capacitance in the readout circuit are known as *readout noise*.

The two main attractions of CCDs as astronomical detectors are their high quantum efficiency (20–40%) and low readout noise (some ten times lower than in electron beam tubes). They also have a large dynamic range (typically 10^4 to 1) and a broad spectral response. Their total area is comparable to that of television devices and they have a similar pixel size. Intensification is not often used, as it reduces both the quantum efficiency and the spectral response.

The low readout noise makes the CCD very suitable for imaging of faint galaxies and the excellent linearity enables accurate sky subtraction. The main problem arises from inhomogeneities in the response caused by manufacturing defects in individual electrodes. Smooth variations can be removed by flat-fielding, but there usually remain some spot defects. Some spurious stellar images or sharp spectral lines may also be produced by stray cosmic rays passing through the detector.

Finally, silicon diodes have such high quantum efficiency that various attempts have been made to use diode arrays, in which each diode has a separate charge-sensitive amplifier. Unfortunately, they have a high readout noise, which makes them unsuitable for observing faint sources. However, their high well capacity (up to 10^7 electrons per pixel) gives them a much larger dynamic range (10^5–10^6 to 1) than is possible even with a CCD. This makes them well suited to obtaining a high signal-to-noise ratio (see Section 3.5) in spectroscopy of bright sources. Examples are the Digicon and Reticon arrays used mainly for spectroscopy.

3.4 Analysers

I have discussed optical detectors extensively because of the dramatic improvements in sensitivity since the 1970s. However, the third component in a telescope system, the analyser, plays an equally important role in determining what information can be obtained from the incoming light. There are four major types of analyser:

(a) photometers, for brightness measurements;
(b) spectrometers, for spectral measurements;
(c) polarimeters, for measuring the degree and angle of polarization;
(d) interferometers, for very-high-resolution angular and spectral measurements.

A photometer simply measures total flux in a well-defined wavelength band, and nowadays usually consists of a photomultiplier or a CCD preceded by a filter (see Section 3.4.3). For measuring many stars, or the surface brightness distribution of galaxies, photographic photometry has the usual areal advantage, but requires careful calibration and is less accurate. I will discuss the accuracy of photoelectric photometry in Section 3.5.

Polarization is not a dominant feature of most optical sources so I will concentrate on discussing spectroscopy and interferometry.

3.4.1 Slit spectrometer

An astronomical slit spectrometer is basically the same as would be found in any physics laboratory. The essential features are shown in Fig. 3.10. An entrance aperture, usually a slit, is placed in the focal plane of the telescope and the diverging beam from the slit is made parallel by a collimating lens. The parallel light is passed through a dispersing element and then the dispersed spectrum is focused onto a detector. In the nineteenth century the dispersing element was a prism, or a train of prisms, and this persisted in long-established observatories well into the twentieth century. Nowadays, the almost universal choice is a transmission or reflection grating, which is much more convenient to use. A grating has the property of producing many spectra, at different angles to the normal and with a different dispersion for each: the central spectrum is undispersed, and the dispersion then increases with the 'order' of the spectrum. Generally, only the first- and second-order spectra are bright enough to be used, but it is possible to concentrate most of the light into a particular order by cutting the ruled grooves on the grating asymmetrically; such a grating is said to be *blazed*. In modern, large telescopes a range of gratings is available, and the dispersion of the spectrograph can be changed in a matter of minutes by replacing one grating by another. The precise wavelength range can be chosen by rotating the grating about its axis to change the angle of reflection.

For high-dispersion spectra, a compact solution is the echelle grating,

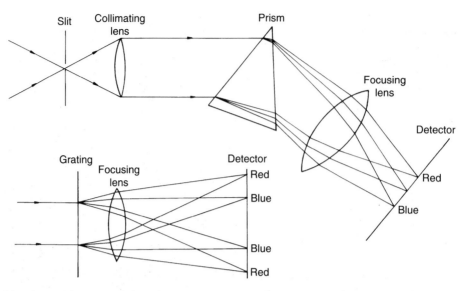

Fig. 3.10. The essential optical components of an astronomical slit spectrometer. The main diagram shows the full light-path from the telescope, the inset showing just the variation when there is a grating rather than a prism in the collimated beam. The slit is placed in the focal plane of the telescope, usually at the Cassegrain focus, so that the image of the object of study is focused onto the slit and as much light as possible passes through. The diverging beam from the slit is made parallel by a collimating lens and is then directed onto a prism or grating. If a prism is used, it is set at minimum deviation for a wavelength near the centre of the spectral range being studied; this minimizes astigmatism. The dispersed light is then brought to a focus again on the detector. As shown in the diagram, the shortest (bluest) wavelengths are dispersed most. For a transmission grating, as shown in the inset, the reverse is true; this is because a longer path difference is needed to produce an interference maximum at longer wavelengths. In practice, reflection gratings are normally used, because they are easier to blaze. However, transmission gratings are sometimes used in a grating–prism combination known as a 'grism': a transmission grating is placed on the front face of the prism and the system is designed so that the deviation of the prism is equal and opposite to the deviation of one of the orders of the grating; the beam has no net deviation, but is still dispersed, as can be seen by imagining a combination of the prism in the diagram with the upper order of the grating, which is deviated in the opposite direction but dispersed in the same direction.

which is blazed to concentrate the light into high orders (typically 10–100). These orders overlap and must be separated by a cross-dispersion grating, yielding a two-dimensional array of short lengths of high-dispersion spectra in adjacent, overlapping wavelength ranges (Fig. 3.11). Because the high dispersion is achieved by going to high order rather than by having a long-focal-length camera, an echelle spectrograph is not particularly large, so

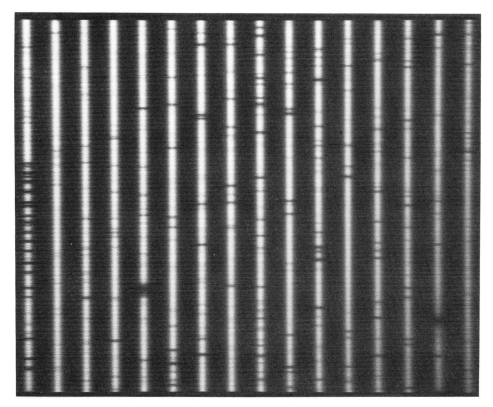

Fig. 3.11. An echelle spectrum of a giant star taken with the Anglo-Australian Telescope. Each vertical strip is part of one order of the spectrum, from order 83 on the far left (central wavelength $\lambda_c = 687.3$ nm) to order 98 on the right ($\lambda_c = 581.8$ nm). Wavelengths decrease from bottom to top along each order, and from left to right with increasing order number. There is no overlap in wavelength because of the small size of the CCD detector. The strong feature in order 87 is the Hα Balmer line of hydrogen, while the many features in order 83 arise from absorption by oxygen molecules in the Earth's atmosphere. (Courtesy Anglo-Australian Observatory.)

high-dispersion spectroscopy can be carried out at the Cassegrain focus of a telescope. The compact design also makes it particularly suitable for use in satellites.

When taking spectra of stars or other point-like sources, the image focused on the slit is small, which makes the spectrum so narrow that it is difficult to pick out features. This was inconvenient for photographic spectra, which used to be examined visually under a microscope. In order to make the spectrum easier to interpret, the image was often trailed backwards and forwards along the slit, by moving the telescope slightly during the exposure,

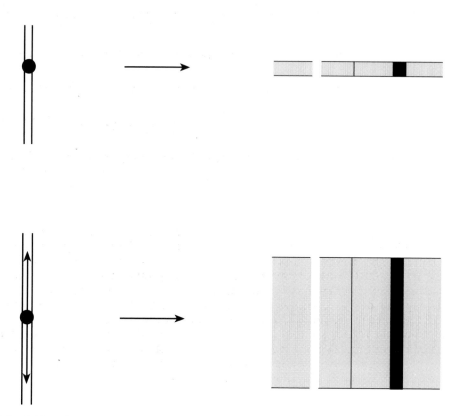

Fig. 3.12. The upper diagram shows a much-magnified sketch of an untrailed spectrum, which is the same width as the stellar image, typically a few arc seconds or less (the size of the seeing disc). Two strong lines are shown, the left-hand one in absorption, the right-hand one in emission. Weak features would be very difficult to detect on such a narrow strip of spectrum. The lower diagram shows the effect of moving the image backwards and forwards along the slit during the exposure, producing a so-called 'trailed' spectrum. The presence of a weak feature can often be more easily verified by seeing whether it is present right across the spectrum, perpendicular to the dispersion. The very narrow emission feature shown between the two strong lines is more easily visible on the lower diagram.

to widen the spectrum (Fig. 3.12). This procedure is rarely† necessary on electronic detectors, where the data are recorded digitally and all the analysis and interpretation is done from the final processed spectrum, which

† In exceptionally good seeing, when the image size is smaller than a pixel, trailing may be used to avoid spurious effects arising from non-uniform sensitivity across the pixel.

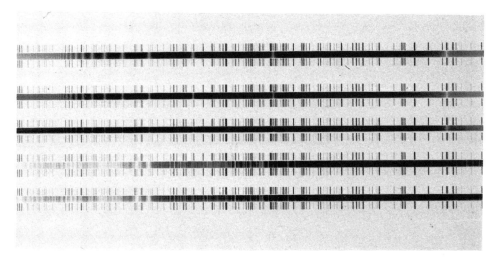

Fig. 3.13. Photographic spectra of five stars, showing in each case the object spectrum in the middle and a comparison arc spectrum on either side; wavelength increases from left to right. The prints are negatives, so the emission lines in the arc spectra appear dark and the absorption lines in the stellar spectra appear as light stripes against the dark continuum. The bottom spectrum is of the giant star Arcturus; the strong pair of lines about a quarter of the way from the left-hand end in the bottom two spectra are the H and K lines of ionized calcium. The top two spectra clearly show the Balmer series of hydrogen, starting with the Hβ line (also visible in the middle spectrum) near the right-hand side and getting closer together towards shorter wavelengths; near the left-hand side they merge into the Balmer absorption continuum (cf. Figs. 8.5, 8.6). (Courtesy D.J. Stickland.)

is produced as a plot of flux against wavelength, rather than from the image on the detector.

So that the wavelengths of the spectral features can be measured accurately, it is also necessary to record an emission-line comparison spectrum. This comparison spectrum comes, for example, from a gas discharge tube of argon or neon or from a copper or iron arc (chosen because they have many lines in the visible), which is mounted close to the spectrometer so that the light can be fed through the same slit. For photography, the comparison spectrum is usually recorded on the same plate as the object spectrum, with one image on each side of the astronomical spectrum (Fig. 3.13) to allow for the slit not being aligned exactly normal to the dispersion. With electronic detectors, the spectrum is recorded separately, because it is sufficient to define an accurate mapping between the known wavelengths of the comparison lines and the pixel positions on the detector. This advantage of having precisely defined pixel positions is particularly useful for the echelle

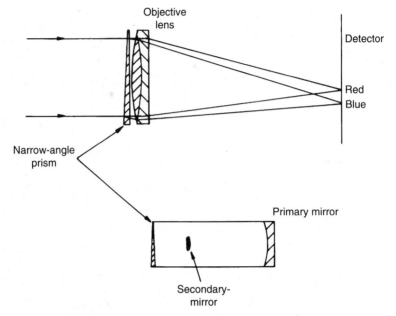

Fig. 3.14. An objective prism spectrometer: a narrow-angle prism is placed in the parallel beam in front of the objective and produces a spectrum directly in the focal plane of the telescope. Because there is no slit, a spectrum is formed of every object in the field of view of the telescope. A low-dispersion transmission grating can also be used. In a reflecting telescope, the dispersing element is placed outside the secondary mirror, as in the lower diagram. The angle of the prism is exaggerated in the drawing; an actual objective prism cannot easily be distinguished by the eye from a circular sheet of plane glass.

spectra mentioned earlier, where it would be very difficult to fit a comparison spectrum onto an already crowded detector.

Spectra of extended objects are not trailed, even on photographic plates, because one wants to see how the spectrum varies across the object, that is, along the slit. In this case, the width of the spectrum corresponds to the angular size of the object along the slit direction and different points across the spectrum correspond to different points in the source. This gives information on the variation of, for example, line-strength and Doppler shift across the source.

3.4.2 Slitless spectrograph

In normal use, a slit spectrograph looks at only one object at a time. If a long slit is used, it may be possible to put a comparison star on the slit at the same time as the target star, but the chance of three or more stars of

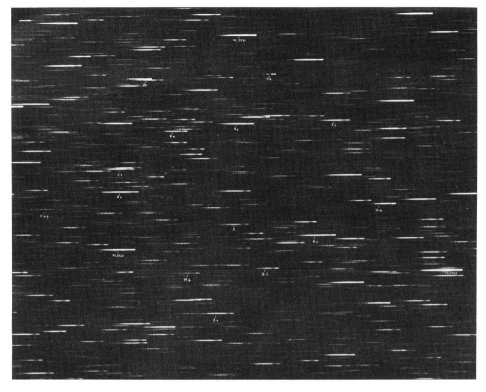

Fig. 3.15. This Hamburg Observatory photograph of stars in the constellation of Cygnus was taken through an 8° objective prism, which has turned every stellar image into a tiny, low-dispersion spectrum. (Royal Astronomical Society.)

interest being lined up along a slit is negligible. If the slit and collimator are omitted and the dispersing element is put in the parallel beam in *front* of the telescope objective (Fig. 3.14) all the images in the field of view of the telescope are dispersed. If a narrow-angle prism or a low-dispersion grating is used, the resulting image is a star field in which each stellar image is replaced by an untrailed, low-dispersion spectrum (Fig. 3.15).

Although it is clearly difficult to obtain detailed information from any one spectrum produced in this way, the technique is very useful for making a rapid, low-dispersion survey of the sky to look for interesting objects that can then be studied in more detail. In particular, it is easy to pick up emission-line objects and objects with unusual continuum spectra, such as planetary nebulae and quasars.

Another use of slitless spectra is to obtain two-dimensional spectral information about single, extended emission-line objects such as hot gas clouds. A slitless spectrum then consists of a number of images of the object at the

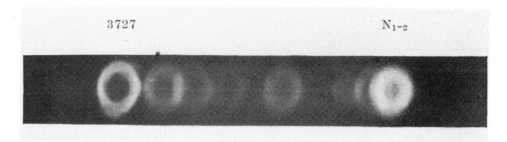

3727   N₁₋₂

Fig. 3.16. A slitless spectrum of an object whose spectrum is almost entirely made up of emission lines, such as this planetary nebula (the Ring Nebula in Lyra), consists of a separate image of the nebula at each emission line wavelength. This reveals at a glance the different spatial distribution of the different ions that correspond to these lines. (Royal Astronomical Society.)

wavelengths of the emission lines, displaced from one another by the dispersion (Fig. 3.16). It is then possible to see at a glance the different extents of the emission from different chemical elements or different ionization states. This technique is only useful for objects with a small number of spectral features, or there is an unacceptable overlapping of images, a problem which also occurs with the spectra of star fields.

Comparison spectra cannot be used with slitless spectrographs, so they are not easy to use for quantitative work. Some of the advantages of slit and slitless spectroscopy can, however, now be combined by making use of fibre optics. A fibre-optic tube is a very narrow, flexible, solid glass fibre whose diameter is generally between 50 and 400 μm, depending on the image scale of the telescope. Light shone in at one end of the fibre propagates along the tube by total internal reflection off the glass–air interface, so that essentially no light escapes through the walls (in practice, a reflective coating is usually put on the outside of the tube) and the light-path can be bent round corners so long as they are not too sharp.

The flexibility of the fibres means that they can be used to image a two-dimensional star field onto a one-dimensional slit (Fig. 3.17). Originally, a metal plate was placed in the focal plane of the telescope to act as a mask. Holes were drilled in this mask at the exact positions of the stars of interest and then an optical fibre was attached to each hole. At their other ends, the optical fibres were lined up along a conventional slit and the light fed through the spectrograph in the usual way. The resultant image consisted of a set of parallel simultaneous spectra of all the stars, together with comparison spectra. More than a hundred stars or galaxies can now be measured at one time, which means that great care has to be taken to

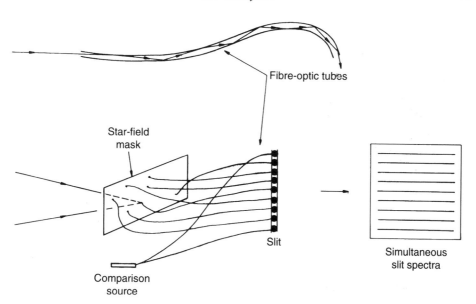

Fig. 3.17. Fibre-optic spectroscopy. The upper diagram shows the light-path through a typical fibre-optic tube, which is some $50\,\mu$m in diameter. A mask is placed in the focal plane of the telescope, with holes at the positions of the stars of interest. The light from these stars is led along fibre-optic tubes to a conventional slit, so that the detector at the end of the spectrograph records simultaneously slit spectra of all the stars in the star field mask. The spectrum of a comparison arc can easily be added on either side of the set of star spectra.

identify correctly which star is where in the sequence on the slit! (An early instrument was known as Medusa, after the snake-haired Gorgon of Greek mythology.) The main difficulties with the technique are making the holes in the mask in precisely the right places and obtaining a good interface between the fibres and the mask. It is also very inconvenient having to construct a new mask for every star field. More recent versions have circumvented this difficulty by having a support frame which enables each fibre to be moved independently to a position in the focal plane corresponding to a target of interest. With computer control of the positioning of the fibres, a set-up for a new field can now be done during an observing run, whereas previously the masks had to be drilled well ahead of time.

3.4.3 *Interference filters*

Astronomers often want to measure the brightness of a star or galaxy in a restricted waveband, for example to determine the colour by comparing observations in different wavebands (see Chapter 8). If the bandpass required

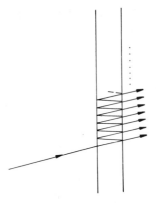

Fig. 3.18. The principle of an interference filter. Light incident at an angle on a glass plate with accurately parallel faces suffers multiple internal reflections before emerging as a set of parallel rays of decreasing intensity, all at the same angle to the plate. These rays will interfere with one another, giving maximum light if the path difference between successive rays is an exact number of wavelengths. In practical use, the angle of incidence is usually 90°('normal' incidence – see Fig. 3.19).

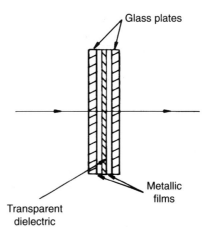

Fig. 3.19. A practical interference filter, used at normal incidence. The glass plates act merely as support for the transparent dielectric, within which the internal reflection occurs. The highly reflective metal films ensure that successive rays are not too different in brightness. Internal reflection in the glass plates is small and has a negligible effect.

is reasonably large (50–100 nm, say), coloured glass or dyed gelatin filters are adequate and have the advantage of being inexpensive. For narrow bandpasses, less than about 5 nm, say, a better solution is to use interference.

The principle of an interference filter is to use multiple reflections between two plane parallel surfaces (Fig. 3.18). Part of the incident ray is transmitted and part is internally reflected twice and then part emerges parallel to the incident ray and part is internally reflected again. Thus a single incident ray gives rise to a set of parallel and successively fainter rays. If the path differences between successive rays are integral multiples of a wavelength, the rays will interfere constructively to produce a bright beam; otherwise they will tend to cancel each other out and give darkness. A filter is usually placed so that the light is at normal incidence and then the calculation is very simple. Fig. 3.19 shows the construction of a typical filter. The glass plates play no optical role, merely acting as mechanical support for the filter. The surface of one plate is coated in turn with a thin evaporated film of high reflectivity, a thin layer of a transparent dielectric and another metal film, the other plate then being added for protection. If the dielectric layer has refractive index n, and is of thickness d, the optical path difference between successive rays at normal incidence is just $2nd$. Maxima then occur for:

$$2nd = m\lambda, \ m = 1, 2 \dots, \tag{3.3}$$

that is, at wavelengths

$$\lambda_m = \frac{2nd}{m}. \tag{3.4}$$

These maxima are separated by the amount

$$\Delta\lambda_m = \lambda_m - \lambda_{m+1} = \frac{2nd}{m(m+1)}, \tag{3.5}$$

showing that they get closer together for large m (Fig. 3.20).

Interference filters are normally used for looking at a particular spectral feature, so the properties of the filter need to be chosen to ensure that the maximum transmission is at the wavelength of interest, for example the Hβ line at $\lambda_m = 486$ nm. We also need the adjacent maxima $\lambda_{m\pm1}$ to be well separated from λ_m; this means that m must be small and thus, since $n \simeq 1$, we must have d not more than a few wavelengths. It is because of this need for a very thin dielectric layer that the films need to be laid down by evaporation in a vacuum chamber. The widths of the maxima are determined by the reflectivity of the metal films: the more reflective the films, the narrower are the maxima and the less transmission there is between maxima.

If there are two maxima within the spectral range of the detector, there is a danger of contamination of the signal by a leak through a neighbouring maximum; in that case, the unwanted maximum can easily be cut out by using a suitably coloured broad-band glass filter as the protective glass plate.

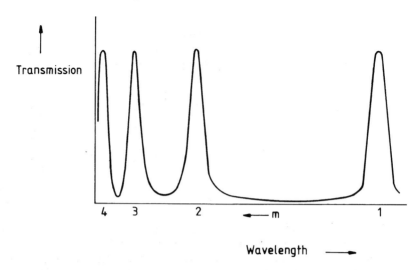

Fig. 3.20. Transmission of an interference filter as a function of wavelength. Higher-order maxima (larger *m*) occur at shorter wavelength and are closer together.

3.4.4 *Michelson stellar interferometer*

Another, more restricted, application of interferometry is to the measurement of the angular separations of close double stars and the angular diameters of single stars. The simplest example, and the only one that I will discuss in any detail for optical telescopes, is the *amplitude* interferometer designed by Michelson in about 1920. Other more complicated designs use the *intensities* of the interfering beams rather than their amplitudes; I will discuss them briefly in the next section.

Michelson's stellar interferometer should not be confused with the Michelson interferometer commonly found in physics laboratories, which works on a rather different principle. The stellar interferometer works on the same principle as Young's double-slit experiment (Fig. 3.21). In the laboratory, light from a point source is collimated and the parallel light falls on a screen with two parallel slits separated by a distance *d*. Each slit acts as a line source and a cylindrical wavefront is emitted from each slit. These wavefronts interfere with one another and the resulting interference pattern is focused onto a screen or other detector. Ignoring the finite widths of the slits, the path difference between two rays travelling at angle θ to the axis is $d \sin \theta$. When this path difference is an integral number of wavelengths, the amplitudes add, giving bright maxima. Thus:

$$d \sin \theta = m\lambda \text{ for maxima, } m = 0, 1, 2 \ldots \quad (3.6)$$

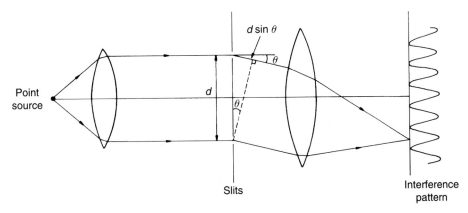

Fig. 3.21. Young's double-slit experiment. Collimated light from a point source passes through two parallel slits separated by a distance d. If parallel rays from the two slits, at an angle θ to the axis, have a path difference $d\sin\theta$ equal to an exact number of wavelengths then they add to produce a maximum in the interference pattern (equation (3.6)). The centre of the pattern is a maximum.

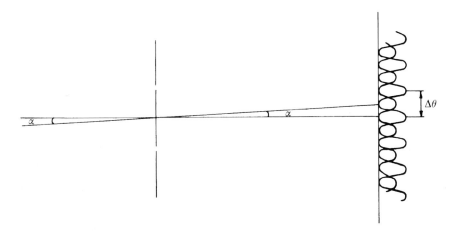

Fig. 3.22. Fringe patterns produced by two independent point sources separated by an angle α. In this diagram, the off-axis source is fainter and so produces a weaker set of maxima. The angle α has been drawn to be equal to $\frac{1}{2}\Delta\theta$ so that the central maximum of the pattern from the weaker source falls exactly on the first minimum of the pattern from the stronger source, giving minimum visibility of the overall pattern (equation (3.7)).

and the pattern has a central bright fringe with further maxima at a spacing $\Delta\theta \simeq \lambda/d$ for small θ.

If there are *two* point sources, separated by a small angle α, each produces an independent fringe pattern but the central maxima are displaced from

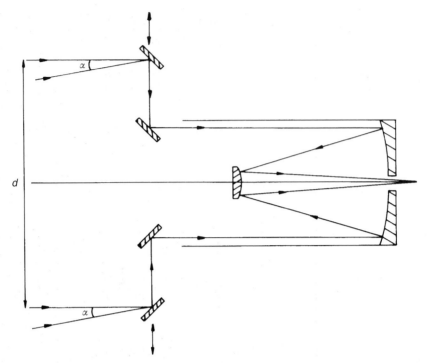

Fig. 3.23. The optical components of the Michelson stellar interferometer. Light rays from two sources separated by an angle α are directed into a telescope via four small plane mirrors. The outer two can be moved apart symmetrically, giving a variable separation *d*. The telescope simply acts to bring the light to a focus.

one another by the angle α (Fig. 3.22). The two fringe patterns will tend to cancel out, causing the visibility of the fringes to decrease with increasing α until the central maximum of one pattern falls exactly on the first minimum of the other. The visibility of the fringes is then a minimum and will be zero if the sources are the same brightness, that is, the fringes will disappear completely. The condition for minimum visibility is that

$$\alpha = \frac{1}{2}\Delta\theta = \frac{1}{2}\frac{\lambda}{d}. \tag{3.7}$$

Note that the light from the two sources is not coherent, so the cancellation of the fringe patterns is *not* an interference phenomenon; there are simply two overlapping patterns of light and dark.

Michelson used this principle to measure separations α of double stars by varying *d* until the fringe pattern reached its minimum visibility, α then being given by equation (3.7). He increased the effective resolution of the 2.5 m Hooker reflector at Mount Wilson considerably by placing in front of it a

Fig. 3.24. The Michelson stellar interferometer, mounted on the 2.5 m Hooker reflector. The two outer mirrors could be moved apart along the girder. (Royal Astronomical Society.)

long girder supporting four small plane mirrors (Fig. 3.23 and Fig. 3.24). The two outer mirrors could be moved apart symmetrically and acted effectively as slits of a variable separation d which could be several times the diameter of the telescope mirror.

The same instrument was used to measure angular diameters of single stars. In this case, there is a set of independent interference fringes arising from each point across the diameter of the star and the fringes vanish when the patterns from opposite edges of the star just coincide (Fig. 3.25), that is, when the diameter corresponds to λ/d; the whole separation between the maxima is then filled in uniformly. Another difference from the case of two point sources is that we must allow for the fact that the source is circular. This introduces the same factor as for diffraction at a circular aperture (Section 2.3 – the Airy disc) and we have, finally:

$$angular\ diameter\ = 1.22\frac{\lambda}{d}, \tag{3.8}$$

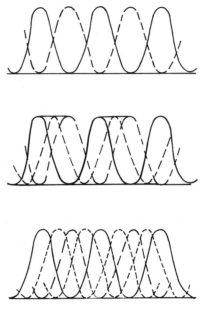

Fig. 3.25. The top diagram shows the superposition of the interference patterns from two point sources of equal brightness. When the maxima of one pattern lie exactly on the minima of the other then, by symmetry, the visibility of the overall pattern is a minimum. Suppose now that the two point sources are the opposite ends of the diameter of a star, so that every point on the diameter between them also acts as a point source, giving its own interference pattern (for simplicity, in this diagram we ignore contributions from points not on the diameter – the discussion here is really for a one-dimensional star!). In this case, the space between the two maxima is completely filled by the maxima of these new patterns, but only on one side; this is indicated by the horizontal bars in the middle diagram, which are a mapping into the interference pattern of the diameter of the star. The outer curves represent the patterns from the two ends of the diameter and the middle curve the pattern from the mid-point of the diameter. The overall pattern does not now have its minimum possible visibility, which will instead occur when the whole space between the maxima of one pattern is filled in by the maxima of the patterns arising along the diameter. This will happen when the maxima from the points at the opposite ends of the diameter just coincide, instead of being interleaved. This situation is shown schematically in the bottom diagram, where the solid line represents the (coincident) patterns from the ends of the diameter, the long-dashed line represents the pattern from the centre of the star's disc, and the short-dashed lines represent the patterns from points one-quarter and three-quarters of the way along the diameter.

where d is the separation at which fringes disappear. The limiting resolution is about 0.01 arcsec, so Michelson's instrument could only be used for a few nearby giant stars. A new version of this instrument, with much higher resolution, is currently being developed in Cambridge (COAST –

the Cambridge Optical Aperture Synthesis Telescope), based on experience gained in radio astronomy (Chapter 4).

3.4.5 *Intensity and speckle interferometry*

When a Michelson stellar interferometer is used to measure the separation of double stars, the light from the two stars is not coherent (neither is the light from different parts of a single stellar disc) and the superposition of the fringe patterns is not an interference phenomenon. However, each separate fringe pattern is formed by interference and it is important that the two path-lengths for the light from a *given* source (one star in a double-star system, or one point on a stellar disc) do not differ by more than the coherence length, l, defined by

$$l = c/\Delta v = \lambda^2/\Delta\lambda,$$

where Δv and $\Delta\lambda$ are the frequency and wavelength bandwidths of the radiation. At optical wavelengths, $\lambda \simeq 500$ nm and a typical bandwidth of 1 nm gives a coherence length $l \simeq 0.25$ mm. Thus, unless the light is made very highly monochromatic (in which case the stars would be undetectably faint), it is important to move the outer mirrors in the Michelson apparatus (Fig. 3.23) with very great precision. For this reason, it proved extremely difficult to construct a Michelson stellar interferometer any larger than Michelson's original instrument.

An ingenious solution to this problem, the *intensity interferometer*, was invented in 1949 by R. Hanbury Brown. Originally used as a radio in-terferometer, the instrument was also developed by Hanbury Brown and R.Q. Twiss for optical use and has had its main success in measuring stellar diameters. As the name suggests, the instrument depends on interference but measures correlations between the *intensities* of signals from two receivers rather than relying on amplitude interference. There are two advantages of this technique, in which the measured signal is the product of the intensities recorded at the two receivers. One advantage is that any noise components are uncorrelated and cancel out (see discussion following equation (4.21)). However, the main reason for the success of the intensity interferometer is that the effective coherence length of the radiation is much longer, which means that equal path-lengths from the two receivers need not be main-tained so accurately. The longer coherence length arises as follows. Because the measured signal at each receiver involves the *square* of the sum of the amplitudes from the two sources, it depends not only on the frequencies of the light from the two sources but also on the sum and difference of these frequencies. A filter is applied so that the instrument only records the low-frequency beat signal corresponding to the frequency difference of the light from the two sources. A typical frequency difference of 100 MHz corresponds to a coherence length of 3 m, allowing path differences of several tens of centimetres without affecting the working of the instrument.

The first optical intensity interferometer was built by Hanbury Brown at Narrabri in Australia in the 1960s. It consisted of two 6.5 m concave reflectors mounted on a circular railway track of diameter 188 m. The reflectors consisted of a paraboloidal mosaic of small hexagonal mirrors. They were not accurately figured, because they acted only to collect light and direct it to photomultipliers mounted at the prime foci; the system was therefore relatively cheap. The large baseline enabled an angular resolution of 0.0005 arcsec to be achieved, but the dependence on the beat signal required the light to be nearly monochromatic and the light losses in the narrow-band filter made the system rather insensitive, despite the large collecting area. It was therefore only used to measure angular diameters for 32 of the very brightest stars, and also to measure some binary separations.

More recently, a quite different way of achieving high angular resolution in the optical waveband has been developed. Known as speckle interferometry, it uses very high time resolution to overcome the blurring effects of scintillation (Section 1.4) and achieve diffraction-limited resolution from ground-based telescopes. An exposure of a few milliseconds freezes the atmospheric motion and might be expected to produce a single image. However, the atmosphere varies spatially as well as temporally and at any instant several hundred turbulent cells lie between the source and the telescope. Each cell produces a separate image of the source at a slightly different point in the focal plane, and the instantaneous picture is a distribution of small spots, over an area roughly equal to the seeing disc, with a concentration of spots towards the centre. The distribution of spots looks like freckles, or like the speckles on an egg, and is known as a *speckle pattern*.

Because there are many turbulent cells, producing a random distribution of phase delays, at least some pairs of cells have similar phase delays and the light from these cells interferes to produce, not a simple image of the source but an interference pattern. Thus each spot in the speckle pattern is actually an intensity peak in a complex interference pattern, the smallest speckles being roughly the size of the Airy disc corresponding to the full aperture of the telescope. By Fourier analysing the speckle pattern it is possible to reconstruct the structure of the original source.

The need for very short time exposures requires the use of large telescopes and high gain detectors, and the method is currently restricted to fairly bright sources. However, the limiting magnitude of +9 or +10 is already several magnitudes fainter than was possible with the intensity interferometer and many more sources can be studied.

The speckle technique has the additional advantage that it can be used with existing telescopes, although this does mean that the resolution is lower (about 0.02 arcsec for the 5 m Hale telescope on Mount Palomar) because no single telescope has an aperture anywhere near the 188 m separation of the receivers in the intensity interferometer.

3.4.6 Area spectroscopy

With the development of fast, powerful mini-computers which can be attached directly to instruments on the telescope, many observations are now possible in ways that could not be achieved before. An example is the gathering of spectral and image information simultaneously, which generates a three-dimensional 'data cube' of intensities in two space dimensions and one wavelength dimension. I will briefly describe two systems developed for use with the 3.9 m Anglo-Australian Telescope.

The first of these is Taurus (so-called because of the bull-like shape of the instrument), which is an imaging Fabry–Perot interferometer. The data cube is generated by imaging the sky directly at a definite wavelength and then repeatedly scanning in wavelength through a defined region of the spectrum and adding the intensities from successive scans (to average out variations in the sky transparency) until a sufficiently strong signal has been recorded. The wavelength is chosen in a way analogous to that used in the interference filter described earlier: two accurately parallel glass plates, silvered on their inner surfaces, are held apart and multiple reflections in this air gap cause interference that allows transmission only at a definite set of wavelengths. For light incident at angle θ to the normal to this 'Fabry–Perot etalon', the wavelengths are given by

$$2d \cos \theta = m\lambda \qquad (3.9)$$

(cf. equation (3.3)). The scanning in wavelength is done by gradually altering the spacing d of the plates of the etalon (about 100 discrete steps in wavelength are normally taken).

The other system, known as ASPECT (for *area spect*roscopy), uses a conventional long-slit spectrograph, with an IPCS detector, and at any instant records spectral information along one spatial dimension. The third dimension of the data cube is in this case built up by repeatedly scanning the slit across the sky in a direction normal to the slit. Intensities from successive scans are again added up in the computer until a sufficiently strong signal has been recorded.

In both these cases, as in many other examples in modern astronomy, computer control is essential to enable fast and accurately phased recording and handling of a very large quantity of data (a typical observing run may accumulate up to several gigabytes of data).

3.5 Photometric accuracy

No physical observation is exact and it is always important to know what the uncertainties are. I will illustrate this by considering what limits the accuracy with which astronomers can measure the brightness of an astronomical source.

The two main sources of uncertainty in measuring the brightness of a source are the *sky brightness* and *noise*. It is the noise that ultimately limits the accuracy, but the sky background is a serious problem for faint sources and must be carefully measured.

3.5.1 Sky background

For many reasons, the night sky is never completely dark, nor is it uniformly bright. The main contributors to sky brightness are:

1. *Scattered moonlight and sunlight* :

   • When the Moon is full, the night sky is very bright and accurate photometry is essentially impossible. Observations of faint sources are best made near New Moon, when the Moon is absent from the sky at night. This is a major consideration when scheduling observing time on large telescopes, and three kinds of night are available:

     dark time: no Moon (about one week near New Moon);
     grey time: Moon absent for half the night (second half near First Quarter, first half near Third Quarter);
     bright time: Moon visible nearly all night (about one week near Full Moon).

   • The Sun also makes a contribution at night, because of sunlight scattered round the edge of the Earth, especially during twilight; at latitudes more than about 48° N or S there is 'astronomical twilight' all night at midsummer (exercise 7.5).

- In the daytime, of course, the Sun makes optical observations impossible. This is not true at other wavelengths, at which there is little radiation from the Sun; radio astronomers can work 24 hours per day and some infrared observations can be obtained during the day.

2. *Airglow*: The upper atmosphere of the Earth is continually being bombarded by hard radiation or energetic particles, mainly from the Sun. The particles are diverted by the Earth's magnetic field so that they reach the night-side as well as the day-side of the Earth. This bombardment excites and ionizes atoms and causes them to emit radiation in particular spectral lines. These night-sky emission lines are easily recognizable in spectra, but not in photometry. Auroral emission makes astronomical observations almost impossible near the magnetic poles of the Earth.

3. *Light pollution* : There is an increasing problem from man-made light sources, as is well known to amateur astronomers trying to observe from sites near towns and cities.

   - *Large cities* are a major source of light at night, mainly from mercury and sodium street lights, which are rarely shielded and direct nearly as much light upwards as onto the roads. Scattering of light in the atmosphere can affect observatories up to 150 km away.
   - Other man-made problems include *artificial satellites*. These appear as bright sources moving across the field of view. There are too many to avoid (or even predict) and most have unknown (and variable) brightness. Near major air routes, aircraft lights can be another hazard.

The spread of these and other sources of light (gas flares from oil fields are the brightest terrestrial light sources seen from Earth satellites) is a major problem for large observatories, especially long-established ones such as those in California. Modern observatories are sited as far as possible from sources of light pollution, but cannot avoid them entirely. Several national astronomical societies, especially the American Astronomical Society, are trying to combat this problem by a process of public education. In Tucson, Arizona, which is only about 20 miles from the Kitt Peak National Observatory and several other major observatories, considerable success has been achieved by demonstrating to lighting engineers and the local authorities that it is actually cheaper to design street lighting that is shielded, so that all the light is directed downwards where it is needed and none of the light is wasted by lighting up the sky.

These sources are a problem because we are observing through an atmosphere, which scatters the light over the whole sky. Above the atmosphere, sources such as the Sun and Moon affect only their immediate surroundings and are far less of a nuisance. However, there still remain three sources of background light that are just as serious for satellite observations as for ground-based ones:

4. *Zodiacal light*: This is sunlight which has been scattered by dust particles in the plane of the solar system (fine debris from comets, etc.). It forms a band around the *ecliptic* (see Chapter 7), which is marked by the constellations of the zodiac.
5. On a larger scale, starlight is scattered by interstellar dust, forming a diffuse background over the whole sky. The scattering is stronger for blue light than for red, so this effect produces a wavelength-dependent background, and also makes distant stars appear redder than they really are: 'interstellar reddening'.
6. There is, finally, a general diffuse background that comes mainly from faint, unresolved stars and galaxies.

This list of six items refers principally to (ground-based) optical telescopes, although infrared and ultraviolet observations are also affected. Ground-based radio telescopes are less affected by natural background, but there are two terrestrial sources of importance. Firstly, radio observations are affected by variations in atmospheric transmission, in the same way as are optical observations (Chapter 1), although radio astronomers can observe through cloud (with some loss of sensitivity). (It is also worth noting here that the ground itself is potentially a source of background radiation: the ground has a temperature of about 300 K and radiates roughly like a black body with a peak wavelength in the infrared (at 10 μm – see Section 5.1). Because the Planck spectrum drops sharply towards short wavelengths, but has a slow decline (the Rayleigh–Jeans tail) towards long wavelengths, the ground is a negligible source in the optical but is still a significant emitter at radio wavelengths. However, good antenna design (Section 4.1) usually reduces this background to an unimportant level.)

Secondly, there is a man-made problem that is, if anything, more serious than light pollution: TV and radio transmissions are creeping into wavebands formerly reserved for radio astronomy, despite strenuous representations at international level by radio astronomers. This is an increasing problem, for

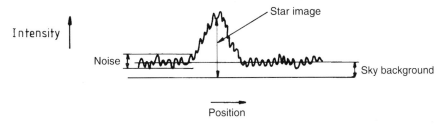

Fig. 3.26. A schematic tracing of the intensity of a photographic image of a star. The star image stands clearly above the sky background but both are blurred by random fluctuations, or noise.

several reasons. The steady increase of TV broadcasting demands wider bandwidths and more channels, and is spreading to satellite broadcasting, from which it is difficult to hide a radio telescope. Satellites are increasingly used for telecommunications as well as broadcasting, and computer links are also demanding ever-increasing bandwidths. On the astronomical side, the growth of observations over a large range of redshift (radio galaxies and QSOs – see Section 11.5) requires broader bands to be available just as the available bands are shrinking.

All these sources of radiation contribute to the measured signal from an astronomical source and therefore contaminate the signal we want to detect, say from a star or a faint galaxy. In the visual, the total surface brightness of the night sky (see Section 11.2.1 for a fuller discussion of surface brightness) is typically about $21 \, \mathrm{mag} \, (\mathrm{arcsec})^{-2}$ (in the absence of the Moon; the Full Moon makes it eight or nine magnitudes brighter). If the sky brightness could be measured *exactly*, it could simply be subtracted off, leaving the signal we want. But no measurement is ever exact; there are always errors, both systematic and random. Systematic errors are particularly difficult to allow for; they are usually specific to a particular measurement and are not easy to discuss in general. Here I will discuss only random errors, known collectively as noise.

3.5.2 *Noise*

Noise may arise in the instrument, for example from thermal motions in the detector, it may be produced in the Earth's atmosphere by random fluctuations in density (see Section 1.5) or it may be intrinsic to the source: because light arrives as discrete photon events, there are always small fluctuations in the arrival rate ('photon noise') which are most noticeable for faint sources. Fig. 3.26 shows the appearance of noise on a tracing of a photographic

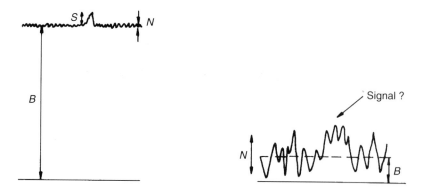

Fig. 3.27. Two extreme examples of noise. In the left-hand diagram, the signal is very weak compared to the background, but is easily detected because the signal-to-noise ratio is large: $S \ll B$ but $S/N \gg 1$. In the right-hand diagram, the signal is comparable in intensity to the background, but its very existence is in doubt because the signal-to-noise ratio is of order one: $S \simeq B$ but $S/N \simeq 1$.

plate. If S is the intensity of the signal above the background and N is the amplitude of the noise intensity, then, roughly, we need $S/N > 1$ for the signal to be detected above the noise. So the crucial quantity that limits accuracy and determines the faintest object that can be detected is not the strength of the signal relative to the background, as one might at first think, but the strength relative to the noise: the *signal-to-noise ratio*, S/N. This is illustrated in Fig. 3.27, which shows that it is perfectly possible in principle to detect a source that is considerably weaker than the background. In practice, detection of a signal with any confidence requires $S/N \simeq 3$ (a so-called '3-sigma detection' – see below).

It is possible to quantify this discussion and estimate how the signal-to-noise ratio depends on the exposure time. To illustrate this, I will just consider random noise in a single-channel, photon-counting detector, although similar considerations apply to photography and to electronic imaging devices. I will neglect instrumental noise, such as dark current and readout noise; for a good detector, these will make a small contribution except for very faint sources.

For *random* noise, in a signal of n counts the noise contributes a mean error σ_n where

$$\sigma_n \approx n^{\frac{1}{2}} \tag{3.10}$$

(this assumes that the photon flux follows a Poisson distribution). Suppose that we use a photomultiplier tube to measure the total number of counts received per second through a circular aperture (see Fig. 3.28) of a size large

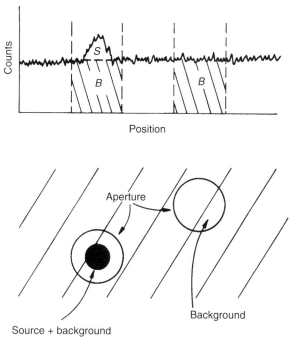

Fig. 3.28. The measured signal always includes the background. The vertical dashed lines in the upper diagram, and the circles in the lower diagram, represent the aperture through which the measurements are made (see text).

enough to include all of the source (for a point source, the aperture would need to be a factor of a few larger than the seeing disc). When the aperture is centred on the source, the measured signal includes not only the source but also the background. If

S = total photon count rate for source,
B = total photon count rate for background,

then the measured signal through the aperture is M, where:

$$M = S + B. \tag{3.11}$$

To find the background, it is necessary to measure the total counts per second through the *same* aperture on a blank piece of sky close to the source. The aperture should not be too large, or there will be very little difference between the signals measured on the source and on the sky: the total background counts depend on the size of the aperture, but the total source counts do not (for apertures larger than the image of the source).

In practice, observations are made by integrating for a time t, spending a fraction f observing the source ('on-source') and a fraction $1 - f$ observing the background ('off-source'). Then:

$$total\ counts\ in\ time\ t\ = Mft + B(1 - f)t. \tag{3.12}$$

We want to know four things:

1. What is S?
2. What is the uncertainty (error) in S?
3. How does this error depend on t?
4. What value of f minimizes the error in S?

In unit time in each mode (on- and off-source) we accumulate M and B counts respectively, with errors:

$$\sigma_M \approx M^{\frac{1}{2}},\ \sigma_B \approx B^{\frac{1}{2}}. \tag{3.13}$$

Integrating over longer times improves the accuracy. For example, in time ft we accumulate Mft counts, with error $(Mft)^{\frac{1}{2}}$. Then our estimate of M is given by

$$estimate\ of\ M = \frac{1}{ft}\left(Mft \pm (Mft)^{\frac{1}{2}}\right) = M \pm \left(\frac{M}{ft}\right)^{\frac{1}{2}} \tag{3.14}$$

and so the uncertainty in M is reduced by the square root of the exposure time: to halve the error, we need to integrate for four times as long. Thus in observing time t we have:

$$\sigma_{M,t} \approx \left(\frac{M}{ft}\right)^{\frac{1}{2}};\ \sigma_{B,t} \approx \left(\frac{B}{(1 - f)t}\right)^{\frac{1}{2}}. \tag{3.15}$$

Now $S = M - B$, so the error in S is given by

$$\sigma_S^2 = \sigma_M^2 + \sigma_B^2 \tag{3.16}$$

(this again assumes a Poisson distribution of photons) or, after time t,

$$\sigma_{S,t}^2 = \sigma_{M,t}^2 + \sigma_{B,t}^2 = \frac{M}{ft} + \frac{B}{(1 - f)t}. \tag{3.17}$$

We can therefore minimize the error in S for a given t by choosing f such that $\partial\sigma^2/\partial f = 0$. Now:

$$\frac{\partial\sigma_{S,t}^2}{\partial f} = -\frac{M}{f^2 t} + \frac{B}{(1 - f)^2 t} \tag{3.18}$$

and σ^2 is minimized if

$$\frac{1 - f}{f} = \left(\frac{B}{M}\right)^{\frac{1}{2}}. \tag{3.19}$$

In the limits of very strong and very weak signals, this result corresponds to what common sense would tell us. For a *very strong* signal, $B \ll M$ and $f \approx 1$. Thus we can spend most of the time measuring $M(\approx S)$ because the background correction is small and need not be determined accurately. For a *very weak* signal, $M \approx B$ and $f \approx \frac{1}{2}$. In this case we need to spend as much time observing the background as the source, since $S = M - B \approx 0$ and the subtraction needs to be very accurate.

The quantity σ is what I have previously denoted by N, so the signal-to-noise ratio is given by:

$$\left(\frac{S}{N}\right)^2 \equiv \left(\frac{S}{\sigma_{S,t}}\right)^2 = \frac{(M-B)^2}{M/ft + B/(1-f)t} = \frac{(M-B)^2 ft}{M + fB/(1-f)}. \tag{3.20}$$

We then see that for the optimum f given by equation (3.19) we can work out the total integration time needed to attain a *given* S/N: that is the ultimate aim, since we need $S/N > 3$, say, to be confident of detecting a source.

Solving equation (3.19) for f and substituting into equation (3.20) gives

$$t = \frac{M(1 + (B/M)^{1/2})^2}{(M-B)^2} \left(\frac{S}{N}\right)^2. \tag{3.21}$$

This shows that for a given source the time taken to reach a given S/N increases with the *square* of S/N, so it is normal practice (where possible) to monitor the signal as the photon count builds up and to stop integrating at a smallish S/N (about 5, say). For very strong signals, $B \ll M$ and the time to reach a given S/N is just proportional to $1/M$. However, for very weak signals, $S(= M - B) \ll M$ and $M \approx B \approx$ constant, giving $t \propto B/S^2$ for a given S/N. It is therefore very time-consuming to search for weak signals amongst the noise, especially in the presence of a bright background.

The idea of signal-to-noise ratio is fundamental to all astronomical observations and the calculation I have just performed is typical of many others. We can now use the concept of signal-to-noise ratio to make a different definition of what is meant by the quantum efficiency of a detector. Our previous definition refers simply to the response of the detector: what fraction of the photons cause a response? The technical term is the Responsive Quantum Efficiency (RQE). However, detectors degrade the S/N of the incident light in *two* ways: they fail to use all the incident photons and they also introduce extra noise. The most useful figure of merit for a detector is then the Detective Quantum Efficiency (DQE), which can be expressed as:

$$DQE = \left[\frac{(S/N)_{\text{out}}}{(S/N)_{\text{in}}}\right]^2 < 1 \qquad (3.22)$$

where $(S/N)_{\text{out}}$ is the signal-to-noise ratio at the output of the detector and $(S/N)_{\text{in}}$ is the value for the light incident on the detector. The larger the DQE, the better the detector. The square of S/N appears because (from equation (3.21)) the measured signal in a given time is (for strong signals) proportional to $(S/N)^2$. The RQE of a detector is defined strictly as the number of photons recorded by the detector divided by the number that would have been recorded by a perfect detector; the DQE allows for any loss of detected photons and for the different RQE of different stages of a multi-stage detector, and is therefore less than the RQE for a given detector. However, in comparing two detectors it is possible for one to be slower at detecting photons, thus having a lower RQE, but to degrade the signal less and thus to have a higher DQE.

3.6 Choosing an observing site

However efficient the detectors, a telescope is useless unless it can see clear sky for a significant fraction of the year. As telescopes and instruments have become more efficient, there has been a corresponding drive to find the best possible observing sites, to make the best use of expensive equipment.

The criteria for a good observing site depend on wavelength, and I will only discuss optical sites. However, it is worth noting that one of the main aims with radio telescopes is to avoid man-made transmissions. If it is not possible to put the telescope in a remote desert (for example, the VLA in New Mexico), the best site may be in a deep valley, whose walls can act as a shield; the 100 m telescope at Effelsberg near Bonn is an example. Note also that radio waves can penetrate cloud, and that the best observing conditions for some purposes may even be a completely overcast sky with a thin drizzle falling, since the atmosphere is often at its quietest under these conditions.

Excellent sites for optical telescopes have at least four important properties:

(a) clear sky (no cloud above the site);
(b) good seeing (minimum turbulence);

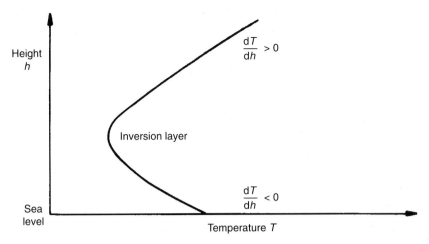

Fig. 3.29. A temperature inversion in the atmosphere. Convection will occur only below the inversion layer, and clouds will tend to gather at the temperature minimum.

(c) dark sky (minimum light pollution);
(d) little water vapour above the site (this is particularly important for infrared work).

The last two conditions require remote sites on high mountains, but there are plenty of these that have heavy cloud cover and/or poor seeing. The best mountain sites are those that lie above a *temperature inversion* (Fig. 3.29). Usually the air temperature decreases with height, but under some conditions the temperature reaches a minimum and then starts to increase with height. Two effects are important. First of all, clouds tend to form at the temperature minimum, because cold air can hold less water vapour and, as the air rises and cools, the water precipitates out as droplets. Second, the air below the inversion layer, with a negative temperature gradient, is unstable to convection (it is a layer heated from below, just like a beaker of water heated by a flame), while the layer above, with a positive gradient, is stable. In the convective layer, any clouds that form are mixed to the top, while in the layer above there is no mixing and turbulence is minimized.

Turbulence can also be caused by violent winds, and will be minimized if the prevailing wind is *laminar*: flowing in smooth layers. This will be true if the wind approaches the site over a smooth surface. The largest smooth surfaces on the Earth are the oceans, so the best sites are those where the wind approaches the site over an ocean. In practice, this happens in two ways. There are places where mountain chains face an onshore

Fig. 3.30. Mount Teide on the island of Tenerife in the Canary islands, seen from the air. The smooth layer of cloud marks the inversion layer: the mountain peak is well above it, as is the observatory of Izaña on the long ridge to the right of the peak. (Photograph by the author.)

wind, for example the coastal ranges in California and Chile; the famous Californian sites (Mt Palomar and Mt Wilson) are now seriously affected by light pollution, but the various Chilean sites are among the best in the world. They also have the advantage of being extremely dry. The other kind of site is on an ocean island with high mountains; the best-known examples are La Palma and Tenerife in the Canary Islands and Mauna Kea in the Hawaiian Islands. These island sites are almost permanently above the inversion layer (Fig. 3.30).

Of course, local structures, such as the dome itself, may also cause turbulence. One solution to dome turbulence is to blow a *steady* stream of air across the primary mirror, to prevent convection.

Many other factors enter into the choice of a particular site, and modern sites have been chosen only after extensive site-testing over a period of many years, to find the best possible site in terms of seeing and clear skies. None the less, the above simple ideas form the basis for initial selection of sites for testing.

3.7 Summary

3.7.1 Detectors

A photographic image is formed in two stages. First, a latent image is formed by incident photons causing electrons to migrate around a silver halide crystal until they combine with silver ions to form a small, stable cluster of silver atoms. These groups of atoms then act as catalysts for the transformation of the whole grain into pure silver when the photograph is developed. The pattern of developed grains forms the visible image.

This is a very inefficient process (usually only about 0.1% of the photons lead to a developed grain), and the strength of the resultant image is related in a very complicated, non-linear way to the number of incident photons. The only reason that photography still plays a major role in professional astronomy is that photographic plates can be much larger than any current type of electronic imaging device and are therefore vital for sky surveys. (For amateur astronomers, there is the additional advantage that photography is much cheaper and simpler than electronic recording.)

Electronic recording makes use of the photoelectric and photoconductive effects. In the photoelectric effect, incident photons cause electrons to be emitted from a semi-conductor which acts as the cathode of a photoelectric cell. However, there is an energy threshold for the photoelectric effect, and less energetic photons (for example in the infrared) simply free electrons from the lattice and increase the effective conductivity of the material. In both cases, one photon affects one electron, and there is an exact linear relation between the incident photon flux and the resulting photocurrent.

This linearity is the major advantage of photoelectric devices, the other two important advantages being a larger dynamic range and a much higher quantum efficiency.

For single-object detection, the photocurrent is amplified in a photomultiplier tube, which uses high voltages to accelerate electrons down the tube. Collisions with successive dynodes generate more electrons at each stage, yielding a burst of up to 10^7 electrons at the anode for each incident photon.

More commonly nowadays, an imaging detector is used. In these devices electrons are again accelerated down a tube, but are focused magnetically onto a target at the end of the tube. This gives a linear response, and a high-resolution image over a reasonable field of view, with good quantum efficiency. The basic principle of focusing high-energy electrons onto some target is a feature of all image intensifiers, which still form a part of several electronic imaging systems. The target in the IPCS (Image Photon Counting System) is a phosphor screen, which is then scanned by a television camera.

The essential advantage of television recording is that the image is scanned repeatedly in a raster scan, which converts a two-dimensional image into a one-dimensional string of numbers giving the intensity of the image at a series of points along the scan. Data in this form can be handled immediately by computer. The other major solid-state device, the CCD, has the same advantage, because it is read out in the equivalent of a raster scan, but has a somewhat different way of recording the signal. Instead of electrons being accelerated down a tube, each incident photon simply frees an electron from the lattice of a semi-conductor which consists of a grid of capacitors, and it falls into the nearest potential well. A visible image is thus transformed into a charge distribution, which is read out at the end of the exposure.

3.7.2 *Instruments*

Instruments are used to analyse the brightness of a source (photometry), the extent to which the light is polarized (polarimetry) and the spectrum of the light (spectrometry). I concentrated on the spectrum, and on the use of interferometry for making high-resolution measurements.

In astronomical slit spectrometers the dispersing element is usually a reflection grating, enabling a compact design in which the dispersion and wavelength range can easily be changed. The use of a slit enables accurate wavelength calibration and repeatable wavelength resolution. However, for survey work, it is sometimes useful to place the dispersing element in front of the main mirror and dispense with the slit. Each object in the field is then imaged as a small spectrum, and many thousands of low-dispersion spectra can be obtained simultaneously on a large Schmidt plate. Slitless spectra can also be used to obtain two-dimensional information about the location of emission lines in extended objects such as planetary nebulae. Some of the advantages of both techniques are combined in fibre-optic spectroscopy, in which the images of a hundred or more galaxies or stars are fed through fibre-optic tubes onto a single slit, so that many slit spectra can be obtained simultaneously.

For high resolution in wavelength and space, interference techniques provide a powerful tool. Multiple reflections between accurately parallel plates are used both in simple interference filters, which select narrow wavelength regions for accurate photometry, and for the Fabry–Perot etalon used to obtain high spectral resolution in the Taurus imaging interferometer. High spatial resolution was obtained in Michelson's stellar interferometer by using the principle of Young's double-slit interferometer, in which a fringe pattern is produced by the addition of two beams with slightly different path-lengths.

For two point sources, or an extended source, the visibility of the pattern decreases as the separation of the sources, or the angular size of the source, increases, enabling the angular size to be determined. In Hanbury Brown's intensity interferometer, the intensities of the two signals were combined, rather than their amplitudes, and much smaller separations could be measured. The modern technique of speckle interferometry uses very high time resolution and Fourier de-convolution to extract diffraction-limited resolution from large telescopes. The angular resolution is less than was achieved by Michelson or Hanbury Brown, but sharp images can be obtained for much fainter sources.

For extended sources, new techniques enable almost simultaneous recording of spectra at every point on a given area of the sky, by scanning either in wavelength (Taurus) or in space (Aspect) to build up a three-dimensional data cube of intensities in two space dimensions and one wavelength dimension. This has proved to be a very powerful tool for studying the dynamics of gas in active galaxies (Section 11.5).

I discussed the accuracy with which the brightness of a source can be measured. There is always some background from the sky, from scattered moonlight and sunlight, from emission in the upper atmosphere and from man-made light pollution, as well as from further afield in the form of scattered light from interplanetary and interstellar dust and of light from unresolved stars and galaxies. However, this background would not be a problem if it could be measured accurately and subtracted off the total signal. The real limit to accuracy comes from the inevitable uncertainty in all measurements, and I discussed the effect of random errors, or *noise*. The crucial quantity is the ratio of signal to noise: this ratio must exceed 2 or 3 if a signal is to be detectable with confidence. For a given signal, the ratio increases only as the square root of the exposure time, so it can require a long exposure to obtain a sufficient signal-to-noise ratio for a weak source. When applying for telescope time on a large telescope it is essential to demonstrate that the proposed observations are feasible by providing an estimate of the signal-to-noise ratio that will be achieved.

Finally, I reviewed briefly how to choose the site for a large telescope so as to maximize its output.

Exercises

3.1   The photographic plates taken by the UK Schmidt telescope have an emulsion area of 356 mm × 356 mm and the effective pixel size is

$5\,\mu m \times 5\,\mu m$. How many pixels are there (a) on the whole plate; (b) per square mm?

3.2   The smallest CCD used on the Cassegrain spectrograph of the Isaac Newton Telescope (INT) on La Palma has a size of $385\,\text{pixels} \times 578$ pixels, each pixel being $22\,\mu m \times 22\,\mu m$. Compare (a) the total number of pixels and (b) the total area with the values for the Schmidt plate in exercise 3.1.

       Repeat this calculation for the large 2048 pixels $\times$ 2048 pixels CCD, with $15\mu m \times 15\mu m$ pixels, used for prime focus photography on the INT.

3.3   A photomultiplier has ten dynodes, at each of which an incident electron causes four electrons to be emitted. Find the gain of the photomultiplier. Why do you think the gain increases when the voltage difference between dynodes is increased?

3.4   An interference filter is designed to have a maximum with a central wavelength of 486 nm. If the filter operates in the second order ($m = 2$) and the dielectric layer has a refractive index $n = 1.35$, find the thickness of the dielectric layer and the wavelengths of each of the neighbouring maxima. How would you ensure that the neighbouring maxima were not transmitted?

3.5   A Michelson stellar interferometer is used at 550 nm to determine the apparent diameter of a star. If the fringe pattern disappears when the adjustable mirrors have a separation of 5 m, find the star's angular diameter in arcsec.

3.6   A photon-counting detector is used to measure a faint source against a background whose count rate is $16\,\text{s}^{-1}$. Assuming that the noise is random, and that equal time is spent observing on and off the source, find the total integration time needed to detect a source whose strength is 1% of the background if the criterion for detection is that the signal-to-noise ratio is 2.

4

Radio telescopes and techniques

4.1 Concepts and definitions

In the preceding two chapters I have introduced many of the basic concepts of observational astronomy, using optical techniques to illustrate them. In this chapter I will concentrate on the differences that arise when we move to the radio window.

The fundamental differences between radio and optical telescopes all stem from the very different wavelengths of the radiation to be detected:

$$\frac{\lambda_{\text{radio}}}{\lambda_{\text{optical}}} \approx \frac{\text{several decimetres}}{\text{several hundred nanometres}} \approx 10^5\text{--}10^6. \qquad (4.1)$$

This is a typical value; the range in ratio is much larger, since the radio window is very wide, covering more than three decades in wavelength. By contrast the optical window covers only about half a decade (a factor of 3).

This large ratio has two major consequences:

1. *Resolution and imaging.* At a wavelength of 1 cm, the resolvable angle for a telescope of aperture D is (cf. equations (2.9), (2.10))

$$\alpha \simeq 40\frac{1\,\text{m}}{D} \text{ arcmin}; \qquad (4.2)$$

   thus for the same resolution as a 1 m optical telescope we would need $D \simeq 20\,\text{km}$! For longer wavelengths, an even larger aperture would be needed (larger than the Earth's diameter at 10 m wavelength). This is clearly quite impracticable for a single telescope, so the resolution of a radio telescope is limited by the largest practicable D, rather than by seeing as for optical telescopes: radio telescopes are generally *diffraction-limited*, with *beamwidth* α given by equation (4.2).

   This has forced radio astronomers into a variety of ways of using

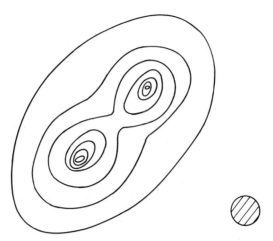

Fig. 4.1. Radio map (contours of equal intensity) of two neighbouring but resolved sources (schematic). The hatched circle represents the beamwidth of the telescope.

arrays of telescopes to obtain higher resolution and, as you will see in Section 4.4, radio angular resolution is now routinely *better* than optical resolution. Optical astronomers are only now beginning to catch up, with their use of interference techniques and active optics to remove the effects of seeing. Radio astronomers have the great advantage that their diffraction-limited observations are much cleaner than ground-based optical telescopes can ever obtain, and interference techniques are therefore much more straightforward to use. Only in space can optical telescopes hope to obtain data of comparable quality.

For a typical radio telescope, the operating wavelength is comparable in size to at least some dimension of the instrument, and the simple geometrical optics used in discussing optical telescopes are no longer strictly valid: diffraction effectively produces additional aberrations. The incident radiation is diffracted by the receiver itself and it is difficult to form undistorted images with a small, single radio telescope. Images of the sky at radio wavelengths can nowadays be obtained using arrays of telescopes (see Section 4.3), or arrays of feed horns (see below) at the focus of a large, single telescope, but with the first, simple, radio telescopes images were constructed by *mapping* : the intensity was measured at a number of positions in a grid on the sky and then a map was obtained by drawing contours of equal intensity. The resolution of such a map is limited by the (diffraction) beamwidth, which is usually shown in a corner of the map. Fig. 4.1 shows a schematic map of two neighbouring but resolved sources.

2. *Detection.* The method of detecting radio waves is fundamentally different from that of detecting optical photons. Radio waves are detected, as in an ordinary domestic radio receiver, by their alternating electromagnetic field. The radiation excites an alternating field in the detector, which is detected electronically as an AC voltage. Because this different mode of detection uses the wavelike properties of photons rather than their particle-like properties, it is possible to obtain information that is not usually available optically. A wave can be described by two properties: the *amplitude*, which is a measure of the strength of the wave, and the *phase*, which measures which part of the wave is passing at a given moment. For a wave of a single frequency ω, the voltage V at time t can be written as

$$V = V_0 \sin(\omega t - \phi), \tag{4.3}$$

where V_0 is the amplitude and ϕ is the phase. The importance of the phase is that it provides information about how waves add together. If two waves have the same phase, the total amplitude is just the sum of the amplitudes; however, if they are out of phase by π, then the signals will have the opposite sign (because $\sin(x + \pi) = -\sin x$), and the total signal will be the difference of the amplitudes (see exercise 4.1). As you will see in the next two sections, it is this kind of phase information that is needed for interferometry, and it is the fact that it is immediately available for most radio signals that makes interferometry such a natural tool for radio astronomers.

The parts of a radio telescope are given somewhat different names from those of an optical telescope, and there is not an exact correspondence to the three-way split into telescope, detector and analyser that I used for optical instruments. For present purposes, I will split radio systems into just two essential parts:

1. *Antenna.* This is equivalent to the optical telescope. It chooses the direction of observation, collects the radiation and converts it to an AC signal.

2. *Receiver.* This is equivalent to a combination of the detector and analyser of an optical telescope. It chooses the signal frequency and bandwidth, using the equivalent of optical filters to define the spectral resolution, processes the signal and records it. For spectroscopic observations, the receiver has many channels, one for each resolved

frequency. Although the detector and analyser are still conceptually distinct, all the processing is done electronically, using circuitry similar to that used in the radio communications industry. I will not discuss receivers in detail. I will say a little more about spectrometers in Section 4.5.

The antenna of a radio telescope differs from an optical telescope in an important respect: it is much more sensitive to polarization. We can understand this by considering a very simple antenna, the single dipole aerial. A half-wave dipole consists of a straight metal rod, half a wavelength long, with the cable to the receiver attached at the centre of the rod. Because the dipole defines a definite direction in space, it responds only to electromagnetic waves which have their electric vectors parallel to the dipole. Most antennae share this property to some extent, so in general a simple antenna is sensitive to the polarization of the incident radiation and records only part of the radiation. Although in practice the whole signal is recorded, by using crossed antennae that are sensitive to the two orthogonal components of the signal, it is also true that radio sources tend to emit more strongly polarized radiation than optical ones. I will therefore discuss briefly the nature of polarized radiation.

A *linearly* polarized wave is the easiest to understand: the plane of polarization, defined as the plane in which the electric vector vibrates, remains fixed in space. Individual photons, emitted by atomic processes in which the atom or electron involved has a definite dipole moment, are linearly polarized. If the dipole moments are randomly oriented in space, as in a hot gas with no external electromagnetic field, the net signal is a random sum of linear polarizations and the wave is said to be unpolarized. If all the atomic dipoles are lined up, as in a dipole transmitter, the wave itself is linearly polarized. In some circumstances, the wave may be *circularly* polarized; in that case the electric vector steadily rotates. Such a wave can be decomposed into two linearly polarized waves of the same frequency and amplitude which are a quarter of a cycle out of phase. Any unpolarized wave can also be decomposed into components, either linearly or circularly polarized, but these components are uncorrelated in phase. Polarized radiation arises when phase correlations are imposed by the nature of the emission process or by interactions with the medium between the source and the receiver. For astronomical sources, the correlations are usually introduced by the presence of large-scale magnetic fields. Many radio emission processes involve magnetic fields, so many radio sources show strong polarization.

I now need to introduce some definitions, and try to relate the rather

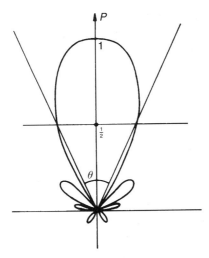

Fig. 4.2. A schematic polar diagram of the power pattern of a simple radio dish. The resolution of the dish is determined by the width of the main peak; θ is the half-power beamwidth (HPBW), sometimes known as the full width at half-maximum (FWHM).

different terminology to what you have already learned about optical astronomy. Radio astronomers usually work in terms of the *flux density*, that is, the flux per unit frequency, measured, for example, in $W\,m^{-2}\,Hz^{-1}$, instead of just the flux. Let S_ν be the flux density of the measured source at frequency ν. The total *power* p_ν measured by the antenna is then related to the flux density by

$$p_\nu = A\,S_\nu, \tag{4.4}$$

where A is called the *effective area* of the telescope. For a simple dish aerial, A is approximately equal to the physical area of the dish, but in general A may be very different from the physical area of the antenna, as can easily be understood by thinking of a dipole aerial. It may also depend on frequency. It is a measure of the sensitivity of the telescope or antenna.

Because radio aerials are directional, the effective area of an antenna varies with the direction of the source relative to the axis of the antenna, the axis being in the direction in which the response is strongest and the effective area is largest. The directional response of an antenna is expressed in terms of the *power pattern P*, where

$$P = A/A_{\max}. \tag{4.5}$$

A schematic polar diagram of a power pattern is shown in Fig. 4.2. The angular resolution of the antenna is determined by the width of the largest peak, known as the *main beam*. The beam is characterized by its width at the point where $P = \frac{1}{2}$, known as the *half-power beamwidth*, often abbreviated as HPBW. The HPBW depends in a complicated way on the design of the aerial, but for a simple parabolic dish equation (4.2), with D equal to the diameter of the parabola, gives a reasonable approximation to the HPBW. The main beam can be thought of as the central peak (Airy disc) of the diffraction pattern of a point source. Because a radio telescope is diffraction-limited, the diffraction rings surrounding the central maximum (Fig. 2.1) will also be visible; that is, the antenna has a non-zero response outside the main beam. These subsidiary peaks in the power pattern are known as *sidelobes*, and are in principle a source of spurious signals and ambiguities: a strong source in a sidelobe may be confused with a weak source in the main beam. The telescope may even be able to detect signals from behind the antenna. In practice, modern antenna design has reduced the response of the sidelobes to an almost negligible fraction of the main beam.

The flux density is only a useful concept for a point source. For an extended source we are interested in the distribution with solid angle and we define the *brightness distribution B_ν*, which is the flux density per unit solid angle and is measured, for example, in $W\,m^{-2}\,Hz^{-1}\,ster^{-1}$. In that case the total power received by the antenna is

$$p_\nu = A_{\max} \int_{4\pi} B_\nu P \, d\Omega \qquad (4.6)$$

where the integration is over all solid angles. In practice, most of the contribution to the integral comes from the main beam. If the beam has a circular cross-section and the HPBW is θ, we may estimate the integral by approximating the power pattern P by a rectangular response of height 1 and total width θ. Then

$$p_\nu \approx A_{\max} \overline{B}_\nu \pi \theta^2 / 4 \qquad (4.7)$$

where $\overline{B}_\nu$ is the average brightness distribution within the beam and I have used the fact that $\theta \ll 1$ radian. If the source is larger than θ in angular size, then it will be resolved by the beam, that is, it will be possible to distinguish it from a point source (see Fig. 4.1).

Even the very strongest astronomical radio sources produce signals at Earth that are weak in comparison with man-made signals. For this reason, radio astronomers describe the flux density in terms of an appropriately small unit, named after one of the pioneers of radio astronomy, Karl Jansky:

$$1 \text{ Jansky (Jy)} = 10^{-26}\,W\,m^{-2}\,Hz^{-1}. \qquad (4.8)$$

Extreme solar radio bursts may reach 10^8–10^9 Jy, and the strongest other sources range up to 10^4 Jy, but the great majority of sources are weaker than a few Jy; modern receivers can detect fluxes of less than one thousandth of a Jansky (one milli-Jansky, or mJy).

The weakness of most radio sources has two important implications. First, radio astronomy is very vulnerable to interference from man-made signals (see Section 3.5). With the explosion of telecommunications this is posing a serious problem and radio astronomers need to exert continual pressure at international level to preserve astronomically important wavebands free from commercial broadcasting.

The other implication is that receiver noise must be kept as low as possible. Receiver noise is mainly due to thermal motions within the electronic circuitry, and can always be expressed in terms of the thermal noise power from a resistor at some temperature. A theorem due to H. Nyquist tells us that the thermal noise power per unit frequency from a resistor at temperature T_N is

$$p_N = k T_N \qquad (4.9)$$

where k is Boltzmann's constant. At room temperature ($T_N \simeq 300\,\text{K}$) $p_N \simeq 4 \times 10^5$ Jy, so it is clear that reducing the receiver noise temperature T_N is very important but is unlikely to be sufficient in general to reduce the noise power to less than the power from the astronomical source.

There are therefore two requirements for an astronomical radio receiver: the noise power should be as low as possible and should also be accurately constant so that it can be treated as a constant background and subtracted off the receiver output to obtain the astronomical signal. Any astronomical signal must be amplified before being recorded, so a typical radio receiver consists of an amplifier and a detector. Various low-noise amplifiers have been developed to keep T_N as low as possible. One of the most effective is a maser amplifier, in which the incident wave is amplified by its stimulating downward electron transitions from some upper energy level in, for example, a ruby crystal. The lower energy state is kept depopulated by using an external energy source, at a higher frequency than the wave being amplified, to pump the electrons to a higher state. At low temperature there are few thermally excited downward transitions and the system is almost noise-free. Another common device is the parametric amplifier, so-called because it consists of an electrical circuit whose parameters are varied in a cyclic manner. For example, the value of a capacitor may be varied; this requires external energy to be supplied, which can be fed into amplifying the input signal if there is a suitable resonance between the frequency of varying the capacitance and the frequency of the input signal. Capacitors have essentially no internal resistance, so this system also has very low noise. None the less, even though amplifiers are often operated at liquid helium temperatures ($\sim 4\,\mathrm{K}$), noise temperatures mostly remain in the range 10–100 K.

Because noise power is always important, it is often convenient to express the astronomical power in terms of temperature as well. If we define the *antenna temperature* T_A by

$$p_\nu = kT_A \qquad (4.10)$$

then Nyquist's theorem tells us that T_A is the temperature of the resistor which, attached to the antenna, would give the same signal. Note that T_A has *no* relation to any temperature in the astronomical source.

The noise temperature is an intrinsic property of the receiver design and is relatively easy to keep constant. However, the gain G of the amplifier is very large (up to 10^{12}) and so must be very stable. The detector receives a signal $G(T_A + T_N)$ and a background of GT_N is therefore subtracted. If the gain changes by a small amount ΔG, but the same background is subtracted, the signal T_A will appear to change by an amount ΔT_A given by

$$G\Delta T_A = \Delta G(T_A + T_N) \qquad (4.11)$$

or

$$\frac{\Delta T_A}{T_A} = \frac{\Delta G}{G}\left(1 + \frac{T_N}{T_A}\right), \qquad (4.12)$$

which shows that any fractional change in G shows up as a fluctuation in the signal that is larger by the factor $1 + T_N/T_A$, which can be much larger than 1 for a weak source. It is therefore important to keep a constant check on the gain stability. This can be done, for example, by some sort of switching system, in which the receiver is continually being calibrated by switching the input between the antenna and a constant calibrating noise source.

At low radio frequencies, stable amplifiers are readily available, but at the high frequencies used in millimetre-wave astronomy stability and high gain both become problems. A very common solution to this problem is to use *heterodyne techniques*, in which the observed signal is mixed with a reference signal at a nearly equal frequency, produced by a stable *local oscillator*. The output of the *mixer* contains oscillatory components with frequencies equal to the sum and difference of the frequencies of the two input signals (see exercise 4.2). By the use of filters, the signal with the difference frequency can be selected. If the local oscillator frequency is close enough to that of the signal from the astronomical source, the difference frequency can be made low enough for it to be amplified by a standard radio frequency amplifier. Exercise 4.2 also shows that the amplitude of the signal at the difference frequency is directly proportional to that of the input high-frequency signal from the source, so this technique also preserves the linearity of the receiver. These techniques are crucial to the success of all the molecular-line studies that are now made in the millimetre region (Sections 6.1.7 and 12.3).

The antenna temperature defined in equation (4.10) is a measure of the total power reaching the antenna. For an extended source, we want to know the brightness distribution and it is convenient to express this in terms of another temperature, the *brightness temperature* T_B, which is the equivalent black-body temperature of the source and is defined as the temperature of the black-body source that would give the same measured brightness distribution. A black body has a Planck spectrum, with intensity ($W\,m^{-2}\,Hz^{-1}\,ster^{-1}$ in SI units):

$$B_v(\text{bb}) = \frac{2hv^3}{c^2}\frac{1}{e^{hv/kT}-1}.\tag{4.13}$$

However, at radio wavelengths $hv \ll kT$ (even at short radio wavelengths this remains true, provided $T \gg 15\,K$, which is true for most radio sources), so we can use the *Rayleigh–Jeans approximation*, which is most conveniently expressed in terms of wavelength:

$$B_v(\text{R–J}) = 2kT/\lambda^2.\tag{4.14}$$

Thus the brightness temperature of a source is given by:

$$B_v(\text{source}) = 2kT_B/\lambda^2.\tag{4.15}$$

If the source *is* a black body, then T_B is its temperature. For non-thermal sources, T_B is just a convenient measure of the signal strength, which varies with wavelength in general.

4.2 Filled-aperture telescopes

Radio telescopes are of two types: filled and unfilled aperture. Filled-aperture telescopes are more closely related to optical telescopes, so I will discuss them first.

The closest analogue to an optical reflecting telescope is the *parabolic dish* (Fig. 4.3(a)). This consists of a parabolic reflector, usually of circular cross-section, which focuses the radio waves onto an antenna, usually a simple dipole or a horn-shaped waveguide, called the *feed*. The signal detected by the feed antenna is transmitted by cable to a central control room. The figure of the reflector has to be accurate to within a wavelength, but at radio wavelengths this allows the use of a mosaic of metal plates at short

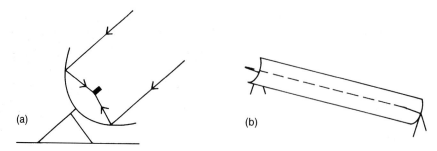

Fig. 4.3. (a) A simple parabolic dish receiver, of circular cross-section, with a prime focus feed. (b) A parabolic cylinder receiver, steerable in altitude, showing a line feed.

wavelengths and an open wire mesh (with mesh spacing ≪ wavelength) at longer wavelengths.

A prime focus feed, as shown in Fig. 4.3(a), is common, but has the disadvantage that the feed must be relatively close to the dish if it is not to pick up a signal from the ground. The focal ratio is then $f/D < 1$ and the system suffers badly from coma. This can be overcome by replacing the feed by a secondary (hyperbolic) reflector which focuses the radio waves back onto a feed placed at the centre of the main reflector (the analogue of an optical Cassegrain system, except that the focus is often in front of the main reflector). Most stray radiation is then from the 'cold' sky rather than from the 'hot' ground. The feed is also in a mechanically stable position and is more easily accessible, especially in large telescopes. For a small telescope, there will often be just a single feed at the focus. For larger telescopes, imaging is made possible by having an array of small feeds in the focal plane.

Small parabolic dishes are usually mounted equatorially but most large steerable dishes, such as the 76 m Jodrell Bank telescope (Fig. 2.10) or the 64 m Parkes telescope (Fig. 4.4), have an alt–azimuth mounting, which is far simpler mechanically. The largest fully steerable telescope at present is the 100 m telescope of the Max Planck Institute for Radio Astronomy at Effelsberg near Bonn (Fig. 4.5).

Because of the cost of fully steerable dishes, many large radio telescopes have only limited mechanical movement, being either transit instruments (steerable in elevation only) or completely fixed. A simple example of a transit telescope is the *parabolic cylinder* (Fig. 4.3(b)), in which the reflector is a segment of a cylinder with parabolic cross-section. The focus is at a line rather than a point, so the feed antenna consists of a line of connected dipoles (line feed). The cylinder is usually aligned east–west and sources are

Fig. 4.4. The 64 m (210-foot) fully steerable radio telescope of the Australian National Radio Astronomy Observatory at Parkes, New South Wales. (Photograph by the author.)

observed as the rotation of the Earth causes them to drift through the beam of the telescope.

An example of a fixed telescope is the 305 m (1000-foot) spherical reflector at Arecibo in Puerto Rico, where advantage has been taken of a large natural bowl in the ground which is near enough spherical that all the supports of the spherical wire mesh reflector are fairly short. The feed is suspended by cables running between three large pylons at the edge of the dish, and is accessible only by a precarious journey along a swaying walkway (Fig. 4.6).

The power patterns of these telescopes reflect their mechanical symmetries and are typical of many others. The parabolic dish is symmetrical about its axis and has a three-dimensional power pattern obtained by rotating Fig. 4.2 about its axis. This is known as a *pencil beam* and is shown in Fig. 4.7(a).

The parabolic cylinder is extended along its axis, so the resolution in directions parallel to the long axis is better than in directions normal to the axis. This produces a flattened beam, known as a *fan beam* (Fig. 4.7(b)) because its shape resembles that of an open fan, in which the larger dimension reflects the resolution of an individual dipole in the feed (the effective

Fig. 4.5. The 100 m Effelsberg radio telescope near Bonn. It is situated in a valley to shield it from man-made radiation. (Max-Planck-Institut für Radioastronomie, Bonn.)

aperture is then the short axis of the cylinder) and the smaller dimension reflects the better resolution obtained when all the dipoles act together (in that direction, the effective aperture is the long axis of the cylinder).

This principle of improving resolution by having a line of antennae can be extended to two dimensions by constructing the telescopes out of many small units arranged in a two-dimensional array and all feeding their signals into a single receiver. Such a *filled-aperture array* is analogous to the optical multi-mirror telescope and is shown schematically in Fig. 4.8. The antennae may be small parabolic dishes, dipoles or some other simple aerial. If they

Fig. 4.6. The 305 m spherical radio dish at Arecibo in Puerto Rico. The instruments are at the prime focus feed, suspended from three tall pylons. (The Arecibo Observatory is part of the National Astronomy and Ionosphere Center, which is operated by Cornell University under a cooperative agreement with the National Science Foundation.)

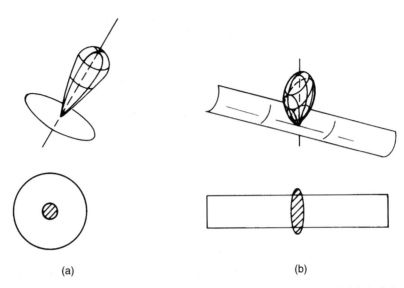

Fig. 4.7. Schematic power patterns. (a) Pencil beam for a paraboloidal dish. (b) Fan beam for a parabolic cylinder. The upper diagrams show the beams in three dimensions, while the lower diagrams are projections along the axis of the beam. Note that in the fan beam the narrower dimension (better resolution) corresponds to the longer axis of the receiver.

are all steerable dishes, then they are all aligned and steered so that they all track the source together.

The individual elements of the array are all at different distances from the receiver. If the cables that feed the signals to the receiver were also of different lengths then the fact that the signal passes along the cable at a finite speed would mean that the same signal would arrive at the receiver at different times from different dishes. In the optical multi-mirror telescope it is necessary to avoid this problem by having the mirrors symmetrically placed about a central receiver. In the radio case the signals are transmitted along cables and it is possible instead to compensate for the different time-delays simply by arranging that all the cable connections are the same length, so that all the signals arrive in phase; the same effect may be achieved electrically, by introducing a phase delay into the circuit.

The same principle can be extended to achieve limited *electrical steering* for arrays of fixed elements. I will illustrate this for the simple case of a three-element linear array of spacing D (Fig. 4.9). Suppose that when the cable lengths from the three elements are all the same the maximum response of the array is in the vertical direction; all the signals from a source vertically overhead then arrive in phase. If we now want to simulate a

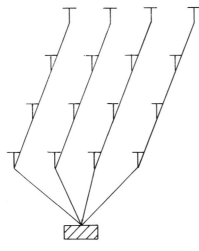

Fig. 4.8. A filled-aperture array of 16 dipoles, arranged in four rows and linked by cables to a single receiver.

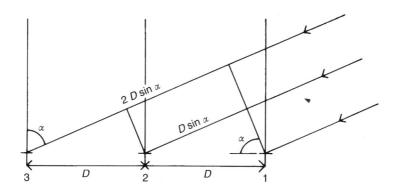

Fig. 4.9. The principles of electrical steering. Signals arriving from an angle α to the vertical reach element 1 first, and have to travel extra distances $D \sin \alpha$ and $2D \sin \alpha$ to reach elements 2 and 3. These three signals can be made to appear to arrive in phase by introducing suitable delays.

maximum response in the direction α from the vertical, we need to introduce progressive phase shifts into the signals arriving at adjacent elements so that now signals from direction α arrive at the receiver in phase. In Fig. 4.9, the signals arriving at elements 2 and 3 from direction α have to travel, respectively, $D \sin \alpha$ and $2D \sin \alpha$ farther than the signals arriving at element 1. This can be compensated for by adding extra lengths of cable to elements 2 and 1, of lengths $D \sin \alpha$ and $2D \sin \alpha$, respectively, or by electrically

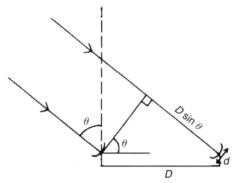

Fig. 4.10. Principles of the two-element interferometer. Two parallel beams arrive with a path difference $D \sin \theta$ and interfere constructively when this difference is an integral number of wavelengths.

introducing phase delays of ϕ and 2ϕ, respectively, where

$$D \sin \alpha = \frac{\phi}{2\pi} \lambda \qquad (4.16)$$

(λ = wavelength of signal). Then all the signals from direction α will arrive at the receiver in phase at the same moment. This is equivalent to tilting the power pattern by an angle α and so achieves the same effect as mechanically steering the telescope. (Note that, except at very long wavelengths, $D \sin \alpha$ will in general be many wavelengths, so the phase changes needed will be many cycles: $\phi = 2n\pi + \phi_0$, where $n \gg 1$ and $0 \le \phi_0 < 2\pi$. The separation D must be known very precisely if an accurate value is to be obtained for ϕ_0: exercise 4.3.)

4.3 Unfilled-aperture telescopes

An array with a small number of well-separated elements (spacing $\gg$ size of element) is analogous to an optical interferometer. It makes use of the fact that the resolution of a large filled-aperture telescope depends only on its largest dimension and can be retained even when large parts of the aperture are removed. This is a cost-effective way of obtaining high resolution, at the expense of losing sensitivity because of a decrease in the effective area.

The simplest example, which is analogous to Michelson's optical stellar interferometer, is the two-element interferometer shown schematically in Fig. 4.10. Two elements of largest dimension d are separated by a distance $D \gg d$. A wave front approaching from an angle θ to the vertical has farther to travel to one of the elements, the path difference being $D \sin \theta$.

If the receiver is at the same distance from both elements, this path

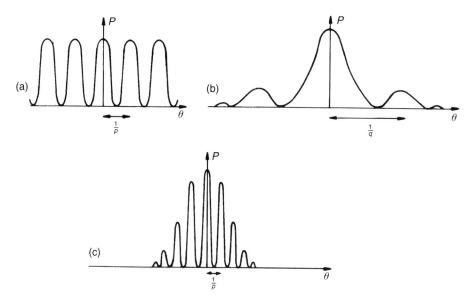

Fig. 4.11. (a) Fringe pattern of a simple two-element interferometer, of spacing $D = p\lambda$. (b) Response pattern of each element, of size $d = q\lambda$. (c) Net response pattern of the system, for the case $q \ll p$.

difference is registered at the receiver as interference between the two signals. If θ is varied, fringes will be detected, with maxima at

$$D \sin \theta = n\lambda, \ n = 0, 1, 2, ... \tag{4.17}$$

As in the last section, the separation corresponds in general to a phase difference $\phi = 2\pi D \sin \theta / \lambda = \phi_0 + 2n\pi$, where n is an integer chosen so that ϕ_0 is between 0 and 2π. Equation (4.17) can thus be rewritten as $\phi_0 = 0$. Note that only the value of ϕ_0 (the 'principal value' of the phase) is important; the waves will add up if the phase difference between them is zero, regardless of the value of n. This emphasizes the importance of being able to record phase information.

If θ is small (i.e. the source is near the plane of symmetry of the interferometer), then the separation of the bright fringes is

$$\Delta\theta \simeq \frac{\lambda}{D} \simeq \frac{1}{p}, \tag{4.18}$$

where

$$D = p\lambda \tag{4.19}$$

(note that p is *not* necessarily an integer; it is just convenient to express D

in wavelength units). The response pattern of a two-element interferometer for small θ is therefore as shown in Fig. 4.11(a). But each element has its own response pattern, with unequal sidelobes, as shown in Fig. 4.11(b). The spacing of the sidelobes is determined by q, where

$$d = q\lambda, \tag{4.20}$$

and so $q \ll p$ (q is also not necessarily an integer). The net response pattern of the system is the product, which is shown in Fig. 4.11(c) and consists of a series of fringes of spacing $1/p$, corresponding to the resolution of the *pair* of elements, with decreasing amplitude on a scale $1/q$ corresponding to the resolution of an *individual* element; the central maximum is the 'main beam' of the interferometer, the other fringes giving the sidelobe response, which is much more significant here than for a single telescope. The overall angular resolution of the interferometer corresponds to the spacing D, a factor D/d times the resolution of one element, although the sensitivity is only twice that of a single element of size d.

We can think of this interference pattern as fixed in the sky, and then as the Earth rotates the radio source being observed drifts across the pattern. Timing the maximum signal gives the position of the source. An extended source will cover several fringes simultaneously, and so will effectively blur the fringes, as in the Michelson case. The visibility of the fringe pattern as a function of time can then be used to deduce the size and structure of the source.

The same principle can be extended to multi-element interferometers, either in one dimension (gratings) or two dimensions (arrays). Again, the response pattern of the interferometer can be thought of as a stationary pattern of fringes through which the source drifts, but the pattern is now more complicated because there are many different spacings between the elements.

What I have said so far would apply equally well to the filled array discussed in the last section, the main quantitative difference being that the fringes (Fig. 4.11(a)) become much narrower relative to their separation, and fewer in number, as the number of elements increases, so the filled array has a less confusing response pattern, consisting almost entirely of the central maximum. The overall resolution of the instrument is, however, still determined simply by the largest separation in the array.

It is expensive to increase the resolution of a filled-aperture telescope, since the effective area (and of course the sensitivity) increases as the square of the resolution, and the cost may well increase faster than that (for much the same reasons as for optical telescopes – see Section 2.6). However, if

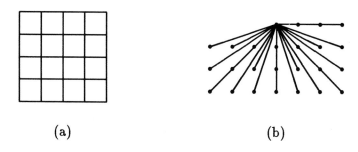

(a) (b)

Fig. 4.12. Principle of aperture synthesis. (a) Filled aperture consisting of 16 elemental areas and 120 spacings. (b) Vectors indicating the 24 distinct spacings present in the filled aperture. (After Fomalont and Wright, 1974.)

finding the structure of known sources is the main consideration, rather than looking for fainter sources, then the important feature to retain is the fringe pattern. This pattern depends on the spacings present in the array, the largest spacings giving information about structure on the smallest resolvable angular scales but the smaller spacings also giving valuable simultaneous information about structure on larger scales. This can be seen by thinking about a source drifting through the pattern: if only the central maximum were present, the whole source would have to cross the beam before we knew its size; because there are also sidelobes present, the overall size of the source can be deduced from the number of sidelobes filled simultaneously by the source. Since each spacing contributes its own sidelobe pattern, all the spacings present in the array contribute some information about the structure of a source.

In a two-dimensional array, each spacing has a length and a direction, and can be represented by a vector **S**. Many of these vectors are identical, because a given spacing can be produced by several pairs of elements in the array. This is illustrated in Fig. 4.12(a) for a 4×4 array. In fact such a 16-element array has 120 spacings ($n(n-1)/2$), of which only 24 are distinct (Fig. 4.12(b)). Clearly, the fringe pattern depends only on the *distinct* spacings, and if we can reproduce these spacings with less than the complete array then we can have the same fringe pattern at a lower cost. The minimum size of array that reproduces all the distinct spacings in the filled array is called a *skeleton* array and is an example of an unfilled-aperture telescope.

More technically, the spacing vector **S** can be described in terms of its horizontal components (u,v); conventionally, u points east and v points north and both are measured in units of the wavelength. The visibility of the

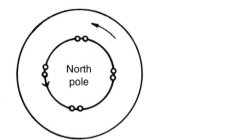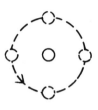

Fig. 4.13. Earth rotation synthesis. A two-element interferometer aligned east–west rotates in space as the Earth rotates, so that one element effectively traces out a circle around the other, synthesizing a ring-shaped aperture for each spacing.

fringes is then a function $V(u,v)$, which is a complicated integral involving the surface brightness distribution over the source; the integral also involves both the response of an individual element of the interferometer array and, crucially, the phase information recorded at each element. (Mathematically, it is essentially the Fourier transform of the brightness distribution, with a given (u,v) corresponding to a single Fourier component.) To recover the brightness distribution, it is necessary to invert the integral by multiplying $V(u,v)$ by a factor which again includes the phase information and then integrating over the uv-plane. Observing at many spacings is thus equivalent to sampling the visibility function at many points in the uv-plane. The better the coverage of the uv-plane, the more accurately can the integral be evaluated and the more information can be obtained about the structure of the brightness distribution of the source.

A snapshot with an unfilled-aperture interferometer will give a first impression of the structure. However, if the source being observed is not varying with time, the coverage of the uv-plane need not be obtained from a single observation. Instead, we can build up the coverage *sequentially* with a set of moveable elements, a technique known as *aperture synthesis*. The moveable elements are commonly small parabolic dishes mounted on railway tracks, which can be deployed to provide the optimum coverage of the uv-plane. Without the constraint of having to have all the elements of the array present at once, it is possible to cover a larger range of spacings and produce a very detailed map of the source. Although there are additional costs and various technical problems associated with having moving elements, aperture synthesis is a very useful way of obtaining high resolution at moderate cost.

The minimum cost is achieved by having a single track with one fixed and one moveable element, giving a variable spacing in a single direction. A two-dimensional aperture can still be synthesized by using *Earth-rotation synthesis*. Fig. 4.13 shows that a two-element system aligned east–west rotates through all distinct orientations with respect to the sky in 12 hours as the Earth rotates. At a given spacing, this rotation synthesizes a ring-shaped aperture and so, by using a range of spacings, a complete filled circle can be synthesized. The main disadvantage of this technique is that it is very time-consuming to obtain sufficient coverage of the uv-plane: a complete synthesis may take up to a month! All practical systems in fact use more than two elements, even if they use the Earth's rotation.

Well-known examples of aperture synthesis instruments are the 5 km telescope in Cambridge in England (where Earth-rotation synthesis was pioneered), the 1.6 km telescope at Westerbork in the Netherlands and the Very Large Array (VLA) in New Mexico, which consists of 27 25 m paraboloids arranged in a Y-shaped pattern along three arms, each 21 km long. The maximum spacing in the VLA is 36 km, enabling a map with a resolution of from 0.13 arcsec to 2.0 arcsec (depending on wavelength) to be obtained. A sensitivity of better than 0.1 mJy can be obtained with 8–12 hours integration time, and in some circumstances 1 mJy sensitivity can be achieved in just five minutes of observing time, enabling slowly varying sources to be studied. The distribution of dishes along three arms at 120° to one another allows good coverage of the uv-plane even for short exposures.

A major problem with aperture synthesis telescopes, and to a lesser extent with all unfilled-aperture instruments, is that the signal-to-noise ratio is very low. To overcome this problem it is usual to use *correlation receivers*, which record only those parts of the input which appear in all elements of the array and reject any uncorrelated parts as noise. This is achieved by arranging that the receiver output is proportional to the average *product* of the output voltages from the components, instead of to their sum. Suppose that there are two components; then the output voltages V_1, V_2 can be written as

$$V_i = V_{iS} + V_{iN}, \; i = 1, 2 \tag{4.21}$$

where the subscripts S, N denote signal and noise respectively. The signal contributions are highly correlated with one another, because they come from the same source, but the noise contributions are random and are completely uncorrelated, either with each other or with the signal. Thus

$$
\begin{aligned}
\text{\textit{receiver output}} \quad &\propto \quad \overline{V_1 V_2} \\
&= \quad \overline{V_{1S} V_{2S}} + \overline{V_{1S} V_{2N}} + \overline{V_{1N} V_{2S}} + \overline{V_{1N} V_{2N}} \\
&= \quad \overline{V_{1S} V_{2S}} \tag{4.22}
\end{aligned}
$$

where the averages are taken over many cycles of the AC voltages, since the other three terms average to zero (there is a low-frequency AC noise output, but no DC output). This technique, which can be used for any interferometer, greatly reduces the sensitivity to noise.

4.4 Very long baseline interferometry

Because of the limited angular resolution available to a single radio telescope, radio astronomers were soon driven to concentrate on interferometers as a means of obtaining the sort of resolution routinely available to optical astronomers. Initially, this was important simply for obtaining positions precisely enough for identification of radio sources with their optical counterparts. As radio telescopes became more sensitive, higher resolution became essential to avoid confusion caused by several (or many) sources being simultaneously in the telescope beam – at a given resolution, the limit to sensitivity is often in practice set by confusion, and telescopes are said to be 'confusion-limited'. So the drive for increasing sensitivity was inevitably accompanied by attempts to obtain higher resolution.

There are obvious practical limits to the size of interferometer array that can use cables to link the elements to the receiver. For larger arrays, the links are themselves made by radio signals, as in the Multi-Element Radio-Linked Interferometer Network (MERLIN) based on Jodrell Bank but with five outstations extending south and west into the Welsh borders with a maximum separation of 134 km, and an outstation in Cambridge at a distance of 220 km.

The longest baselines now in use are comparable with the radius of the Earth, involving, for example, telescopes in California and Europe, or California and Australia. At such distances, even radio links are ineffective and instead the signals are recorded on magnetic tapes, using very accurately synchronized clocks and very stable oscillators to provide precise time and phase markers. The tapes are subsequently analysed together, using a computer to search for correlations that correspond to interference fringes. Structures on scales of 10^{-3}–10^{-4} arcsec can be deduced from the fringe pattern, and positional accuracies of 10^{-3} arcsec are possible, providing a more precise grid of positions than is yet available optically. It is ironic that radio astronomers, driven to interferometry by the low resolution of single telescopes at radio wavelengths, have now surpassed optical resolutions.

The very longest baseline used in radio astronomy to date is the Earth–Moon distance. When the Moon passes in front of a radio source (a lunar *occultation*), the signal is diffracted by the edge of the Moon and a diffraction pattern of the source sweeps across the observer. If this pattern is recorded by a single radio telescope, the size and structure of the source can be deduced, and also its position, because the motion of the Moon is known very precisely and the time of occultation gives a very precise position; corrections have to be made for the non-circular shape of the lunar limb which arises from the presence of mountains, craters or other irregularities. Historically, this technique is important because in 1963 it was used to give the first precise position of the nearest quasar, 3C273 (Section 11.5).

4.5 Radio spectroscopy

Spectroscopy plays a rather different role in radio astronomy from the one it plays in optical astronomy, for two reasons. Firstly, a far higher proportion of strong radio emission is continuum radiation. When the Dutch astronomer J.H. Oort learned about the first radio observations, he was quick to recognize the potential importance of radio spectral lines and he encouraged H.C. van de Hulst to consider possible line-emission mechanisms. None the less, for some 20 years the only important spectral line in radio astronomy was the 21 cm line of neutral hydrogen predicted by van de Hulst in 1944 and first observed in 1951.

The second difference is that instead of using a grating to disperse the incoming radiation in wavelength the signal is measured separately in a number of discrete frequency intervals (channels) using the equivalent of a set of optical filters. Since all radio antennae are sensitive only over a certain frequency band, all radio observations are to some extent spectroscopic.

Within the broad frequency band to which the antenna is sensitive, the receiver of a radio telescope can be tuned to select a narrow band centred at any desired frequency. The simplest way to observe a single spectral line is therefore to choose a bandwidth small compared with the linewidth and to scan the central frequency of the band through the line profile. Calibration is achieved by switching rapidly in frequency between the line and the neighbouring continuum and recording only the difference in signal. However, use of a narrow band reduces the sensitivity of the telescope, so this is a time-consuming procedure for weak sources. A much better solution is to use a multi-channel receiver which can detect signals simultaneously in a large number of adjacent narrow frequency bands. Frequency switching is still necessary for calibration.

With the steady improvement in telescope sensitivity, many more lines from other atoms and molecules have become detectable and it is clear that the radio spectrum is in fact very dense with lines. This makes frequency switching a less satisfactory form of calibration and comparison is often made instead with a neighbouring piece of sky or with a terrestrial calibrating source (known as a 'load'). Sky comparison is almost impossible at 21 cm, since most of the sky is filled with emitting hydrogen, so frequency switching is still regularly used for neutral hydrogen observations.

Multi-channel receivers require a separate amplifier and detector for each channel and it is not particularly easy to change the spectral resolution. A common modern alternative is to replace the electronic circuitry by a fast on-line computer which does the spectral analysis by digital techniques and is much more flexible. A third technique, which is widely used in molecular-line studies, uses a transducer to transform the electromagnetic signal into ultrasonic waves. The variations of density, and thus of refractive index, caused by the passage of the acoustic waves form a sort of three-dimensional diffraction grating through which laser light is passed (the process is analogous to the Bragg diffraction of X-rays by a crystal lattice, and the device is known as a Bragg cell). The ultrasound pattern is reconstructed from its effect on the laser, and the original radio spectrum is then deduced.

Spectral-line measurement involves narrow bandwidths, and can easily be handled by a single receiver. However, there is a limit to the bandwidth over which a given receiver can operate and for line-searches or for measurements of continuum sources over a very large range of frequency it is necessary to use several different receivers, and thus to change the equipment at the feed. For a multi-element array this would be both laborious and costly, so single-dish receivers are favoured for such work.

In modern radio astronomy, spectrometry plays just as prominent a role as the original continuum observations. As well as continuing observations of neutral hydrogen, there are two other particularly important areas of study. Ionized hydrogen can be observed by the radio *recombination lines* emitted when electrons are captured into highly excited levels of hydrogen, and cascade down through levels with principal quantum numbers of several hundred (e.g. the 166α line, corresponding to a transition from $n = 167$ to $n = 166$). The other main constituent of the interstellar medium, molecular gas, also radiates primarily in the radio, especially in the millimetre range; the most useful line is the 2.6 mm rotational line of CO, which can be used to survey the distribution of this very common molecule throughout our own and other galaxies (Section 6.1.7).

4.6 Summary

Radio wavelengths are typically a million times larger than optical wavelengths. This means that a single-dish radio telescope has poor angular resolution and cannot easily form images. It also means that detection techniques are very different, using the wave properties of photons rather than their particle properties. In particular, radio astronomers are able to record the phase of the incoming wave as well as its intensity, a fact that is vital in the application of interferometry. Another difference is that polarization plays a larger role than in optical observations. Not only are many sources intrinsically polarized in the radio but the antenna of a radio telescope is usually sensitive to polarization and a single antenna will only detect part of the incident radiation; crossed antennae are used to record the whole signal.

The effective area of a radio telescope is the ratio of the total power recorded by the antenna to the flux density of the source. It is a measure of the response of the antenna, and it varies with the direction of the source relative to the axis of the antenna. The shape of the variation is known as the power pattern and is equivalent to the diffraction pattern of a point source; the maximum response is generally along the symmetry axis of the antenna, and is the equivalent of the Airy disc. There are sidelobe responses at various angles from the axis, corresponding to diffraction rings, and careful antenna design is needed to minimize their confusing effect on the signal. An extended source has its brightness distribution spread out over the power pattern, and is resolved if its size is greater than the half-power beamwidth of the antenna.

Astronomical sources are generally very weak radio sources, and the noise power of the receiver is always important and must be accurately measured and subtracted from the total signal. Because the main source of noise is thermal motions in the receiver electronics, the noise is most usefully measured in terms of temperature, and it is often convenient to express the signal in terms of temperature also: the antenna temperature for a point source or the brightness temperature for an extended source.

Radio telescopes come in many shapes and sizes, but all have either filled or unfilled apertures. Single-dish telescopes (for example, paraboloids, parabolic cylinders or spherical dishes) are all filled-aperture, and are similar to optical telescopes in overall design, with prime or Cassegrain foci. Large dishes may have limited movement. To extend the size of a telescope without prohibitive cost, many small telescopes may be arranged in a two-dimensional array, analogous to the multi-mirror or mosaic optical telescopes. If the array itself cannot be steered mechanically to follow a source as the Earth rotates, limited electrical steering may be possible by arranging suitable phase delays in the cables from the elements of the array to the central receiver.

Even filled-aperture telescopes have limited angular resolution and, to obtain better resolution, radio astronomers make use of interferometric techniques similar to those used in the Michelson stellar interferometer. The angular resolution is determined by the largest separation between a pair of elements in an array, even if the array is unfilled and has large gaps in it.

In a filled array, there are many identical spacings. The interference pattern is determined essentially by the distinct spacings, so it is possible to remove some elements of the array without changing the information in the fringe pattern or the resolution of the array. This is the principle of using an unfilled aperture. An even cheaper solution is to reproduce all the spacings sequentially with a set of moveable elements instead of having them all present simultaneously; this is aperture synthesis. Often the Earth's rotation is used as part of the synthesis of a ring-shaped aperture.

Of course, the sensistivity of an unfilled array depends only on the total effective area of the individual elements, and is generally small. To prevent the signal-to-noise ratio also being small, correlation receivers are used, in which the receiver output is the average product of the output voltages from the components. Averaging over many cycles of the AC voltages enhances the signal but greatly reduces the contribution from the uncorrelated noise.

The highest resolution radio telescopes have components separated by such large distances that they can no longer be linked by cables. In some cases (for example, MERLIN) radio links are possible; in others, where the baselines are between continents (VLBI), the signals are recorded separately at the different sites and combined later by computer. Such observations yield angular resolutions considerably better than can yet be achieved optically.

Radio spectroscopy does not use a dispersing element similar to an optical diffraction grating. A receiver can be tuned to receive radiation in a particular narrow range, which can then be switched continuously to provide spectral information. Alternatively, simultaneous spectral information can be obtained by using a multi-channel receiver with many adjacent frequency bands, or by using digital techniques to analyse the signal.

Exercises

4.1   Write down a general expression for the sum of two waves of frequency ω, with different amplitudes and phases (see equation (4.3)). Show that if their amplitudes have the same value A, the amplitude of the sum has the values $2A$ and 0 when the difference of phase has the values 0 and π, respectively.

4.2 Suppose that an astronomical signal of the form $V_{ast} = V_1 \sin(\omega_1 t - \phi_1)$ is mixed with a signal from a local oscillator whose form is $V_{lo} = V_0 \sin(\omega_0 t - \phi_0)$ and that the output from the mixer is $V_{out} = A(V_{ast} + V_{lo})^2$. Show that the output from the mixer contains one term which is proportional to $V_1 \sin[(\omega_1 + \omega_0)t - \phi]$ and another which is proportional to $V_1 \sin[(\omega_1 - \omega_0)t - \psi]$, where expressions should be given for ϕ and ψ in terms of ϕ_1 and ϕ_0.

4.3 A two-element interferometer operating at a wavelength of 1 m has a separation of 100 m. It is required to introduce a phase delay ϕ given by equation (4.16) (see also the note following that equation). Show that if $\alpha = 30°$ then an error of 0.1 m in measuring the separation will produce an error of 0.1π in ϕ_0, where $\phi_0 = \phi - 2n\pi$ is between 0 and 2π and is the actual phase delay to be applied. (Thus a 0.1% error in the separation is amplified to a 5% error $((0.1\pi/2\pi) \times 100)$ in the phase because the separation is many wavelengths.)

4.4 The Arecibo radio telescope in Puerto Rico is a 305 m diameter filled-aperture spherical dish, while the Very Large Array in New Mexico has nine 25 m diameter parabolic dishes on each of three arms (27 dishes in all), radiating out at equal 120° angles to one another; each arm is 21 km long. Calculate the ratios of
    (a) the total collecting areas,
    (b) the angular resolutions,
of the two telescopes. Which telescope would you choose for observing (i) faint sources; (ii) structure on a small angular scale?

4.5 An extended source with a uniform brightness distribution has a brightness temperature of 10^4 K when observed at a wavelength of 0.21 m. Estimate the antenna temperature that would be measured if this source were observed with a radio telescope of half-power beamwidth 10 arcmin and maximum effective area 100 m^2. [Hint: in equation (4.7) θ is measured in radians. 1 radian = 3438 arcmin.]

5

Observing at other wavelengths

I have concentrated so far on optical and radio observations, because they are ground-based and involve the majority of astronomers and telescopes. However, many of the new and exciting results in modern astronomy have been obtained by exploring the sky for the first time at other wavelengths that are inaccessible, or accessible only with difficulty, to ground-based observers. Since the 1950s the use of balloons, rockets and satellites has opened up the entire electromagnetic spectrum to observation and has revealed some quite unexpected sights; I will review some of them in Chapter 6.

The techniques involved in observing at some of these other wavelengths are very different from those I have discussed so far and I will now look at some of these differences.

5.1 Infrared observations

The infrared is the only other spectral region that can be even partially explored from the ground. In the range 1–25 μm there are various narrow windows between molecular absorption bands, and in the far infrared beyond 300 μm there is gradually decreasing atmospheric absorption. The far infrared band merges into the microwave radio region at about 1 mm, and is sometimes known as the sub-mm band. The near infrared region up to about 25 μm has been well explored from the ground, but the far infrared has been less well studied, partly because it is in a transition region between infrared and radio techniques where observations are technically quite difficult. Several mm-wave telescopes, such as the UK's James Clerk Maxwell Telescope (JCMT) in Hawaii, are now rapidly filling in the gaps.

In the near infrared, various techniques are available. The earliest infrared measurements were made using uncooled thermal bolometers. These are instruments which absorb radiation on a blackened surface and convert it to heat. The temperature is then measured by a change in some temperature-dependent property of the material, such as the electrical conductivity. The quantum efficiency of such a device is very high, but it generally has poor temperature resolution and is strongly affected by ambient temperature fluctuations. Both of these disadvantages have been overcome in the cooled germanium bolometer, developed by F.J. Low in 1961. The use of a gallium-doped germanium semi-conductor produces a strongly temperature-sensitive electrical conductivity, which greatly improves the sensitivity. The main improvement, however, is the strong reduction in thermal noise by operating at liquid helium temperatures ($T \sim 1$–2 K). This requires the use of vacuum techniques, so infrared astronomers need to be cryogenics experts. Both silicon and germanium are currently used as bolometers and with a suitable absorbing surface detectors can be made that are sensitive into the mm wavelength region.

At wavelengths less than about 5 μm, photoconductive and photovoltaic detectors are the most commonly used. The photoconductive devices use the principles discussed in Section 3.3.3. Photovoltaic detectors are similar, except that what is measured is the potential difference produced across the detector by charge separation; common materials are lead sulphide (PbS) and indium antimonide (InSb). These devices are also cooled, but it is usually sufficient to use liquid nitrogen ($T \sim 77$ K), which does not need a vacuum and is much easier (and safer) to handle than is liquid helium. Other doped semi-conductors, such as Si:Ga and Ge:Ga, can be used in a similar way at considerably longer wavelengths, up to several hundred μm. New materials, with a variety of spectral responses, are constantly being developed.

The main difficulty in making infrared observations is that there is always a strong thermal background. For black bodies of temperature T the emission has a peak at λ_{max} where

$$\lambda_{max} \, T \approx 3000 \, \mu\text{m K} \qquad (5.1)$$

and so the telescope and its surroundings, which are at about 300 K, emit strongly in the infrared, with a peak at about 10 μm. This background varies both in space and in time, because of constantly changing conditions (e.g. air movements and movements of the telescope), and at 10 μm the background thermal emission may be up to a million or more times stronger than the emission from the source being observed. A comparable situation in the optical would be to observe in the daytime with a luminous telescope surrounded by flickering lights! Infrared observations are therefore dominated by the need for accurate subtraction of a constantly changing background. To do this, it is necessary to observe the background as well as the source constantly: this is usually done by switching rapidly backwards and forwards between the source and a nearby patch of blank sky, a procedure known as 'chopping'. It is not feasible to swing the whole telescope to and fro quickly and accurately, and instead the telescope beam is usually wobbled on and off the source using an oscillating mirror. This mirror may be a small plane mirror oscillating in position in the focal plane, or for a Cassegrain system it may be the secondary mirror oscillating in its pointing direction. The amplitude of the offset on the sky is typically less than 1 arcmin and the oscillation frequency may be anything between 1 and 1000 Hz, but is usually between 10 and 100 Hz.

Atmospheric absorption is another major problem for infrared astronomers, one of the major absorbers being water vapour. Infrared telescopes are therefore sited as far as possible on high mountains, above the main cloud-forming layers. An example is the 3.8 m UK InfraRed Telescope (UKIRT) on the 4300 m (14 000 feet) extinct volcanic peak of Mauna Kea in Hawaii.

Even at these altitudes, the region between about 25 and 300 μm is blocked by the atmosphere (mainly by carbon dioxide and water vapour). Unmanned balloons can float for long periods at altitudes of about 40 km, which is above most of the atmosphere. Aircraft can only ascend to about 20 km, but have the advantage that they can carry an observer. For many years, infrared observers in the USA have used the Kuiper Infrared Observatory, permanently mounted in an aircraft, to make high-altitude observations for several hours at a time. Rockets can reach completely above the atmosphere, but only for much shorter flights. All these techniques have been used and have given tantalising glimpses of the infrared sky.

A major development was the launch in 1983 of the InfraRed Astronomy Satellite (IRAS), jointly sponsored by the USA, the Netherlands and the UK. Because of being in vacuum, it was possible to cool the entire telescope to liquid helium temperatures without the terrestrial problem of condensation on the mirrors. This drastically reduced the background (by a factor of 10^{12}!) and made the detection of faint sources much easier, although the Earth's radiation belts produced some problems. The disadvantage of cooling the telescope was the limited life of the satellite; the liquid helium supply lasted for only some 300 days. None the less this was a very successful flight, surveying the whole sky and resulting in a catalogue of some 10^6 sources. The sensitivity is about 1 Jy and observations were made in four broad bands centred on 12, 24, 60 and 100 μm. There was also a low-resolution spectrometer covering the range 7–24 μm. Some of the results are given in Chapter 6.

5.2 Ultraviolet observations

Optical detection techniques can be adopted with little change in the ultraviolet part of the spectrum. The only reason that ultraviolet astronomy is distinct from optical astronomy is that the ozone layer in the Earth's atmosphere completely blocks any radiation shortward of about 300 nm and no ground-based observations are possible.

Conventionally, the ultraviolet region has been taken to extend from the atmospheric cut-off at about 300 nm to about 90 nm. The reason for the short wavelength boundary is that at wavelengths less than 91.2 nm there is strong Lyman continuum absorption in the interstellar medium: the photons are absorbed by ionizing neutral hydrogen from the ground state. This limits observations at $\lambda < 91.2$ nm to directions in which there is little neutral hydrogen. The interstellar fog does not begin to clear again until the X-ray region at about 10 nm.

Until recently, it was believed that the interstellar hydrogen formed a fairly uniform veil and that observations in the extreme ultraviolet (EUV) would be limited to objects within about 100 pc of the Sun; there are very few hot, luminous stars within that distance, so it was thought that the Sun itself would be the only important target. One of the surprises sprung by the ROSAT X-ray satellite, which made the first all-sky survey in the EUV soon after its launch in 1990, was that the hydrogen distribution is in fact very patchy, and in some directions it is possible to see a very long way, revealing many more sources than were expected and simultaneously providing a map of the local gas distribution. These observations are being followed up by the EUVE satellite, launched in 1992, which is using a 0.4 m telescope in the range 7–70 nm (ROSAT covered only 10–20 nm). X-ray techniques are needed in this range, and the telescope uses grazing-incidence optics (Section 5.3.3).

Balloons and rockets have been used as well as satellites to explore the ultraviolet spectrum. Brief rocket flights began as early as 1946. The earliest balloon flights were the series of Stratoscope I and Stratoscope II flights initiated by astronomers at Princeton University in 1957. These were mainly designed for optical astronomy in conditions of good seeing, and serious ultraviolet astronomy had to await the launch of satellites such as the American Orbiting Astronomical Observatories (OAO), OAO-2 launched in 1968 and the highly successful OAO-3 launched in 1972 and named Copernicus in honour of the 500th anniversary (in 1973) of the birth of that famous Polish astronomer.

By far the most successful ultraviolet satellite to date has been the International Ultraviolet Explorer (IUE), a joint venture by NASA, ESA (the

European Space Agency) and the British Science and Engineering Research Council (SERC), which was launched in 1978 for a two or three year mission and is still in active use as this book goes to press (September 1994). One reason for its huge success is that it is operated in the same way as a large ground-based telescope, with astronomers from many institutions visiting the two ground stations in the USA and Spain and using the telescope interactively via a radio link for their own observing programmes. This has enabled ultraviolet spectroscopy to become a routine tool in many investigations, such as studies of cataclysmic variables (Chapter 9) and active galaxies (Chapter 11), and this mode of operation has become a model for later satellites, such as the Hubble Space Telescope (which also has significant ultraviolet capability) and ROSAT.

5.3 X-ray and γ-ray observations

The division between X-rays and γ-rays is rather arbitrary and is usually based on the photon energies. Because it is convenient to be able to say that positron–electron annihilation always leads to two γ-rays, we may take γ-rays to be photons whose energies are greater than the electron rest-mass, which is about 0.5 MeV (the eV is defined in Section 3.2). We therefore define:

$$\gamma\text{-rays:} \qquad\qquad \lambda \;<\; 0.002\,\text{nm}$$
$$E \;>\; 0.5\,\text{MeV},$$

$$\text{hard X-rays:} \quad 0.002\,\text{nm} \;<\; \lambda \;<\; 1\,\text{nm}$$
$$0.5\,\text{MeV} \;>\; E \;>\; 1\,\text{keV},$$

$$\text{soft X-rays:} \qquad 1\,\text{nm} \;<\; \lambda \;<\; 10\,\text{nm}$$
$$1\,\text{keV} \;>\; E.$$

A useful approximate relation for converting between wavelength λ and energy E is $\lambda E \approx 1\,\text{nm keV}$. The soft X-rays merge into the EUV and the notional limit of 10 nm is simply where interstellar absorption starts to become serious.

X-ray astronomy started soon after World War 2 with observations of the Sun using V-2 rockets. The first non-solar X-ray source, known as Sco X-1 because it is the brightest X-ray source in the constellation Scorpius, was discovered by a rocket flight in 1962. Various other rocket flights detected in all about 30 sources, but the X-ray sky was not properly explored until the 1970s, when a series of satellites was launched that transformed X-ray astronomy. The first proper sky survey, in the energy range 2–20 keV, was undertaken by the American Uhuru satellite and led to a catalogue of 161 sources. Uhuru was launched in 1970 from the coast of Kenya, on Kenyan Independence Day, and its name is the Swahili word for 'freedom'; for X-ray astronomy it meant freedom from the limitations of rocket flights.

Uhuru was followed by Copernicus (see Section 5.2), which carried British X-ray detectors, and then by the British Ariel 5 and Ariel 6, the Dutch ANS satellite and a series of NASA satellites culminating in the Einstein Observatory, launched in 1978 and carrying the first X-ray imaging telescope. More recently, there have been the European Exosat and ROSAT, and the Japanese Tenma and Ginga, so X-ray astronomy has been well-served by satellites.

γ-ray astronomy has not developed at such a pace, but the SAS-II and COS-B satellites, both launched in the 1970s, completed very useful surveys of the sky and found many interesting sources. Later satellites include CGRO, the Compton Gamma-Ray Observatory, launched in 1991. The US military series of Vela satellites, designed to monitor violations of the nuclear test ban treaty, also contributed to astronomy by discovering mysterious γ-ray bursts whose origin is still uncertain. Balloon flights have also played an important role in γ-ray astronomy.

The detectors carried by the satellites are very different from the ones I have discussed so far, mainly because the individual photons have very high energy and there are correspondingly fewer of them for a given energy flux. All X-ray and γ-ray detectors are therefore photon-counting devices, and fluxes are usually quoted in counts per second. For example, Uhuru detected sources with strengths between about 1 and 2000 ct s^{-1}. In the spectral range 2–6 keV, and for an effective area of 840 cm^2, one Uhuru count corresponds to an energy flux of 1.7×10^{-14} W m^{-2}. Because of the enormously wide bandwidths used in X-ray astronomy (2–6 keV is equivalent to 1.2×10^{18} Hz), it does not make sense to convert this to Jy to compare with radio telescope sensitivities.

Photographic film will detect X-rays (indeed, one of the first applications of X-rays was in medical photography) and can be used in rocket or balloon flights where the detector will be recovered. However, it is clearly not possible to use film in unmanned satellites. I do not have the space here to describe other X-ray and γ-ray detectors in detail, so I will just outline some of the main techniques, which are related to methods of particle detection.

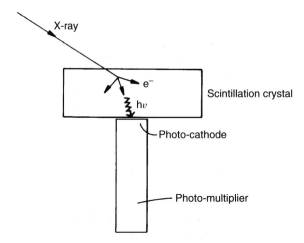

Fig. 5.1. The principle of a scintillation detector. An incident X-ray releases electrons, which are recaptured by impurity atoms, releasing a flash of visible light which is detected by a photomultiplier.

5.3.1 X-ray detectors

For energies less than about 20 keV, the main type of detector is a *proportional counter*. This consists of a gas-filled tube with a wire down the middle held at a high positive potential (~ 2000 V) with respect to the walls of the tube. A photon entering the tube ionizes a gas atom and the free electron has enough energy to create more free electrons before it accelerates towards the central anode wire. As in a photomultiplier, this acceleration rapidly causes a cascade of more free electrons, causing a detectable pulse to be recorded in the external circuitry when the electrons reach the anode wire. Each incident photon produces an initial number of ion–electron pairs that is proportional to the photon's energy and the tube voltage is chosen so that the number of electrons produced in the cascade is directly proportional to the number of ion–electron pairs produced by the photon. The counter thus not only registers the arrival of the photon but can measure its energy. Practical counters are usually rectangular boxes filled with an inert gas and with one side made of a thin X-ray-transparent material such as beryllium or mylar film. Counters can be made position-sensitive, for example by having an orthogonal grid of anode wires within the box.

For energies above about 20 keV, X-rays tend to pass straight through a gas-filled counter without being absorbed, so various solid-state detectors are used. The oldest of these is the scintillation detector shown schematically in Fig. 5.1. The incident X-ray ionizes an atom in the crystal lattice, producing a high-energy free electron which can excite many other electrons

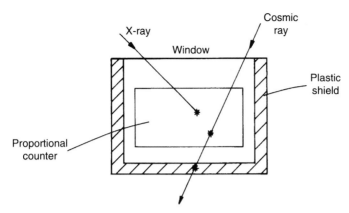

Fig. 5.2. Schematic anti-coincidence device. Incident X-rays are absorbed in the proportional counter, while the higher-energy cosmic rays merely lose a little energy and pass through, causing another event in the plastic shielding. For X-ray detection, events that also trigger the plastic scintillator are rejected.

into the conduction band. The key difference between this detector and the proportional counter is the presence of impurity atoms in the lattice, which can capture some of these electrons and produce flashes of visible light – 'scintillation'. These photons are detected by a conventional photomultiplier attached to one face of the crystal. The scintillation crystals are usually made of sodium iodide or caesium iodide, with thallium as the impurity.

More recently, semi-conductors have been used as high-energy detectors. These are usually p–i–n junction diodes (see Section 3.3.3) made of silicon or germanium, with lithium as the impurity. The incident X-ray releases a high-energy electron in the intrinsic region of the semi-conductor sandwich, and this electron produces a number of electron–hole pairs which separate in an external electric field and are detected as a charge pulse at the junction contacts. For good efficiency in the X-ray region, the intrinsic region needs to be several millimetres thick. Unlike the signal from a proportional counter or the photomultiplier of a scintillation detector, there is no gain and care is needed to avoid noise. However, because the complete detector is solid and small, it is very well adapted to use in satellites. The recent Japanese/US ASCA satellite has a CCD detector in one of its instruments, which has the additional advantage of providing good spectral resolution.

Most X-ray detectors also respond to high-energy particles, such as cosmic rays (Sections 1.1 and 5.4), and it is important to be able to distinguish X-ray and cosmic ray events. This is done by a variety of anti-coincidence devices. For example (Fig. 5.2), the detector may be surrounded by a plastic shield which acts as a scintillation counter for cosmic rays. An X-ray would be absorbed in the detector, so any event which also triggers the scintillator can be rejected as a cosmic ray.

Detection techniques are energy-dependent, so a set of X-ray detectors sensitive in different energy ranges provides crude spectral information. For higher-resolution work, the commonest device is the Bragg crystal spectrometer, in which the regular lattice of a crystal acts as a three-dimensional diffraction grating. The scattering from the lattice atoms also introduces polarization, so crystals can also be used as polarimeters.

5.3.2 *γ-ray detectors*

Low-energy γ-rays (< 10 MeV) can be detected using the same techniques as for hard X-rays, scintillation detectors and semi-conductor diodes, the only difference being that the incident γ-ray may produce an electron–positron pair in the field of a lattice atom rather than being absorbed by the lattice and producing an electron–ion pair. The electron–positron pair may be of such high energy that it passes straight through a scintillation detector, making a second detector necessary.

At higher energies (> 10 MeV), electron–positron pair production is the dominant absorption mechanism and detection techniques concentrate on following the paths of the pairs of high-energy particles. One way of doing this is to use a stack of spark chambers. A spark chamber is basically a rectangular array of small proportional counters operated at such a high voltage that any ionizing particle causes electrical breakdown and produces a spark in the counter. If a high-energy electron passes through a stack of spark chambers, it leaves a three-dimensional track in the form of a spark in each counter through which it passes. Analysis of the tracks of an electron–positron pair provides information about the energy and direction of the incident γ-ray.

The very highest energy γ-rays ($> 10^5$ MeV) can in fact be detected from the ground, although only indirectly. The γ-rays are absorbed in the upper atmosphere, producing a shower of very-high-energy secondary electrons which in turn interact with the atmosphere and produce Čerenkov radiation (see Section 1.2). The difficulty is to distinguish these γ-ray-induced light flashes from the much more numerous events produced by cosmic ray electron showers. However, several groups have begun to obtain useful results.

5.3.3 X-ray and γ-ray imaging

At first sight it might be thought that imaging would not be a problem in the X-ray region; at wavelengths around 1 nm, the size of the Airy disc is about $0.2/D$ arcsec, where D is the aperture in *millimetres*, so even a tiny telescope should give high resolution. However, this argument fails to take account of two important factors. Firstly, to attain this theoretical resolution the telescope mirror would need to be figured to an accuracy of much better than a wavelength, that is, to about 0.1 nm. Secondly, and more importantly, most X-rays are too energetic to be reflected at normal incidence and conventional telescope optics cannot be used except for very soft X-rays, verging on the ultraviolet; even then, a special reflecting coating, with layers of alternating refractive index, is necessary to produce reflection at normal incidence.

Two approaches have been used to solve this imaging problem. The simpler solution is to use some form of mechanical collimator in front of a two-dimensional detector such as a position-sensitive proportional counter. The purpose of the collimator is to restrict the range of directions from which photons are received; this is done by arranging a set of slats or baffles which only allow the detector to 'see' a small region of sky. The baffles are usually made of thin metal plates arranged in an array of rectangular tubes rather like the divisions in an ice-cube tray or in a honeycomb; the typical field of view is a few degrees across, and is determined by the height-to-width ratio of the tubes. (Medical X-rays use a somewhat similar principle, recording not a focused image but a shadow of the human skeleton. However, for medical applications the X-ray source is uniform and the 'collimator' – the human body – is non-uniform, whereas in astronomical applications the collimator has a uniform pattern and is used to observe a non-uniform source.)

For higher-resolution imaging, use is made of the fact that X-rays can be reflected from a polished metal surface if they strike it at *grazing incidence* (less than about 2°). This property was applied to X-ray microscopy by H. Wolter in 1952 and first proposed for astronomical telescopes by R. Giacconi and B.B. Rossi in 1960. A schematic drawing of a grazing-incidence telescope is shown in Fig. 5.3. It consists of a section of a paraboloid of revolution followed by a section of a hyperboloid. The paraboloid alone would form an image but at grazing incidence a single paraboloid suffers severely from coma for off-axis images and is restricted to a useful field of view of less than about 30 arcsec. The hyperboloid reduces the effect of aberrations considerably and allows a resolution of a few arcsec over fields of many arc minutes.

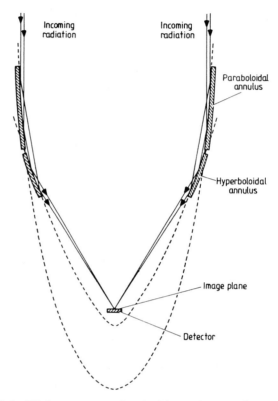

Fig. 5.3. Principles of the Wolter-type grazing-incidence X-ray telescope. X-rays far from the axis are brought to a focus after successive grazing-incidence reflections from cylindrical sections of a paraboloid and a hyperboloid. The reflected angles are greatly exaggerated for clarity of drawing. A central baffle cuts out unfocused on-axis rays. (Diagram by C.R. Kitchin.)

The use of grazing incidence relaxes the requirement of accuracy in figuring the mirrors by a factor of the order of the sine of the grazing angle, though accuracies of several nm are still required. Unfortunately, grazing incidence also drastically reduces the effective area, by a similar factor, so high-resolution imaging telescopes are not very sensitive: the high-resolution telescope on EXOSAT had an effective area of only $10\,cm^2$.

Arcsecond resolution requires detectors with a better resolution than is possible using simple position-sensitive proportional counters. One solution is to use a *microchannel plate*. This consists of a closely packed array of a large number of very narrow hollow metal tubes. An X-ray at a particular point in the image will enter one of these tubes and liberate an electron from the wall. A potential of several thousand volts along the tube accelerates this electron, which then collides with the wall and liberates further electrons. A single X-ray photon can produce a burst of 10^7 electrons or more at the far end of the tube. The mechanical collimation preserves the image, which is recorded by an array of closely spaced fine wires.

γ-rays are too energetic for even grazing-incidence telescopes to be effective and γ-ray images have so far been of lower resolution, relying on mechanical collimation for soft γ-rays and on particle track analysis for hard γ-rays. The instruments on the SAS-II and COS-B satellites had resolutions of $2°$–$5°$; the CGRO satellite has a resolution of $0.4°$ at best.

5.4 Cosmic rays, neutrinos and gravitational waves

Although most of our information about the Universe reaches us in the form of electromagnetic radiation, there are a few other possible information channels, two of which have been seriously exploited so far. A third is being actively explored.

5.4.1 Cosmic rays

Cosmic rays were first definitely detected in 1912 and 1913 in a series of heroic manned flights in open balloons to heights of more than 5 km, which showed clearly that above 1 or 2 km the average ionization increases with height. It is now recognized that the radiation responsible for this ionization, originally believed to be γ-rays, in fact consists of high-energy particles. These particles are mainly protons, electrons and a few light nuclei and have energies ranging from 10^7 eV to 10^{20} eV; the numbers at a given energy E drop off very sharply with energy, $N \propto E^{-2.6}$, making the most energetic cosmic rays extremely rare.

A single cosmic ray entering the top of the atmosphere may on rare occasions reach the ground and be detected directly. More commonly, these primary cosmic rays are absorbed in the upper atmosphere, producing a shower of secondary cosmic rays, mainly electrons (see Section 1.2). A great deal of work on high-energy cosmic rays is still done from the ground, observing these 'extensive air showers' with large arrays of separated detectors, covering many square kilometres. The atmosphere acts as a natural amplifier, producing a million or more particles at ground level for each rare primary cosmic ray. For lower-energy cosmic rays, the fluxes above the atmosphere are larger and satellites are a more effective tool. At energies below about 10^{10} eV, the cosmic rays are significantly affected by the magnetic field streaming out from the Sun in the so-called solar wind, and measurements are difficult to interpret; the 22-year solar activity cycle causes strong modulation of the observed cosmic ray flux.

All the hard X-ray and γ-ray detection techniques that I have described above can also be used for cosmic ray detection – spark chambers, scintillation counters, Čerenkov radiation and so on. For balloon flights or ground observations, where bulk is not a consideration and where the detector can be recovered, other techniques can be used, such as bubble chambers and stacks of nuclear emulsion (very-fine-grained photographic film), both of which give high-resolution records of the ionization tracks of individual cosmic rays. These tracks contain a great deal of information, especially on the energies, masses and charges of the particles, but their analysis is very laborious.

5.4.2 Neutrinos

If current ideas are correct, the whole Universe is bathed in a sea of neutrinos, strange particles with spin but no charge or (probably) mass, and no electric or magnetic moment, predicted by Pauli in 1933 to satisfy conservation laws in particle interactions. Unfortunately, neutrinos are extremely difficult to detect, not because they are absorbed by the Earth's atmosphere but because they are hardly absorbed by anything: the mean free path of a 10 MeV neutrino, even in water, would be 0.1 light years! This makes neutrinos a potentially very important source of information about the very early Universe, since neutrinos produced in the first few minutes of the Big Bang would easily survive without absorption until the present day. Other potential sources of neutrinos are supernova explosions (Sections 9.4.2 and 10.4.5) and violent events at the centres of active galaxies (Section 11.5).

The difficulty with all these neutrino sources is simply that most detectors are not sensitive enough. The first experiment in neutrino astronomy, which was the only one in existence for some 20 years, looked instead for neutrinos from much closer to home: nuclear reactions at the centre of the Sun. The detector consists of a large tank containing 100 000 gallons of cleaning fluid (tetrachloroethene, C_2Cl_4). The heavier stable isotope of chlorine is able to capture neutrinos by the reaction

$$Cl^{37} + \nu \longrightarrow A^{37} + e^-. \tag{5.2}$$

The argon isotope which is produced is a radioactive gas, with a half-life of about 35 days; the gas is flushed out every two or three months and the decay electrons are counted. The limit of sensitivity of the experiment is about 10^{-36} neutrino captures per second per Cl^{37} atom (this is known as one solar neutrino unit, or 1 SNU). R. Davis and his colleagues have been running the detector nearly continuously since 1964 and there is now a clearly significant signal. Although the signal-to-noise ratio (Section 3.5) is low and the detections are both indirect and averaged over a period of months, neutrinos have certainly been detected.

Unfortunately, the result is somewhat embarrassing, since the detector is omni-directional, and the Sun alone is predicted to produce a signal some three times larger than that observed. The detector is located nearly a mile below the ground, in the Homestake goldmine in Lead, South Dakota, to minimize the background signal, which is mainly from cosmic rays. All possible care has been taken with the calibration, and there seems no doubt that the discrepancy is real. In 1988, confirmation came from the independent Kamiokande experiment in Japan, originally set up to look for the decay of the proton. This experiment (described below) used a totally different technique, but produced essentially the same discrepancy.

Over the years since the Davis result first appeared, theoretical models of the Sun have been refined time and again, until now the 'standard' model predicts a signal of (7.9 ± 2.6) SNU, while the average observed flux between 1970 and 1984 was (2.1 ± 0.9) SNU. The first attempts to resolve the discrepancy all concentrated on trying to reduce the neutrino production rate without reducing the luminosity of the Sun (which is fixed by observation). This is possible in principle, because the neutrinos that are detectable by the Davis experiment come from the very centre of the Sun, while the main energy production comes from further out; a small drop in temperature at the very centre could reduce the neutrino flux without a significant effect on the luminosity. In practice, all attempts to date to construct a 'non-standard' model of the Sun have failed some other test of stellar structure.

More recently, attention has switched to the possibility that the discrepancy requires the introduction of new physics into the problem. The current favourite solution appeals to the fact that three types of neutrino are known; the electron, μ and τ neutrinos. If the electron neutrinos, produced in the nuclear reactions in the Sun, were to be able to change into a mixture of all three types between production and detection, there would be a reduction in flux of electron neutrinos at the Earth by about the factor of three required. Is this possible? According to three theoretical physicists, Mikheyev, Smirnov and Wolfenstein, electron neutrinos scattering off electrons inside the Sun *can* be converted into μ and τ neutrinos, but only if at least one of the neutrinos has a mass. This is impossible in standard theories of the electromagnetic and weak forces ('electro-weak' theory), but *is* possible in new theories that seek to unify the electromagnetic, weak and strong interactions. In these 'Grand

Unified Theories' (GUTs), the Mikheyev–Smirnov–Wolfenstein (MSW) effect is expected to operate, making the solar neutrino experiment a possible test for GUTs. The mass required to resolve the discrepancy is very small, less than one fifty-thousandth of the electron mass, and is not in conflict with any experiments to date; indeed, some experiments have already claimed that the neutrino does have a small mass.

How can we test these ideas? One of the difficulties of the Davis experiment is that it detects only high-energy neutrinos from a rare side-chain of the hydrogen fusion reactions that power the Sun; the main proton–proton reaction produces neutrinos whose energy is only half that of the threshold of the experiment. The scattering experiments described below have the advantage of direct detection of individual neutrinos, allowing measurement of arrival times, energies and the direction of the source. However, their threshold is still too high to allow detection of the main-chain neutrinos. In the 1980s new experiments were devised using gallium instead of chlorine. In these experiments, gallium is transformed into germanium by the reaction:

$$\nu_e + {}^{71}Ga \longrightarrow e^- + {}^{71}Ge. \tag{5.3}$$

This isotope of germanium is radioactive, with a half-life of about 11 days, and it is extracted and the decay rate recorded, as in the chlorine experiment (the details are more complicated, because gallium and germanium are both solids, and must be handled in solution). The major advantage of these experiments is that the threshold is low enough for the main-chain neutrinos to be detectable – for the first time, we have an experiment capable of probing directly the main fusion reactions in the Sun.

Two gallium experiments are underway. The GALLEX consortium (Europe/USA/Israel) is based in the Gran Sasso tunnel in Italy, while the SAGE experiment (originally USSR/USA) is at Baksan in the Caucasus mountains. At the time of writing (1993), neither experiment has produced a definitive result. Since the flux of the main-chain neutrinos is directly proportional to the luminosity of the Sun (which depends almost entirely on the energy released in the main chain), the prediction of the neutrino flux is constrained mainly by the observed luminosity and is essentially independent of the details of the solar model. There is a model-independent minimum predicted flux of about 80 SNU. If the main-chain neutrinos are observed to be lower than this prediction, the problem cannot lie in the Sun, and new physics must be required. On the other hand, if the gallium experiments agree with the predicted flux of main-chain neutrinos – and preliminary results suggest that they may – it will still be necessary to understand why there are too few side-chain neutrinos.

Neutrino astronomy was transformed in 1987 with the accidental observation of a neutrino burst from the bright supernova in the Large Magellanic Cloud (Sections 9.4.2, 10.4.5). About ten neutrinos were detected simultaneously in each of two experiments which had originally been set up to look for evidence of proton decay. The experiments are Kamiokande II – the second Kamioka Nucleon Decay Experiment, consisting of a tank of 680 tons of water, about 1 km below the ground in the Kamioka metal mine in the Japanese Alps – and IMB – a similar tank of about 6.8 kilotons of water in the Morton–Thiokol salt mine near Fairport, Ohio in the USA. Both experiments detected the neutrinos by interactions in *water*. There are two important reactions, one being an inverse of neutron decay and the other just involving scattering of neutrinos by electrons:

$$\bar{\nu}_e + p \longrightarrow n + e^+$$

$$\nu + e^- \longrightarrow \nu + e^- \tag{5.4}$$

The electrons and positrons are travelling at such high speeds that they actually exceed the speed of light in water ($v > c/n$, where n = refractive index of water $\simeq 4/3$). They therefore emit pulses of *Čerenkov radiation* (Section 1.2), which are detected by thousands of photomultipliers arranged in a grid on the walls of the tank of water.

These results were much more satisfactory than the solar neutrino experiment, because the observed flux of neutrinos was consistent with theoretical predictions from models of supernova explosions. Since supernovae occur at the final stages of stellar evolution, and the models rely on all the previous stages being more or less well understood, this suggests that the problem in the solar neutrino experiment does indeed lie in the properties of neutrinos rather than in our models of the Sun.

5.4.3 Gravitational waves

According to general relativity, a rapidly changing gravitational field should emit waves, which are essentially oscillations in space–time. If Einstein's theory is correct, these waves will propagate through the Universe with a speed c equal to that of light, and can in principle be detected on the Earth. *Gravitational waves*† have not yet been detected, but their detection would

---

† Note that the term 'gravity waves' should not be used. Gravity waves are internal waves at a density discontinuity in a fluid, and have nothing to do with general relativity. They are responsible, for example, for the regular lines of cloud sometimes seen in a calm atmosphere in the lee of a hill.

be such an important test of general relativity that a great deal of effort is being put into searching for them.

Why should we expect to see any? Perhaps the most likely source to be observed is the decay of the orbital motion in a very close binary star system: if there are two neutron stars in orbit around one another (see below for an example), the rate of energy loss by emission of gravitational waves will be large enough for the size of the orbit to decrease quite rapidly, and the two stars will eventually coalesce in a huge pulse of gravitational radiation. Other possible sources are the collapse of the core of a supernova to form a neutron star, the collapse of a massive object to form a black hole or the decay of cosmic strings. Unlike light, gravitational radiation is essentially quadrupole in nature, which means that spherically symmetric, or even axially symmetric, sources cannot radiate, cutting down the possible sources.

Gravitational wave detectors are very simple in principle: the earliest ones were simply massive metal cylinders, suspended horizontally. The passage of a gravitational wave would cause a ripple in the local gravitational field and set the cylinder vibrating at its natural frequency. The maximum sensitivity would be for a wave travelling at right angles to the axis of the cylinder. Unfortunately, the amplitude of the vibration is expected to be so small – positional changes of less than one part in 10^{18}, and more likely one part in 10^{21} or 10^{22} – that very sophisticated techniques are needed to detect the vibration. The first detectors used piezoelectric crystals, which respond to mechanical strain by developing a potential difference across the crystal which can be detected by an external circuit. It is also necessary to have at least two separated detectors, and to look for coincidences, so as to cut out local sources of vibration, such as earthquakes, heavy road traffic, or even just footsteps in the next room. The need to use large masses (several tons), in order to obtain an even faintly observable signal, also has the effect of restricting the detectable frequencies to the kHz range.

Great excitement was caused in the late 1960s by J. Weber's claim to have detected pulses of gravitational radiation, apparently from the Galactic Centre. The strength of the signal was much larger than expected and many other groups started to build detectors. Although several of these detectors are much more sensitive than Weber's original apparatus, none of them has confirmed his results, which are now attributed to some unrecognized source of systematic error. However, his pioneering work provided the initial impetus for gravitational wave studies, although the general belief is that it will still be some years before detectors are sensitive enough to provide positive results.

Weber's detector was a bar made of aluminium, chosen because of its high Q-value: the Q-value measures the decay rate of the resonant response to a signal at the natural frequency; a high-Q material will oscillate for a long time after a pulse has passed, making the pulse easier to detect. Refinements by Weber's and other groups include very careful shielding from terrestrial disturbances, cooling to liquid helium temperatures to reduce thermal noise and using other material, such as niobium and sapphire, with even higher Q-values.

A different kind of detector is the laser interferometer in which a gravitational wave changes the relative path-lengths in two arms of an interferometer, the tiny changes being measured by fringe shifts. Experiments to detect gravitational waves in this way are being developed in particular by three groups in the USA, the UK and Germany, and Italy and France, with arm lengths of some 3 km. If they continue to receive funding, these experiments now seem to be the most likely to be the first to achieve sufficient sensitivity. Another possible method of detection, which would be sensitive to very-low-frequency gravitational waves, is to monitor the motion of a spacecraft by very accurate Doppler tracking. A passing gravitational wave would distort space–time, producing an apparent abrupt change in the speed of the spacecraft as it responded to the changing gravitational field, like a boat rocking on a water wave.

Although none of these methods has yet produced a positive result, there is indirect evidence that gravitational radiation does exist. The best evidence comes from the so-called 'binary pulsar', known as PSR 1913+16, which is a pulsating radio source, or 'pulsar' (Section 10.4.4). It is a binary system consisting of two neutron stars, highly condensed stars only a few km in diameter (Section 10.4.2), in a very close orbit. Theory predicts that emission of gravitational radiation should cause the two stars to lose energy and to spiral gradually in towards each other; observation shows a decrease in period at the predicted rate. Other effects of general relativity on the orbit have also been observed, such as the precession of the major axis by about $4°$ per year; this was the first system in which it was possible to measure relativistic effects in a strong gravitational field directly.

5.5 Summary

Near infrared astronomy is possible from the ground because there are narrow gaps in the spectrum between the various molecular absorption bands (mainly water vapour and carbon dioxide) which absorb most of the radiation as it passes through the Earth's atmosphere. Between about

$25\,\mu$m and $300\,\mu$m, however, the absorption is total and observations are possible only from space. Detectors use a variety of solid-state techniques, and must be cooled to reduce thermal noise. The major problem with infrared observations is that there is always a strong, and rapidly changing, thermal background from the telescope itself, and from the surroundings for ground-based instruments, and as much time must be spent in measuring the background as in measuring the source.

The ultraviolet part of the spectrum is also inaccessible from the ground, mainly because of absorption by ozone. However, detection techniques are similar to those in the optical region. Ultraviolet observations have become almost routine as a result of the unprecedented success of the IUE satellite, launched in 1978. The extreme ultraviolet, at wavelengths shorter than $91.2\,$nm, has been opened up by the ROSAT and EUVE satellites.

At even shorter wavelengths, observations of X-rays and γ-rays are also possible only from rockets and satellites and detection techniques are very different from those used in the optical. Individual photons have very high energies and behave almost like high-energy particles; as a result, essentially all detectors are photon-counting devices. At energies less than about $20\,$keV, it is possible to use proportional counters, in which the X-ray causes ionization in a gas-filled tube and a current flows that is proportional to the energy of the X-ray. For higher energies, scintillation counters and other solid-state devices are used. Anti-coincidence devices are employed to discriminate between the photons and high-energy particles, such as cosmic rays, which may also trigger the detector. For high-energy γ-rays, detection is possible by following the tracks of electron–positron pairs produced by the absorption of the γ-rays, or by observing Čerenkov radiation from secondary electrons generated in the upper atmosphere.

The main problem with X-ray and γ-ray observations is that the photons are too energetic to be reflected by normal telescopes and imaging is very difficult. The simplest solution is some form of mechanical collimator, which simply restricts the field of view, but the angular resolution is poor. More recent imaging satellites use grazing-incidence reflection and achieve arcsecond resolution, at least for X-rays.

Similar detection techniques are used for cosmic rays, high-energy particles whose energies typically exceed those of γ-ray photons. They were first detected in balloon flights, but much of today's observation is done from the ground, observing showers of secondary electrons, or Čerenkov radiation, produced by absorption of primary cosmic rays in the upper atmosphere.

One of the most difficult particles to detect is the neutrino, although astronomers believe that they are extremely common. The solar neutrino experiment has detected neutrinos in a tank of cleaning fluid, where chlorine is transformed to radioactive argon by capture of a neutrino, and the argon atoms are counted by their radioactive decays. Embarrassingly, the number of neutrinos detected is only one-third as many as are expected from the Sun, a result which seems to be confirmed by independent measurements which use proton capture reactions in water. Either something is wrong with models of the Sun or, as seems more likely at present, the neutrino has a tiny mass and there is the possibility of the electron neutrino produced in the nuclear reactions inside the Sun being changed into a mixture of all three known types of neutrino. New experiments to test these ideas are underway, using gallium instead of chlorine.

A more satisfactory result was obtained from two experiments to detect neutrinos in tanks of water, which serendipitously observed a burst of neutrinos from the 1987 supernova in the Large Magellanic Cloud. In that case, the observed flux of neutrinos was in agreement with supernova models, suggesting that stellar evolution is reasonably well understood.

Gravitational-wave astronomy is still in the embryo stage, with no confirmed detections. The detection of these oscillations in space–time is simple in principle – just the vibration of a large mass as a wave passes – but extremely difficult in practice, with amplitudes smaller than one part in 10^{20} being expected from plausible sources. The most promising form of detector at present is the laser interferometer, which measures tiny changes in the relative lengths of two very long arms of an interferometer. Confidence that a signal will be detected in due course comes partly from strong indirect evidence that gravitational radiation is occurring: orbital changes in the 'binary pulsar' are in excellent agreement with the predictions of general relativity.

Exercises

5.1   Use equation (5.1) to find the temperatures of the thermal sources that would have their maximum emission in the main bands of the electromagnetic spectrum: γ-rays (0.001 nm), X-rays (1 nm), ultraviolet (100 nm), optical (500 nm), infrared (100 μm) and radio (0.1 m).

5.2   An X-ray telescope mirror consists of a nearly cylindrical section, of radius 1 m and length 0.25 m, of a paraboloid. Estimate the total area of the mirror. If X-rays strike the mirror at an average angle of 1°, estimate the effective area for collecting X-rays.

6

Interlude – pictures of the sky

6.1 The sky at many wavelengths

6.1.1 Optical blinkers

Now that we have discussed observational techniques it is time to explore in more detail what these techniques enable us to discover. In this chapter we will look at the sky at many wavelengths in order to obtain an overall impression. We will discover that the view through the narrow optical window gives a very incomplete picture and that exploration of the sky at other wavelengths has revealed many quite new phenomena, as well as giving us fresh insight into the properties of familiar objects.

Since earliest times, people have been aware of the sky, dominated by the Sun during the day and the Moon and stars at night. Calendars were based on the monthly movement of the Moon against the fixed background of the stars, and the seasons were marked by the slower annual motion of the Sun. The regular wanderings of the brighter planets were charted by many ancient civilizations.

Looking beyond our solar system, we now recognize that the sky is dominated by the structure of our own Galaxy, the Milky Way of the ancients, whose flattened disc cuts the sky in half (Fig. 6.1). This disc is made up of myriads of stars and of clouds of gas and dust which hide the centre of the Galaxy. Out of the plane of the Galaxy we see bright, nearby stars, globular star clusters (Fig. 9.12) and external galaxies. On deep photographs we detect faint clusters of galaxies and distant quasars, whose curious optical properties were only discovered after they had first been observed as radio sources.

Our knowledge of the sky at optical wavelengths is much more complete than at other wavelengths. The first serious sky surveys were undertaken almost as soon as the telescope was invented and by the end of the nine-

Fig. 6.1. A mosaic of optical photographs of the Milky Way, re-drawn at the Lund Observatory. The Magellanic Clouds, satellites of our Galaxy, can be seen in the lower right of the picture. (Lund Observatory.)

teenth century there were many catalogues of stars and 'nebulae', as galaxies and gas clouds were both then called. These multiplied throughout the twentieth century, culminating in the systematic surveys undertaken with large Schmidt telescopes (Chapter 2). The Palomar Observatory Sky Survey of the 1950s mapped the whole northern sky to a roughly uniform limit of between 20th and 21st magnitude, the best that was possible with the photographic emulsions then available. In the 1970s and 1980s a similar survey of the southern sky was undertaken jointly by the UK Schmidt in Australia and the European Southern Observatory (ESO) Schmidt in Chile; both are improved copies of the Palomar instrument. With improved optics and better emulsions, the limiting magnitude is considerably fainter, and the resolution is also better. The survey was conducted in two wavebands, blue (in Australia) and red (at ESO). The blue limit, on Kodak IIIa-J emulsion, is at $J = 22.5$ while the red limit, on IIIa-F emulsion, is at $R = 22$. The Palomar Schmidt has now been re-furbished, using the experience gained with the southern Schmidts, and is repeating its survey of the northern sky to a limit at least as faint as the ESO/UK survey. These second-epoch plates will provide invaluable information about variable sources.

Extracting information from these plates is made much easier by the use of automatic plate-measuring machines (Section 3.1). The UK pioneered the use of these machines, of which the best known are the original GALAXY machine (General Automatic Luminosity and XY) developed by Vincent Reddish at the Royal Observatory Edinburgh in about 1970 and its modern counterparts COSMOS in Edinburgh and APM in Cambridge. A single plate contains so much information that its storage in a more immediately accessible form has proved a problem until recently; with the development of optical discs, it is now possible to store all the positions, shapes and brightnesses of all the millions of objects in an almost instantly available format, enabling huge catalogues of stars, galaxies and other objects to be generated almost automatically.

Many objects which were first observed at optical wavelengths have spectra that peak in the visible. This is obviously a selection effect, since we would have been unlikely to have seen objects whose main emission was at other wavelengths, although stars were known whose spectra clearly peaked beyond the visible, either in the ultraviolet or the infrared. However, it did mean that astronomers had very little idea of what to expect when they expanded their view to other wavelengths.

We now know that the Universe contains a vastly greater range of phenomena than our imaginations had forecast. At short wavelengths, the photons are highly energetic and give us information about regions of intense activity and high temperature, some of which are not detectable at all in the visible. At long wavelengths, the low-energy photons come from very cool sources whose existence was previously unsuspected, or from huge regions of relativistic plasma which emit only in the radio. In the remainder of this chapter we will outline some of these discoveries, all made in the second half of the twentieth century, and many of them since 1970, by sky surveys at other wavelengths.

6.1.2 *The ultraviolet sky*

We turn first to a view of the sky which is at first sight little different from what we see in the optical. Fig. 6.2 shows a panorama of the sky at ultraviolet wavelengths. It is still true that there is a concentration of stars to the plane of the Milky Way. However, the concentration is not so extreme, and there is an apparent bunching along the plane. The reason for the first feature is that the hot, luminous stars that radiate most strongly in the ultraviolet are relatively nearby and thus occupy a larger angle in the sky than the optically bright stars; this is because the dust in the plane of the Milky Way strongly absorbs ultraviolet radiation and hides the more distant hot stars. The hot stars are also young, and are concentrated in the spiral arms in which they

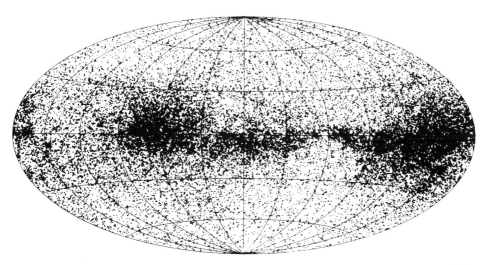

Fig. 6.2. A negative image of the ultraviolet sky, taken by the TD-1A satellite. (Copyright Royal Observatory Edinburgh.)

were born. The regions of high star density mark out two diametrically opposite directions in which we are looking along the length of the local spiral arm within which the Sun is situated.

More subtle differences become apparent when we start to make spectral studies in the ultraviolet. The interstellar gas is cold, and most of the atoms and molecules are in their lowest energy levels, and do not emit very strongly. However, when radiation from distant stars passes through the gas it can be absorbed if the energy of the photon is enough to excite the atom or molecule to its first energy level above the ground state; the spectrum of the star then shows an absorption line arising in the gas between the observer and the star. For a few, relatively rare, species optical photons can effect this excitation, and so-called 'interstellar lines' of sodium, calcium and simple molecules such as CN and CH have been known in stellar spectra since the early years of the twentieth century, recognized by their being much narrower than the lines arising in the star's atmosphere. For the commonest species, however, the first excited state is so far above the ground state that the more energetic ultraviolet photons are needed. Ultraviolet spectra of hot stars then revealed for the first time the true chemical composition of the cool interstellar gas; in particular, the H_2 molecule was observed for the first time on a rocket flight around 1970.

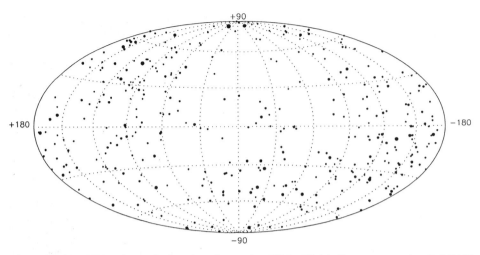

Fig. 6.3. An EUV view of the sky, from the Wide Field Camera on the ROSAT satellite. Note the lack of sources towards the Galactic Centre, where the absorption by neutral hydrogen is greatest. (K.A. Pounds.)

6.1.3 *The extreme ultraviolet (EUV) sky*

With the launch of the ROSAT X-ray satellite in 1990, it became possible for the first time to make observations in the EUV region at wavelengths below 100 nm. This region of the spectrum is technically difficult to observe, because it is on the boundary between optical and X-ray techniques, but perhaps the main reason there had been no previous systematic attempts to explore this wavelength range was because it was believed that little would be seen. As we saw in Section 5.2, interstellar hydrogen absorbs all ultraviolet photons with wavelengths less than 91.2 nm, because such photons have enough energy to ionize the hydrogen. Surprisingly, the interstellar hydrogen turned out to be quite patchy, and EUV observations have not only discovered many nearby white dwarfs and even external galaxies but have been able to map out the local distribution of hydrogen. The results of the all-sky survey are shown in Fig. 6.3, in which the more-or-less uniform distribution is an indication that most of the sources are nearby. There is certainly no hint of clustering to the plane of the Milky Way (which would be a horizontal line in the diagram), and even the local spiral arm seen in the UV is not apparent. There is a dearth of sources towards the Galactic Centre (the centre of the map), where the absorption by neutral hydrogen is greatest.

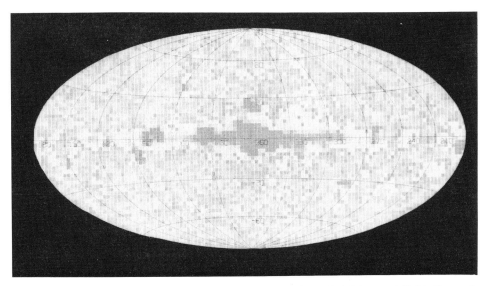

Fig. 6.4. The X-ray sky, seen by HEAO-1. (NASA/Goddard Space Flight Center.)

6.1.4 The X-ray sky

A survey of the sky in X-rays shows a much more uniform distribution of sources than is seen in the optical (Fig. 6.4), but there is still some concentration of sources to the galactic plane. These local sources are mainly X-ray binaries, unknown before the Uhuru and Ariel V satellites of the early 1970s, and supernova remnants. The wispy, ragged supernova shells seen in the optical (Fig. 9.23) turn out to be much more regular in the X-ray (Fig. 6.5, lower picture), showing that much of the gas is shocked to very high temperatures, which emit mainly in the X-ray.

Outside the galactic plane, most of the X-ray sources are very distant active galaxies or clusters of galaxies; the presence of huge masses of hot gas in clusters was first revealed by X-ray observations (Fig. 6.6). However, some nearby globular clusters also contain X-ray sources, which are again X-ray binaries.

There is also a diffuse X-ray background, which may just be the superposition of many faint, unresolved sources.

6.1.5 The γ-ray sky

There are still relatively few γ-ray observations, but it has been known since the early 1960s that the sky is indeed a source of γ-rays. Successive satellites have shown that many known energetic events, from solar flares to pulsars,

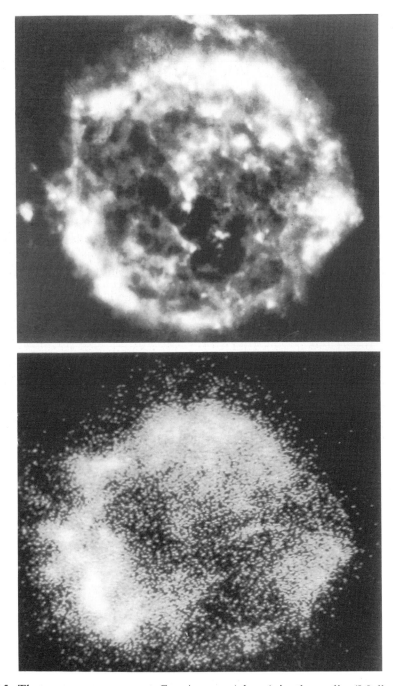

Fig. 6.5. The supernova remnant Cas A, seen (above) in the radio (Mullard Radio Astronomy Observatory) and (below) in X-rays (Einstein Observatory, High Resolution Imager).

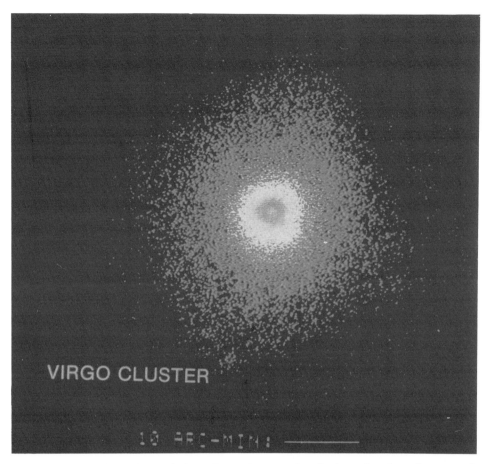

Fig. 6.6. An X-ray view of the Virgo cluster of galaxies, showing the smooth distribution of hot gas around the central galaxy, M87 (Einstein Observatory, Imaging Proportional Counter).

emit copious amounts of γ-radiation, and the 512 keV line from the mutual annihilation of a positron and an electron in cosmic rays near the centre of our Galaxy has been seen. Outside the Galaxy, γ-rays are expected to be emitted in profusion by active galactic nuclei (Section 11.5).

The Compton Gamma-Ray Observatory (CGRO), launched in April 1991, soon began to reveal many new sources. The first all-sky map in γ-rays is shown in Fig. 6.7. The Milky Way dominates, as a strong source of diffuse emission, but several point sources can also be seen. They include four QSOs, extreme examples of active galactic nuclei, and three pulsars. The Crab and Vela pulsars are the only ones to have been detected at optical wavelengths, and were known before the CGRO observations to emit γ-rays. The QSO

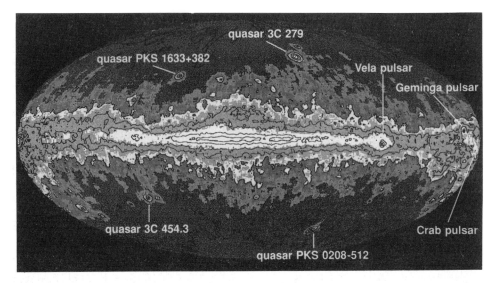

Fig. 6.7. The sky in γ-rays, at energies above 100 MeV, as seen by the EGRET experiment on the Compton Gamma Ray Observatory. (NASA/Goddard Space Flight Center.)

sources are exceptionally powerful: they emit most of their energy in γ-rays and outshine by factors of 10^4–10^5, in γ-rays alone, the total luminosities of normal galaxies at all wavelengths.

The Compton observatory also mapped the distribution of the mysterious γ-ray bursters first revealed by the US Vela satellites. Any one source is seen only once, as a roughly 10 s burst of gamma radiation, and they occur about once a day, coming apparently randomly from all directions in the sky. For many years, the most plausible model had been that violent events near neutron stars were responsible, but this didn't fit with the sky distribution, since neutron stars are concentrated in the galactic disc. It was widely expected that the 20 times more sensitive instruments on CGRO would show that we had just been looking at nearby sources and that the real distribution was indeed flattened in the galactic plane. Surprisingly, the sky distribution remained essentially uniform, suggesting that our solar system really is at the centre of a roughly spherical distribution. The scale of that distribution remains completely unknown, with suggestions ranging from an origin in the Oort cloud of comets, which extends less than half-way to the nearest star, through a population of neutron stars in a hugely extended galactic halo, to neutron star collisions in distant active galaxies. It has even been suggested that the bursts are tracers of incredibly violent events in the very early Universe.

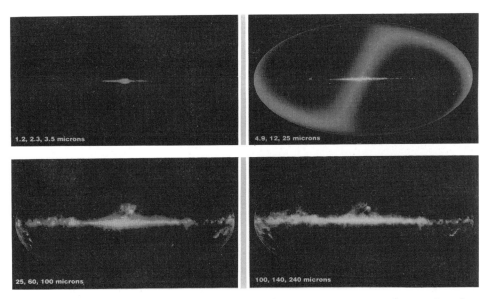

Fig. 6.8. The infrared sky as seen by COBE at a range of wavelengths. (NASA/Goddard Space Flight Center.)

6.1.6 The infrared sky

Most of the planets are strong infrared sources, and it was near infrared observations that revealed that all the outer planets, from Jupiter to Neptune, actually emit more radiation than they absorb from the Sun: they are self-luminous, probably because they are slowly contracting and releasing gravitational energy. As well as these discrete sources, the whole of the ecliptic is marked out by emission from interplanetary dust. In fact, cool dust is one of the principal sources of infrared emission, and the whole sky is covered with 'cirrus' – tenuous clouds of cold interstellar dust ($T \simeq 20\,\mathrm{K}$), obscuring many of the background sources and emitting strongly at wavelengths of a few hundred μm (Fig. 6.8). The full extent of these clouds was first revealed by the IRAS survey (Section 5.1).

The sky map which resulted from the IRAS survey showed again the familiar picture of a concentration of sources to the galactic plane. These galactic sources are clouds of cool gas and dust, often associated with HII regions. Many of them are sites of active star formation, where the newly born star or protostar is still cocooned in an envelope of accreting gas and dust, and is optically invisible (Fig. 6.9).

The COBE satellite gave a more detailed view of the infrared sky, over a greater range of wavelength (Fig. 6.8). At the shortest wavelengths, less than

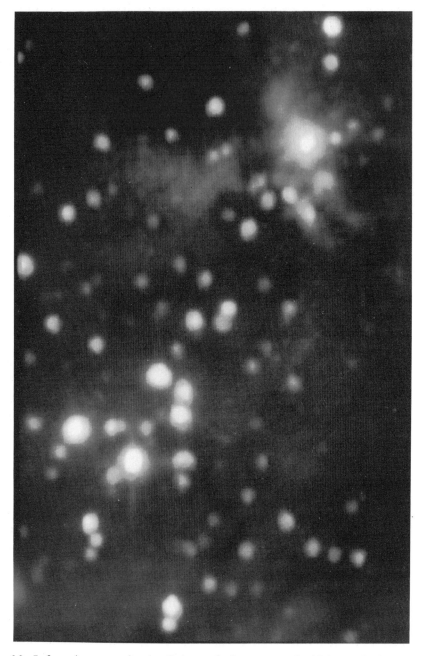

Fig. 6.9. Infrared sources in the Orion nebula, many of which are invisible in the optical. The nebulosity in the upper right of the diagram is the Kleinmann–Low Nebula, which contains a number of protostars and newly formed stars buried in cocoons of dust. (Anglo-Australian Observatory.)

10 μm, starlight dominates and the disc and central bulge of the Galaxy are seen. The disc is still seen at somewhat longer wavelengths, but the most prominent feature at around 10–20 μm is emission from the interplanetary dust in the ecliptic plane. At still longer wavelengths we begin to see the very cool interstellar dust clouds associated with regions of star formation; the major Orion star-forming region can be seen at the extreme right, just below the galactic plane, and the Magellanic Clouds can be picked out further below the plane, about half-way between Orion and the Galactic Centre. At the longest wavelengths, of a few hundred μm, cirrus can be seen in a broad band around the whole sky.

A major surprise of the IRAS survey was the large number of galaxies that have their peak emission in the infrared, with total luminosities of about 10^{12} $L_\odot$. Most of the energy from these galaxies is being absorbed by huge clouds of dust surrounding the nucleus and then re-radiated in the infrared. The most luminous galaxy known at present, IRAS 10214+4724, has $L > 10^{14}$ $L_\odot$; the majority of the radiation is emitted at wavelengths between 1 μm and 1 mm.

Many molecules have vibrational transitions in the infrared, and infrared spectroscopy is an important tool for detecting interstellar and circumstellar molecules. The huge shell around the giant carbon star IRC+10216 is particularly rich in molecules, with more than two dozen detected. Cool dust and gas shells around this and similar red giant stars are also revealed by black-body emission in the infrared, at temperatures of a few hundred degrees.

6.1.7 The microwave sky

The most dramatic discovery in the microwave spectrum was, of course, the cosmic background radiation which is the evidence for the Big Bang (Section 15.3), and the most famous microwave maps of the sky are those from the COBE satellite that were published in newspapers in 1992, showing the 'ripples' in the background that are the seeds of galaxy formation.

However, the bulk of the observations in the microwave region relate not to cosmology but to studies of the interstellar medium (Section 12.3). Most molecules have their rotational transitions in the microwave region of the spectrum and microwave observations have led to the discovery of dozens of molecules since the early 1970s, at a remarkably constant rate of nearly four per year. The most abundant are H_2 and CO, but complex molecules such as alcohol and long-chain molecules such as HC_nN with n up to 11 have also been discovered in giant molecular clouds in the interstellar medium.

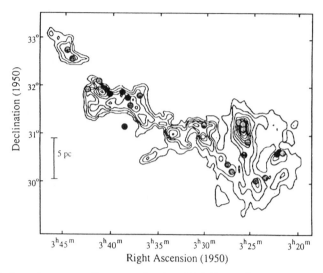

Fig. 6.10. Infrared sources discovered using IRAS are shown as gray and black dots superimposed on a contour map of carbon monoxide. (Reprinted with permission from *The Astrophysical Journal*, 1993.)

Probably the more complex molecules have been formed in high-density gas near the surfaces of stars and then ejected via the stellar winds that are seen in many red giants.

These giant molecular clouds also seem to be stellar nurseries, concealing infrared sources within them which are probably protostars still accreting gas from the clouds that formed them (Section 10.1). Fig. 6.10 shows a number of infrared sources buried in a cloud of carbon monoxide.

Carbon monoxide, with a transition at 2.6 mm, is by far the commonest molecule to be observed in the microwave, because it is a very stable molecule and is therefore very abundant. Maps at 2.6 mm give a good overall impression of the distribution of molecular gas (Fig. 6.11). Molecular hydrogen is far more abundant, but is difficult to observe; because the two atoms are identical, the molecule has no dipole moment and thus no strong emission lines. Also, the main infrared emission line of H_2, at $28\,\mu m$, is unobservable from the ground because it coincides with an atmospheric absorption band. However, it is generally believed that it is collisions with hydrogen molecules that excite CO to the upper level of the 2.6 mm transition (no other species is sufficiently abundant to give enough collisions), so observations of CO also provide a tracer for the distribution of H_2. The 2.6 mm line of CO thus has an importance in galactic studies comparable to that of the 21 cm radio line of atomic hydrogen (see next subsection

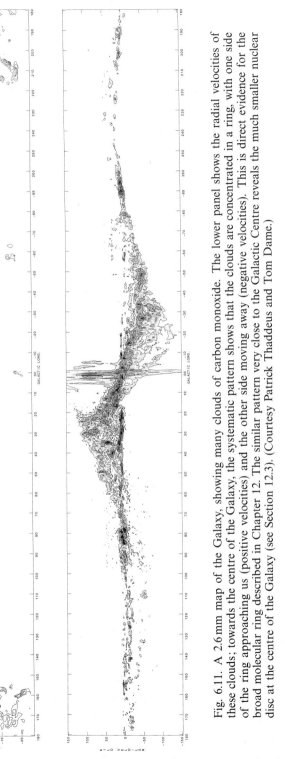

Fig. 6.11. A 2.6 mm map of the Galaxy, showing many clouds of carbon monoxide. The lower panel shows the radial velocities of these clouds; towards the centre of the Galaxy, the systematic pattern shows that the clouds are concentrated in a ring, with one side of the ring approaching us (positive velocities) and the other side moving away (negative velocities). This is direct evidence for the broad molecular ring described in Chapter 12. The similar pattern very close to the Galactic Centre reveals the much smaller nuclear disc at the centre of the Galaxy (see Section 12.3). (Courtesy Patrick Thaddeus and Tom Dame.)

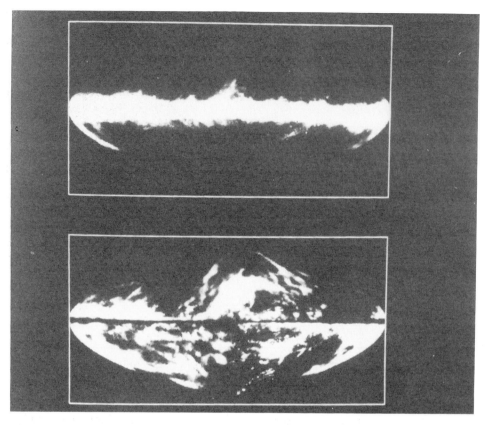

Fig. 6.12. The distribution of neutral atomic hydrogen in the Galaxy (21 cm radio observations). The top diagram shows the concentration to the galactic plane: most of the gas has $|b| < 5°$, where b is the galactic latitude. In the lower diagram, the density has been multiplied by $\sin |b|$ to suppress the contribution from near the plane and reveal the much fainter emission from out of the plane. (Reproduced, with permission, from the *Annual Review of Astronomy and Astrophysics*, Volume 28, ©1990, by Annual Reviews Inc.)

and Section 12.1). Fortunately, the microwave region is not used much for communications and the sky is remarkably free of man-made interference at these wavelengths; the survey which produced the map in Fig. 6.11 was carried out using a small radio dish on the roof of the Physics Department of Columbia University, in the heart of New York city.

6.1.8   The radio sky

The appearance of the radio sky depends strongly on whether it is observed in the 21 cm line of atomic hydrogen or in the continuum.

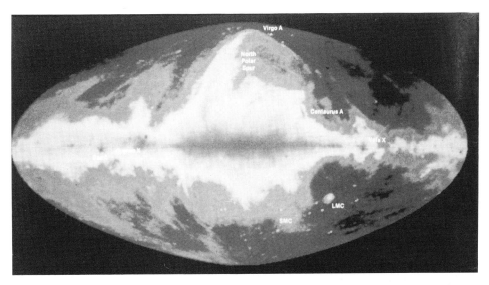

Fig. 6.13. The whole sky at 73 cm. The Galactic Centre is at the centre of the picture. (Max Planck Institut für Radioastronomie, Bonn.)

The 21 cm transition is an unusual one. The lowest energy state of atomic hydrogen is split into two closely spaced levels. The upper level corresponds to the spins of the electron and the hydrogen nucleus being parallel, while the lower level corresponds to their being anti-parallel, which they prefer. Occasional collisions excite some atoms to the upper level, and spontaneous downward transitions produce the 21 cm photons. However, the transition is extremely rare, the upper level having a lifetime of some ten million years. In the laboratory, the transition would never occur, because collisions would de-excite the upper level long before a photon could be emitted. Only in the extreme vacuum of deep space are such collisions rare enough for these hyperfine transitions to occur naturally – and they are only detectable because there are enough hydrogen atoms present to compensate for the extreme rarity of any one of them emitting a photon.

Atomic hydrogen is very widespread throughout the Galaxy, and the 21 cm line is the main tool used to map its distribution. A sky map at 21 cm (Fig. 6.12) at once shows that the hydrogen is concentrated very strongly towards the mid-plane of the Galaxy; it is in fact mostly in a thin disc of thickness only 100 or 200 pc. The disc is not continuous, but is composed of many diffuse clouds, of typical size about 5 pc. Extensions away from the mid-plane are caused by local clouds and spiral arms; some of the filaments and arches in Fig. 6.12 may be gas swept up by supernova explosions. The main disc is so well-defined that it is used as the reference plane for galactic coordinates (Section 7.2).

At 73 cm the sky looks very different (Fig. 6.13), dominated by a huge arch of emission covering half the sky. Here we are observing continuum emission from relativistic electrons moving in the galactic magnetic field. The field is largely parallel to the plane of the Galaxy, but the prominent arch is the North Polar Spur, a very local loop of magnetic field arching over our heads out of the galactic plane, perhaps the shell of a very old remnant from a nearby supernova. The radio view is completely unobscured by dust, and the Galactic Centre can be seen as the brightest region in the galactic plane.

Although these dominant features are very local, emission from relativistic electrons (*synchrotron emission*) is one of the main radio emission mechanisms, occurring in sources as different in size as supernova remnants and double radio galaxies.

6.2 Summary

Despite the different views of the sky at different wavelengths, a recurring theme has been emission from our own Galaxy, which often dominates the sky. Observing at non-optical wavelengths is thus not just a question of getting above the Earth's atmosphere: we also contend with other 'local' sources in our attempt to see the distant universe, and at many wavelengths it is difficult to see faint extra-galactic sources in the galactic plane. Indeed, one of the reasons for the careful survey of the sky by COBE is precisely to determine the local emission from the solar system and the Galaxy with sufficient accuracy to subtract it and reveal the true pattern of emission from the microwave background. Thus observations of the sky at many wavelengths not only reveal a multitude of new sources: they also make it clear that inconvenient absorption is not limited to the Earth's atmosphere!

7

Coordinates and time

7.1 The celestial sphere

From the early attempts to make patterns in the sky by grouping stars into constellations, people have been interested in the positions of the stars and planets. Careful measurements of positions led to Kepler's laws of planetary motion, to the discovery of stellar motions and the rotation of our Galaxy and to the measurement of stellar distances, as well as to many other important discoveries. It is now more important than ever to be able to measure positions of objects in the sky, so as to be able to relate observations made at different wavelengths and to discover whether they are observations of the same object – unless we know, for example, the precise position of a quasar at optical, radio and X-ray wavelengths, we cannot compare the optical, radio and X-ray observations of the quasar and construct its spectrum at all wavelengths. This requires us not only to pinpoint the position of the quasar but also to be able to tie together coordinate systems constructed by optical, radio and X-ray astronomers.

I will discuss the measurement of position in terms of optical observations, but the basic principles, and the various coordinate systems, are the same at all wavelengths. The tying together of coordinate systems defined at different wavelengths is done by successive approximations, first of all identifying bright radio sources as well-known optical sources (for example the Crab nebula) and then gradually refining their positions. I will not discuss the process in any more detail.

Although the objects we observe are all at different distances, we do not in general know these distances, and the position of an object is simply described by its direction. In order to relate the positions of different objects, it is convenient to represent the objects by points on an imaginary sphere of radius much larger than the Earth's radius and centred on the observer.

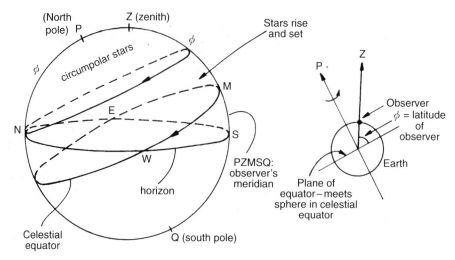

Fig. 7.1. Stars appear to be fixed on the inner surface of a sphere. Positions on this sphere are defined relative to projections of the Earth's equator and rotation axis onto the sky. The sphere is centred on the observer, whose position on the Earth affects the relation between the pole and the horizon.

Each object is taken to be at the point where the direction from the observer to the object intersects this 'celestial sphere'. The rotation of the Earth is represented by the celestial sphere rotating in the opposite direction once a day – so we are back to the ancient view of an Earth-centred Universe and a stationary Earth! In modern parlance, we choose a reference frame fixed on the observer, and look at the apparent motions in that frame.

Since most objects are at extremely large distances, it is often equivalent to take the sphere to be centred at the centre of the Earth (geocentric celestial sphere) or even in some cases at the centre of the Sun (heliocentric celestial sphere). Only for objects within the solar system is it essential to remember that the sphere is centred on the actual observer.

The celestial sphere and its relation to the observer are shown in Fig. 7.1 for an observer in the northern hemisphere. The point directly overhead is known as the *zenith* and the plane normal to that direction defines the *horizon*, the half of the sphere below the horizon being invisible. The plane of the Earth's equator defines another circle on the sphere, the *celestial equator*, and the rotation axis of the Earth projects into the north and south *celestial poles*, P and Q. Directly below the north pole, on the horizon, is the north point N and the equator cuts the horizon in the east and west points, E and W. The circle from P through Z to Q cuts the horizon in S, the south point; this circle is known as the *observer's meridian* and all objects on it are

Fig. 7.2. *Left*: Star trails around the north celestial pole, seen from La Palma, over the domes of the Jacobus Kapteyn and Isaac Newton telescopes. Because the exposure lasted the whole night, traces of both sunset and sunrise can be seen: the domes are reflecting the setting sun (to the left of the picture), while dawn is brightening the horizon to the right. (Royal Greenwich Observatory.) *Right*: Star trails around the south celestial pole, seen over the dome of the Anglo-Australian telescope. Lights around the dome rail are torches of observers, out to look at the weather; those at the base of the dome are car headlights. (Anglo-Australian Observatory.)

due south of the observer. (For an observer in the southern hemisphere, the south point is on the horizon directly below the south pole, and the meridian cuts the horizon in the north point; see exercise 7.1.)

Distances between different points on the sphere are measured by the angle at the centre between the directions to the points (this is equivalent to taking the radius of the sphere as the unit of distance). From Fig. 7.1, it can be seen that $PN = ZM = \phi$, where ϕ is the latitude of the observer: that is, the altitude of the pole above the horizon is the same as the latitude of the observer. This is true for the southern hemisphere also, where the altitude of the south celestial pole above the horizon is the same as the southern latitude of the observer.

As the Earth rotates from W to E, so the celestial sphere appears to rotate from E to W and the stars revolve about the poles on circular paths parallel to the equator (Fig. 7.2). Some stars never set and remain visible

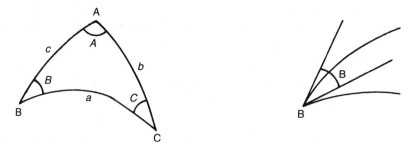

Fig. 7.3. A spherical triangle ABC has sides whose lengths *a*, *b*, *c* are expressed in angular measure. The sides are great circle arcs. The angles *A*, *B*, *C* are defined as the plane angles between the tangents to the corresponding sides.

at night, in some part of the sky, all year; these are called *circumpolar stars* (Fig. 7.1). Stars farther from the pole rise and set and are visible at night only during that part of the year when the Sun is in the opposite part of the sky. Which stars are circumpolar depends on the latitude of the observer; stars within ϕ of the north pole are circumpolar for an observer at north latitude ϕ, and stars within ϕ of the south pole are never seen by such an observer; the reverse is true in the southern hemisphere. A circular gap in the ancient constellation figures, centred on the south pole, shows that the people who named the constellations must have lived at a latitude of about 36° N, probably in the eastern Mediterranean.

The reason for introducing the celestial sphere is two-fold: it forms a convenient pictorial representation of the different directions of astronomical objects, and these directions can easily be related to one another by using the methods of spherical trigonometry. For present purposes, we need only a few basic principles of spherical trigonometry; these are best understood by actual use, and you are strongly recommended to try the exercises at the end of this chapter (more than for any other chapter, for this reason).

We must first distinguish carefully between *great circles* and *small circles*. Great circles on a sphere are those produced by intersection with a plane through the *centre* of the sphere; examples are the celestial equator and the horizon. Lines of longitude on the Earth would also be examples if the Earth were a perfect sphere. Any plane *not* through the centre intersects the sphere in a small circle; examples are lines of latitude on the Earth and the apparent paths on the sky of circumpolar stars. Spherical trigonometry deals almost entirely with great circles.

A great circle arc is the shortest distance between two points on a sphere (so great circle routes are usually used by airlines). We therefore define a *spherical triangle* to be a triangle whose sides are all great circle arcs. An angle between two sides of a spherical triangle is defined as the angle between the *tangents* to the two great circle arcs. Typical spherical triangles are shown in Fig. 7.3. To avoid ambiguities, each side and each angle is taken to be less than 180°. Unlike plane triangles, the sum of the angles has no definite value, but it is always greater than 180°.

With these definitions, the sides and angles of a spherical triangle can be shown to be related by the following formulae, which are the spherical analogues of the cosine and sine formulae of plane trigonometry, and reduce to them in the limit of small sides (exercise 7.2):

$$\cos a = \cos b \cos c + \sin b \sin c \cos A,\qquad(7.1)$$

$$\frac{\sin a}{\sin A} = \frac{\sin b}{\sin B} = \frac{\sin c}{\sin C},\qquad(7.2)$$

where the capital letters refer to the angles of the spherical triangle and the small letters to the sides opposite these angles. Proofs of these formulae can be found in any textbook on spherical trigonometry (see the bibliography) and I will not spend time proving them here. Other, more complicated, formulae are also given in these books, but they are unnecessary for our purposes; almost every simple problem in positional astronomy can be solved by using one or both of equations (7.1) and (7.2).

You have now learned as much as you need to know about spherical trigonometry and I can go on to describe the various coordinate systems that are used to define the positions of astronomical objects.

7.2 Coordinate systems

There are several coordinate systems that can be used to describe positions on the celestial sphere. The simplest of these is based on the horizon (Fig. 7.4). The two coordinates are the *altitude*, a, and the *azimuth*, A. If a great circle arc, ZB, is drawn through the zenith and the object, then a is the distance of the point B above the horizon, measured along the great circle. The distance ZB is called the *zenith distance* z; clearly $z = 90° - a$. The azimuth is the angle at Z between the great circles through the pole and the object; equivalently, it is the distance along the horizon from the north point to the foot of the great circle arc ZB (exercise 7.3). There have been several different conventions for the precise definition of azimuth, and astronomers and navigators have not always agreed upon which one to use; the current astronomical convention is that it is measured eastward

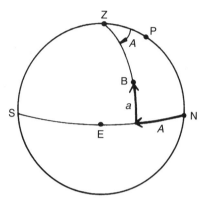

Fig. 7.4. Definitions of azimuth (A) and altitude (a). a is measured vertically from the horizon; the distance $ZB = 90° - a$ is known as the zenith distance (z). A is measured eastward from the north point N in both hemispheres (see text).

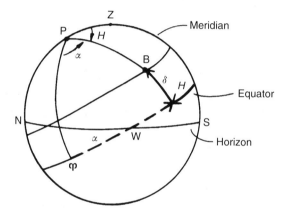

Fig. 7.5. The equatorial coordinate system. Hour angle (H) is measured *westward* from the observer's meridian. Declination (δ) is measured north (+) or south (−) from the equator. Right ascension (α) is measured *eastward* from the first point of Aries (♈; Fig. 7.6). For any object B: $HA\,B + RA\,B = HA\,♈$.

($0° \le A \le 360°$) from the north point in both hemispheres, but you will find different conventions in older books.

   Because the Earth is rotating, an object's altitude and azimuth are constantly changing; this means that they cannot be used as entries in a catalogue of positions, although they are useful concepts for earth-bound observers who need to know at a given instant that the object they want to observe is in an accessible part of the sky and not hidden behind a neighbouring mountain or below the horizon ($a < 0$). A more convenient system for cataloguing purposes is one based on the equator, which rotates with the

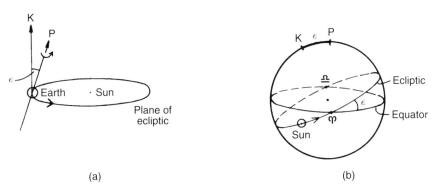

Fig. 7.6. (a) The plane of the Earth's orbit around the Sun projects into the sky as the ecliptic: the apparent path of the Sun through the stars. The rotation axis of the Earth is tilted at an angle ϵ to the normal to its orbital plane. This angle is known as the *obliquity of the ecliptic* and has a value of $23°27'$(very slowly changing with time). (b) The ecliptic and the equator intersect at ♈ (Aries) and ♎ (Libra). The Sun passes through ♈ on March 21, crossing the equator from south to north in the direction of increasing right ascension.

apparent motion of the stars (Fig. 7.5). The obvious analogue of altitude is the *declination, δ*, which is the distance of the object from the equator along the great circle through the pole (PB); clearly $PB = 90° - \delta$. Visible stars may be north or south of the equator; northern declinations are reckoned positive, and southern ones negative.

The analogue of the azimuth is the *hour angle, H*, measured at the pole westward from the observer's meridian (in both hemispheres) ($0° \leq H \leq 360°$). Because of the rotation of the Earth, H increases uniformly with time, going from $0°$ to $360°$ in 24 hours. The hour angle of a particular object is therefore a measure of the time since it crossed the observer's meridian – hence the name. For this reason, it is often measured in hours, minutes and seconds of time rather than in degrees, arcminutes and arcseconds; clearly $1^h \equiv 15°, 1^m \equiv 15', 1^s \equiv 15''$ and so on.

The declination is a fixed coordinate suitable for a catalogue; the hour angle is not, and astronomers choose instead the constant *difference* in hour angle between the object and a fixed reference direction, called the *first point of Aries* and denoted by the zodiacal symbol for the constellation Aries, the Ram: ♈. This difference is called the *right ascension* of the object and is usually abbreviated as RA or given the symbol α. It is measured from ♈ in the opposite direction to H, that is, in the direction of a right-handed screw motion about the direction to the north pole. Like H, it is usually measured in hours rather than degrees ($0^h \leq \alpha \leq 24^h$).

Table 7.1. *Right ascension and declination of the Sun during the year*

	Spring equinox (March 21)	Summer solstice (June 21)	Autumn equinox (Sept. 21)	Winter solstice (Dec. 21)
α	0^h	6^h	12^h	18^h
δ	$0°$	$+23°27'$	$0°$	$-23°27'$

Any fixed reference point on the equator could have been used to define the zero of right ascension. The first point of Aries is chosen by reference to the apparent motion of the Sun which is a reflection of the orbital motion of the Earth around the Sun. The apparent path of the Sun in the sky is known as the *ecliptic* and is the intersection of the plane of the Earth's orbit with the celestial sphere. Because the rotation axis of the Earth is tilted at an angle ϵ to the normal to the plane of the Earth's orbit (Fig. 7.6(a)), the ecliptic is inclined at an angle ϵ to the equator (Fig. 7.6(b)). The angle ϵ is known as the *obliquity of the ecliptic* and is currently about $23°27'$. The ecliptic and the equator intersect at two points, associated with the zodiacal constellations of Aries (the Ram) and Libra (the Scales); the first point of Aries is defined to be the point where the Sun, which moves in the ecliptic in the direction of increasing RA, crosses the equator from south to north. This occurs at the spring equinox, on about March 21, when day and night are of equal length; the right ascension of the Sun then increases through the year, as shown in Table 7.1. The Sun's declination also varies during the year; at the summer and winter solstices (when the Sun 'stands still' in the sky at its maximum declination before starting to move back towards the equator), the Sun is directly overhead at noon at the Tropics of Cancer and Capricorn, respectively, these being the zodiacal constellations (the Crab and the Goat) associated with those parts of the ecliptic where the Sun is at these times (see exercise 7.4).

The Earth's axis of rotation is not fixed in direction, but precesses slowly in space, like a spinning top, about the normal to the Earth's orbit. Because of this, the pole P traces out a small circle of radius ϵ about the pole of the ecliptic, K. This causes the equator to move also, and as a result the first point of Aries is not actually a fixed reference point; it moves gradually backward along the ecliptic, by about 50 arcsec per year, and is in fact currently in the neighbouring constellation of Pisces (the Fishes). This means that even right ascension and declination are not quite fixed coordinates and catalogue positions have to specify the date (e.g. 1950.0 or 2000.0) to which they refer. More precisely, the coordinate frame used for catalogue positions is defined by the positions of the equinox (♈) and equator at a particular date, so astronomers talk about positions referred to (for example) 'the equinox of 1950.0'. Although the pole takes some 26 000 years to make a complete circuit, the effect of precession is much larger than the uncertainties in the positions of objects (Section 7.5), so all positional measurements must be corrected for precession.

Precession only rotates the reference frame, and has no effect on the relative positions of stars. However, the stars are not stationary in space: they are all moving around the centre of the Galaxy (Chapter 12), in different orbits, and the effect is that nearby stars have measurable motions relative to the Sun. The projections of these motions on the sky are known as the stars' *proper motions*, and they *do* cause changes in the relative positions of stars. It is therefore not enough to give a single catalogue position for a star, since it will change *within the reference frame* as a result of proper motion. For stars with significant proper motion, it is necessary to specify both the proper motion and the date of the observation (known as the *epoch*), as well as the catalogue position and the equinox to which it refers! Once the position at a given epoch has been corrected for proper motion, the corrected catalogue position has to be described by both the equinox of the reference frame (e.g. 1950.0) and the epoch of the new position (e.g. 1985.7 for a position which was correct seven-tenths of the way through 1985). If no epoch is given for a position, it is assumed that the epoch is the same as the equinox of the reference frame.

Right ascension and declination are the most commonly used coordinates. However, for solar system astronomy it is often more convenient to refer positions to the mean orbital plane of the system, using *ecliptic coordinates*. Ecliptic latitude β is analogous to declination, but measures distance north or south of the ecliptic. Ecliptic longitude λ is measured from ♈ in the same direction as right ascension but along the ecliptic rather than the equator.

When studying the structure of our Galaxy (Chapter 12), it is convenient to work in terms of *galactic coordinates*, whose reference plane is the disc of our Galaxy. The intersection of this plane with the celestial sphere is known as the *galactic equator*, and is inclined at about 63° to the celestial equator. Galactic longitude l is measured in the same direction as right ascension, but along the galactic equator. The zero of longitude is in the direction of the Galactic Centre, in the constellation of Sagittarius (the Archer); it is defined precisely by taking the galactic longitude of the north celestial pole (at the

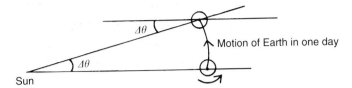

Fig. 7.7. In one day the Earth has an orbital motion of $\Delta\theta \simeq 360°/365\frac{1}{4}$, and so has to rotate by the angle $360° + \Delta\theta$ with respect to the stars in order to rotate once with respect to the Sun. The solar day is therefore longer than the sidereal day by $24^h \times \Delta\theta/2\pi \simeq 4$ minutes.

epoch 1950.0) to be exactly 123°. Galactic latitude b is measured north or south of the galactic equator.

7.3 Time

We have seen that the rotation of the Earth causes a regular change in the apparent positions of stars and that the hour angle of an object can therefore be used as a measure of time. In particular, astronomers use the hour angle of the first point of Aries as a measure of the rotation of the Earth with respect to the stars (more precisely, it measures the rotation of the Earth relative to the equinox, but the difference is small and I will ignore it in this book). This is known as the *local sidereal time* (LST) for an observer:

$$LST = HA\, \text{♈},\qquad(7.3)$$

and is zero when the first point of Aries crosses the observer's meridian.

However, there are two timescales associated with the motions of the Earth. One motion is the rotation of the Earth about its axis and the other is the orbital motion of the Earth around the Sun. The orbital motion of the Earth causes the Sun to appear to move with respect to the stars, which makes the time measured by the hour angle of the Sun, the *local solar time*:

$$local\ solar\ time\ = HA\, \odot,\qquad(7.4)$$

different from the local sidereal time. In a sidereal day the Earth rotates once with respect to the stars. In that time the Earth has moved along its orbit by about $(360/365\frac{1}{4})°$, so the Earth has to rotate by that extra amount before it has rotated once with respect to the Sun (Fig. 7.7). A solar day is therefore longer than a sidereal day by about $24^h/365\frac{1}{4} \simeq 4^m$ or, in solar time:

$$1\ \text{sidereal day} \simeq 23^h 56^m.\qquad(7.5)$$

Another way of expressing this is that the Sun appears to move backwards amongst the stars by about 4 minutes (or about 1°) each day. (The Moon, which orbits the Earth once a month, appears to move backwards amongst the stars by some 13° each day, there being about 13 lunar months in a year.)

A sidereal day has a constant length, apart from small fluctuations in the Earth's rotational rate. However, because the Earth's orbit around the Sun is an ellipse, the Earth's orbital speed varies during the year, causing a solar day to vary in length: the apparent rate of motion of the Sun in the ecliptic is not uniform with respect to the stars. For this reason, and also because the ecliptic is inclined to the equator, the rate of change of the hour angle of the Sun varies from day to day. This *apparent solar time* is what is measured by a sundial, but it is not convenient for most modern purposes because of its variability. Instead, astronomers measure time in terms of a fictitious body called the *mean Sun*, which is supposed to move along the *equator* with a uniform motion in right ascension equal to the mean motion of the real Sun (about 4 minutes of RA per day). The mean Sun then has a uniform rate of change of hour angle as the Earth rotates, and we define:

$$local \; mean \; solar \; time \; (LMST) \; = HA \; of \; mean \; Sun \; \pm 12^h. \qquad (7.6)$$

The 12^h correction is put in to make the mean solar time 0^h at midnight (rather than at midday when the Sun crosses the meridian); the plus sign is used when $0^h \le HAMS \le 12^h$ and the minus sign when $12^h \le HAMS \le 24^h$. Sundial time differs from local mean solar time by up to about a quarter of an hour either way. The two times coincide four times a year – on or about April 16, June 14, September 1 and December 25. A sundial is about 14^m slow in mid-February and 16^m fast in early November.

The time defined by equation (7.6) is a purely local time, because the hour angle depends on the observer and is different at different longitudes. For navigational purposes it has long been convenient to refer all times to a standard longitude, which was chosen by international agreement in 1884 to be the longitude of the old Royal Observatory at Greenwich. Correcting for the observer's longitude λ east ($\lambda > 0$) or west ($\lambda < 0$) of Greenwich, we define a reference or *universal time*:

$$UT \; = HAMS \; - \lambda \pm 12^h, \qquad (7.7)$$

which is actually the local mean solar time at Greenwich.

For local convenience, time within a particular country or geographical region is defined by *time zones*, 15° wide in longitude, within each of which the time is the same and approximates to the local mean solar time at the centre of the zone. The zone time is defined by:

$$zone\ time\ = UT\ + n^h \tag{7.8}$$

where n is constant for a particular zone and is negative for western longitudes and positive for eastern ones; $n = \lambda/15$ where λ (in degrees) is the
longitude at the centre of the zone. In most cases n is taken to be an integer,
so that different zones are an exact number of hours apart. An exception
is India, whose national time corresponds to a time zone with $n = 5\frac{1}{2}$. For
local geographical reasons, the time zone borders may not always be exactly
along lines of longitude; for example, the International Date Line, which
marks the centre of zones ±12, takes large excursions around groups of
Pacific islands so that all the islands in the group agree on the date.

Before the advent of accurate clocks, the rotation of the Earth was the
fundamental timekeeper for all purposes, both astronomical and civil. As
soon as clocks became precise enough to detect fluctuations in the rotation
rate of the Earth, the importance of astronomical observations for civil
timekeeping began to diminish and nowadays *intervals* of time are defined
by atomic clocks, the corresponding time scale being known as UTC –
coordinated universal time. This drifts slowly with respect to UT, because
of fluctuations in the Earth's rotation rate, and because the Earth's rotation
is gradually slowing down, and leap seconds are added from time to time
(usually on January 1 or July 1) to keep UTC in phase with UT. Although
such small corrections have no observable consequences for daily life, they
need to be made, or else eventually (civil) noon would be at (astronomical)
midnight! Thus we still rely on astronomical observations to keep the
zero-point of our timescale correct.

7.4 A practical example: planning an observing trip

You may be feeling that all this is very dull. To try to persuade you that
it is, none the less, extremely useful, let us consider an imaginary observing
trip. Suppose that we want to make observations of the cataclysmic binary
IP Peg, whose coordinates (right ascension and declination, equinox 1950.0)
are: $23^h\ 20^m\ 39^s, + 18° 08' 33''$. The very first question to ask is whether we
need to look for a telescope in the northern or southern hemisphere. Since
the declination is positive, a northern hemisphere telescope would be better
(see exercise 7.18).

The next crucial question is: at what time of year is it best to observe the
star? We want the star to be as high in the sky as possible, so that there is a
minimum of absorption in the Earth's atmosphere. This will be true at the

time of year when the star crosses the observer's meridian near the middle of the night, for the star is nearest the zenith when it is on the meridian. When the star crosses the meridian, it has (by definition of hour angle) hour angle $= 0^h$. From Fig. 7.5 and equation (7.3), we have for any object in the sky

local sidereal time (LST) $=$ *hour angle (HA)* $+$ *right ascension (RA)*.
$$(7.9)$$

Hence the LST when the star crosses the meridian is just the right ascension of the star.

How do we relate the LST to the mean solar time, which is what we need to do to decide whether this time is during the night or not? Now we use the fact that at midnight (mean solar time) the Sun (strictly, the mean sun, but we neglect that correction here) has an hour angle of 12^h. Equation (7.9) applies to the Sun, so the LST at midnight on a particular day is the right ascension of the Sun on that day, plus 12^h. Now we are nearly finished. Table 7.1 shows how the right ascension of the Sun varies throughout the year; it is zero on the spring equinox (roughly March 21) and changes by 2^h per month.

For our cataclysmic binary, the LST for the best observing conditions is about 23^h. From the previous paragraph, we see that this occurs when the RA of the Sun is 11^h, or, more generally:

RA of Sun at best observing time $=$ *RA of object* $- 12^h$.   (7.10)

At 2^h per month, 11^h corresponds to $5\frac{1}{2}$ months, so the best observing conditions would be available around the first or second week in September.

This is the first calculation which astronomers have to make whenever they apply for observing time on a large telescope. There are many more factors to consider as well, for example whether it is possible to observe when the Moon is in the sky, what instrumentation is needed, and what size of telescope to use, but the time of year for the observation is fundamental.

I will just make two other calculations, to illustrate the use of some of the ideas I have introduced in this chapter. One important question, which I have already mentioned briefly, is how high in the sky the star will be at the time of observation. The star has its highest altitude (minimum zenith distance) when it crosses the meridian. Referring to Figs. 7.1 and 7.5, you will see that when a star B is on the meridian it has a zenith distance $z = ZM - \delta$, where δ is its declination and ZM is the zenith distance of the equator where it crosses the meridian. From Fig. 7.1, $ZM = \phi$, where ϕ is the latitude of the observer, so if we know the latitude of our telescope we can calculate the minimum zenith distance very easily from $z_{min} = \phi - \delta$. Note that stars with $\delta = \phi$ pass through the zenith as they cross the meridian.

If the star is to be usefully observed z_{min} must not be too large or absorption in the Earth's atmosphere will seriously affect the observations. A practical limit is about $50°$. To find out how long we can observe a star, we must therefore also calculate how long it remains above a zenith distance of $50°$. Referring again to Fig. 7.5, apply the cosine formula (7.1) to the spherical triangle PZB, in which $PZ = 90° - \phi$, $PB = 90° - \delta$ and $ZB = z$. The angle $Z\hat{P}B = H$, the hour angle of the star at the time of observation, and also (by definition) the time since the star crossed the meridian. If we put $z = 50°$, then H will be the length of time after meridian crossing for which the star will be usefully observable. By symmetry, it will also be observable for the same length of time before meridian crossing, so the total useful observing time will be $2H$, where H is the solution of the equation:

$$\cos 50° = \sin \phi \sin \delta + \cos \phi \cos \delta \cos H. \tag{7.11}$$

Here I have used equation (7.1) together with the relations $\cos(90° - b) = \sin b$ and $\sin(90° - b) = \cos b$ for any angle b. Note that this calculation has assumed that the sky is dark enough throughout this time for the star to be observable! That will usually be true if we have arranged the observing date so that meridian crossing occurs at or near local midnight – otherwise the star may already be above $z = 50°$ at sunset, or still be above $50°$ at sunrise.

For our cataclysmic binary IP Peg, observed from La Palma (latitude $28°45'43''$N), the minimum zenith distance is $10°37'10''$, and the star remains above $z = 50°$ for just over seven hours. On 13 September 1993, when the meridian crossing was at local mean midnight, the Sun was well below the horizon throughout that time.

7.5 Accuracies

We have seen how the idea of a celestial sphere can be used to define the positions of objects, using a variety of coordinate systems. The *relative* positions of individual stars can be measured to about 0.01 arcsec, but the reference frame that defines a coordinate system cannot be tied down to that accuracy, for two reasons. First of all, the observations that define the reference frame are taken over many years, with different instruments, on different sites, and corrections for precession have to be made when combining these observations into a single reference system. The second reason is that the stars that define it move as a result of their proper motion. These two effects, and other smaller effects that I do not have the space to discuss here, make it difficult to define an absolute reference frame very precisely, and in fact the uncertainties grow larger with time as ever larger corrections for proper motion are applied to the stars' originally measured positions.

The most accurate reference frame at the present time is the so-called FK5: the fifth *Fundamental Katalog*, produced in Heidelberg, which is the basis for the epoch 2000.0 positions (Section 7.2), which are now replacing the older 1950.0 positions based on an earlier version of the catalogue. It is an amalgam of catalogues which were compiled between about 1935 and 1951, and the overall accuracy of the FK5 system, derived by looking at the internal consistency between the catalogues, is between 0.01 and 0.03 arcsec (depending on declination) *at the mean epoch of the catalogues.* However, the accuracy of the FK5 system at the present day is much less, because of systematic errors arising from proper motion corrections between the epoch of the catalogues and the present. These vary over the sky, from 0.04 to 0.2 arcsec per century, being worst at extreme southern declinations, and dominate the errors in the present FK5. None the less, over most of the sky the present system is good to better than 0.1 arcsec. Absolute positions of single stars have similar errors, which depend strongly on zenith distance, being least at the zenith. Measurements from individual observatories may differ from the absolute positions by a few tenths of an arcsecond.

Much more accurate positions will be available when the final results are published from HIPPARCOS, an astrometric satellite launched in 1989; preliminary results show that it will produce a reference frame with an overall accuracy of about 0.0015 arcsec (1.5 milliarcsec, or mas), at the mean epoch 1991.5 – nearly a hundred times better than FK5 at that epoch. However, the proper motions have errors of 1 or 2 mas per year, so again we see that after ten years or so the errors in positions will be dominated by those in the proper motions. Thus any reference frame that is based on stars degrades with time and must be continually renewed. Attempts have been made to base a fundamental reference system on objects with negligible proper motion, such as QSOs, but they are scattered too thinly over the sky (except at faint optical magnitudes) to provide a well-defined frame at all points, and it is therefore essential for the foreseeable future to continue observing and refining the positions of stars.

The best radio positions are known from VLBI to better than 1 mas, which is about one-tenth of a beamwidth at 5 GHz. Since the beamwidth depends on frequency, the accuracy of positions depends on the frequency of observation, and is best at high frequency. Positions from smaller arrays, such as MERLIN and VLA, are probably good to 10 mas. Because it is possible to use radio sources with negligible proper motion to define the radio reference frame, the radio frame is somewhat more self-consistent than the optical one, but it cannot be used to define right ascension because there are no radio sources in the solar system that are suitable for fixing the position of the first point of Aries (the plane of the solar system cannot be determined by radio observations). None the less, the correspondence between optical and radio coordinates is good to 0.1–0.2 arcsec, the limit being set by the optical errors.

Relative positions can be determined to better accuracy, at least for neighbouring objects, and here also the radio astronomers now do best, determining the structure of radio sources to milliarcsecond accuracy with very long baseline interferometry.

Positional accuracies for most satellite-based observations are less good. The difficulty here is not the pointing accuracy of the satellite, which depends on optical tracking of reference stars and typically gives absolute directions to better than 0.1 arcsec and relative directions (which depend only on the stability of the tracking sensors) to better than 0.01 arcsec. The limitation arises simply from the low angular resolution of most non-optical telescopes. Although ultraviolet and infrared space telescopes have good resolution and are limited only by the pointing stability, even one of the best X-ray satellites, the Einstein Observatory, achieved a spatial resolution of at best 1 arcsec, and γ-ray positions are still uncertain to tens of arc minutes, the uncertainty increasing towards higher energies.

I have explained that position and time are intimately related because we live on a moving Earth. Satellites are also mobile observatories and accurate timing is a vital part of satellite observations. Absolute timings, from the ground or from a satellite, can be made to about a microsecond, and relative timings to considerably better than a nanosecond using stable oscillators; the limit to the precision is the response time of the electronic timing circuits and in the very best cases can be tens of picoseconds. Accuracies of that kind have made possible the study of many rapidly varying sources, such as pulsars (Section 10.4.4), as well as helping in positional measurements.

In the earliest times, astronomy was almost entirely concerned with positional measurements. Although positional astronomy is not at present a very fashionable area of research, it remains of fundamental importance for all observational work. In the remainder of this book I will not very often use explicitly the ideas that have been introduced in this chapter, but they underlie all the results that I will discuss.

7.6 Summary

Most astronomical positions are defined in terms of coordinates based on the Earth's equator as a reference plane. Declination is measured north and south of that plane and right ascension is measured round the equator from the first point of Aries, which is where the ecliptic crosses the equator from south to north. The hour angle of the first point of Aries measures sidereal time (with respect to the stars) and the hour angle of the Sun measures solar time. These are different because the Earth is orbiting around the Sun. Neither time is completely uniform and for civil purposes atomic clocks are used to measure time intervals, although the phase is still related to astronomical time.

As well as these fundamental definitions, I discussed the use of the celestial sphere to represent stellar positions, and introduced enough spherical trigonometry (spherical triangles, sine and cosine formulae) to enable you to make simple calculations. Some of the ideas were applied in the planning of an imaginary observing trip. Other coordinate systems were mentioned (altitude and azimuth, ecliptic coordinates and galactic coordinates). I explained how precession causes the rotation of the equatorial coordinate frame and how proper motions degrade the accuracy of any reference frame based on stars. The best optical reference frame is currently the one based on observations by the HIPPARCOS satellite.

Positional astronomy is of fundamental importance for relating measurements made at different times, from different places and in different wavelength regions. This chapter has outlined the basic ideas involved, although the practical application is much more complicated.

Exercises

7.1 Draw the diagrams in Fig. 7.1 for an observer in the southern hemisphere.

7.2 By considering the limit of small sides, show that the cosine and sine formulae of spherical trigonometry (equations (7.1) and (7.2)) reduce to the usual formulae for plane trigonometry. You may assume that, for small x, $\sin x \simeq x$ and $\cos x \simeq 1 - x^2/2$ and that cubes and higher powers of small quantities may be neglected.

7.3 Consider two great circles drawn through the zenith, Z, and cutting the horizon at A and B. Show that the (angular) length of the great circle arc AB is equal to the angle at Z between the two great circles. [This result holds quite generally for any two great circles drawn through a 'pole' of a third; the 'poles' of a great circle are the places on the celestial sphere where the normal to the plane defining the great circle meets the sphere.]

7.4 Show, by drawing celestial spheres or otherwise, that at the summer and winter solstices (June 21 and December 21 respectively – see Table 7.1), the Sun is directly overhead at midday for an observer at the Tropics of Cancer and Capricorn (latitudes $+23°27'$ and $-23°27'$ respectively).

7.5 Astronomical twilight is the period of time after sunset during which the Sun is less than $18°$ below the horizon. Neglecting the angular size of the Sun, find the latitude at which astronomical twilight just lasts all night at the summer solstice.

7.6 How much of the visible sky is circumpolar for an observer (a) at the North Pole? (b) on the equator? What fraction of the whole celestial sphere is visible at *some* time for each of these two observers?

7.7 Given that the mean radius of the Earth is 6371 km, and assuming that the Earth is a perfect sphere (not quite true!), calculate the rotation speed of the Earth in $km\,s^{-1}$ at latitude $51°$ N.

7.8 What is the zone time for a place $117°$ E when the hour angle of the mean Sun there is $+3^h$?

7.9 What is the local sidereal time (LST) when the first point of Aries crosses the observer's meridian?

7.10 A star is observed to cross the meridian at an altitude of $+60°$ at a time when the LST is $+5^h$. If the observer is at latitude $+30°$, what are the right ascension and declination of the star? What would its altitude be at this time for an observer at latitude $-30°$?

7.11 Show that, when a star of declination δ sets for an observer of latitude ϕ, its hour angle H is given by

$$\cos H = -\tan\phi\tan\delta.$$

7.12 Suppose that the Sun is on the celestial equator and that the Moon's phase is first quarter. What is the hour angle of the Moon at sunset?

7.13 At local sidereal time (LST) 0^h, a star is observed on the observer's meridian (i.e. due South for a northern observer) at an altitude of $49°$. If the observer's latitude is $51°$ N, what are the declination and right ascension of the star?

7.14 A star has right ascension $+6^h$, declination $+30°$. Find its altitude above the horizon for an observer at latitude $30°$ N when the LST is 9^h. At what LST would it be directly overhead? At what time of year would the star be directly overhead at local midnight (hour angle of

Sun $= 12^h$)? (The Sun's right ascension may be taken to increase uniformly at a rate of 2^h per month; RA($\odot$) $= 0^h$ on March 21.)

7.15   A transit telescope is used to observe a star with right ascension 12^h, declination $+5°$. If the star has an altitude of $65°$ at the time of observation, find the latitude of the telescope. What is the local sidereal time (LST) of the observation? If the observation were made on March 21 (RA of mean Sun $= 0^h$), at longitude zero, what would be the universal time (UT)?

7.16   At latitude $30°$ S, a certain star is observed to have an altitude of $30°$ when it is due west. Find the star's declination. If the hour angle of the first point of Aries is 6^h when the star is due west, find the star's right ascension.

7.17   At what zone time will the Sun be due south at a place of west longitude $100°$ on a date when the equation of time (= hour angle of Sun − hour angle of mean Sun) is $+15$ minutes?

7.18   An observatory is situated at latitude $30°$ north. What is the (southern) declination of a star that has a zenith distance of $50°$ when it crosses the meridian? (This is an example of how to calculate the practical limit for observing the southern sky from the north.)

7.19   What is the hour angle of the Full Moon at local midnight?

8

Magnitude systems and stellar spectra

In the preceding chapter I established a framework on which to build observations. You have already seen, in Chapter 6, an outline of these observations and I will now turn to discuss some of them in more detail.

The most immediately obvious astronomical objects are the stars, and I will begin by discussing their brightnesses and spectra. In succeeding chapters I will describe the structure and evolution of stars before going on to discuss more briefly the great aggregates of stars that astronomers call galaxies and that were only discovered after telescopes became available.

8.1 Stellar magnitudes

As you saw in Chapters 1 and 2, optical astronomers express brightness in terms of magnitudes. This is a historical quirk, arising from the logarithmic response of the eye, and is peculiar to the visible part of the spectrum. In other wavelength regions, as you saw, for example, in Chapter 4, the observed signal is directly recorded as a flux.

However, the use of magnitudes in optical astronomy is likely to continue for some time, so I must now be a little more precise about the definition. Consider a star, at distance d, which has a total power output L (over all wavelengths); L is known as the *luminosity* of the star and is measured in watts. Then the flux F reaching the Earth is:

$$F = \frac{L}{4\pi d^2}.$$ (8.1)

This is the *bolometric* flux, that is, the flux that would be measured by a detector which was equally sensitive at all wavelengths. The corresponding magnitude is known as the apparent bolometric magnitude, m_{bol}, and is defined by

$$m_{bol} = -2.5 \log_{10}(L/4\pi d^2) + \text{constant}$$

(8.2)

$$= -2.5 \log_{10} L + 5 \log_{10} d + \text{constant}.$$

This magnitude measures the *apparent* brightness of a star, which depends both on its intrinsic brightness, or luminosity, and on its distance. It is useful also to define a magnitude that can be used to describe the *intrinsic* brightness; the usual choice is the absolute bolometric magnitude, M_{bol}, which is defined as the bolometric magnitude which the star would have if it were at a standard distance (taken to be 10 parsecs – see Table 1.2):

$$M_{bol} = -2.5 \log_{10} L + 5 \log_{10} 10 + \text{constant}$$

(8.3)

$$= -2.5 \log_{10} L + \text{constant}.$$

The difference between the apparent and absolute bolometric magnitudes is then a measure of distance, known as the *distance modulus*:

$$m_{bol} - M_{bol} = 5 \log_{10}(d/10 \,\text{pc}).$$ (8.4)

However, no-one can ever measure the bolometric flux or magnitude directly. This is partly because of absorption in the Earth's atmosphere, but more fundamentally because no receiver is equally sensitive at all wavelengths and any observation only detects some fraction of the total incident flux. Suppose that $f(\lambda)$ is the flux distribution at the Earth (in $W\,m^{-2}\,nm^{-1}$) and that $s(\lambda)$ is the spectral sensitivity of the receiver, which here is taken to include the whole telescope system. Then, if I neglect the effect of the Earth's atmosphere, which I have discussed earlier (Section 1.2), the measured flux (above the atmosphere) is

$$F = \int_0^\infty f(\lambda)\, s(\lambda)\, d\lambda.$$ (8.5)

Different magnitude systems are then defined by different functions $s(\lambda)$.

There are many magnitude systems in use in astronomy, each determined by a particular combination of filters which allows a particular waveband of interest to be isolated. A typical sensitivity function is shown in Fig. 8.1: the width of the waveband varies from about 2 nm for narrow-band magnitude systems to about 100 nm for broad-band systems. The commonest system is the broad-band *UBV* system introduced by H.L. Johnson and W.W. Morgan in the early 1950s. The central wavelengths and bandwidths of the three

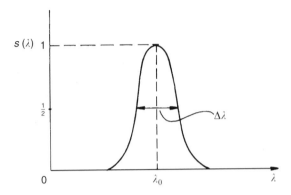

Fig. 8.1. A schematic spectral sensitivity function, characterized by a central wavelength λ_0 and a width $\Delta\lambda$ that may be anything from 1 or 2 nm (narrow band) to 100 nm (broad band).

Table 8.1. *The central wavelengths and bandwidths of the filters of the UBV photometric system. The three filters correspond roughly to the ultraviolet, blue and visual parts of the spectrum.*

	λ_0/nm	$\Delta\lambda$/nm
U	350	70
B	440	100
V	550	90

filters are given in Table 8.1 where, for example,

$$V = m_V = \text{constant} - 2.5\log_{10}\left(\int_0^\infty f(\lambda)s_V(\lambda)\,d\lambda\right). \qquad (8.6)$$

The constant is chosen by assigning a magnitude to a particular star (see Section 2.2 – the visual magnitude scale uses the star λ U Mi).

The original *UBV* system was defined in terms of a particular piece of equipment mounted on a particular telescope. Since then many other observatories have constructed *UBV* filters for use on their own telescopes. It is virtually impossible to construct two completely identical filter–telescope combinations, so each observatory really has its own local instrumental system, which is some approximation to the standard system. It is clearly important to be able to compare observations made with different telescopes, so astronomers need a way of calibrating instrumental systems and thus transferring observations onto a standard system. This is done by observing a set of standard stars, whose magnitudes on the defining system are well

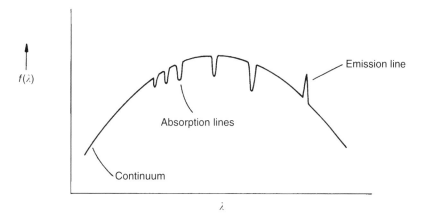

Fig. 8.2. A highly schematic stellar spectrum. The continuum emission is approximately black-body. All stars show superimposed absorption lines, which may be very numerous. A few stars also show emission lines.

determined, and using these to calibrate the local magnitude system. Thus any practical magnitude system is defined by two equally important factors: not only the response function $s(\lambda)$ but also a set of standard stars which are used for calibration.

8.2 Stellar spectra

A schematic stellar spectrum is shown in Fig. 8.2. It consists of a continuum, which is roughly that of a black body, on which is superimposed a considerable number of absorption lines; some stars also show emission lines. A black-body continuum is uniquely specified by the temperature of the black body, as you saw when I discussed the brightness temperature of radio sources. Although stars are not black bodies, their continua are near enough to black-body spectra for it to be useful to define an *effective temperature*, $T_{\rm eff}$, which is the temperature of the black body of the same size as the star that would emit the same total power. For a black body, Stefan's law tells us that the surface emissivity is σT^4, where σ is the Stefan–Boltzmann constant $(5.67 \times 10^{-8}\,{\rm W\,m^{-2}\,K^{-4}})$; since stars are assumed to be spherical, the effective temperature is related to the luminosity, L, and the radius, R, by:

$$L = 4\pi R^2 \sigma T_{\rm eff}^4. \tag{8.7}$$

Defined in this way, the effective temperature is an estimate of the surface temperature of the star. It is usually found from the shape of the contin-

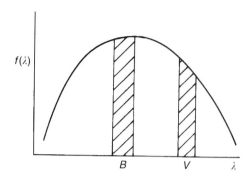

Fig. 8.3. The shape of the continuum in a stellar spectrum can be estimated by measuring the flux in two wavebands (e.g. B and V) and taking the ratio of these fluxes. For a black body, this ratio would be uniquely determined by the temperature.

uum, which can be estimated by measuring the stellar flux in two different wavebands, for example the B and V bands of the UBV system. As can be seen from Fig. 8.3, the ratio of the fluxes in these wavebands enables us to estimate the shape of the continuum and thus to estimate T_{eff} by comparing the shape to that of a black body. The ratio of the fluxes is measured by the difference of the corresponding magnitudes; since stars of different temperatures also have different colours, astronomers call this difference of magnitudes a colour index:

$$colour\ index = B - V. \tag{8.8}$$

A second colour index in the UBV system is $U - B$. Very roughly, the $B - V$ colour index is related to the effective temperature by

$$B - V = 7000\,\mathrm{K}/T_{eff} - 0.56; \tag{8.9}$$

cooler (and therefore redder) stars have larger $B - V$.

For black bodies, there would be a nearly linear relationship between $B - V$ and $U - B$. The departure of real stars from black bodies can be seen in the actual two-colour relationship shown in Fig. 8.4. The major dip away from the black-body line for stars with $B - V$ near zero is largely due to absorption by hydrogen in the Balmer continuum at wavelengths less than 364.6 nm. Fig. 8.5 shows the corresponding electron transitions and Fig. 8.6 shows the shape of the stellar continuum compared to that for a black body; there is clearly considerably less flux in the ultraviolet and blue parts of the spectrum from the star than from a black body. The rest of the dip in the two-colour diagram is due to line absorption by hydrogen (the Balmer series) and by other elements.

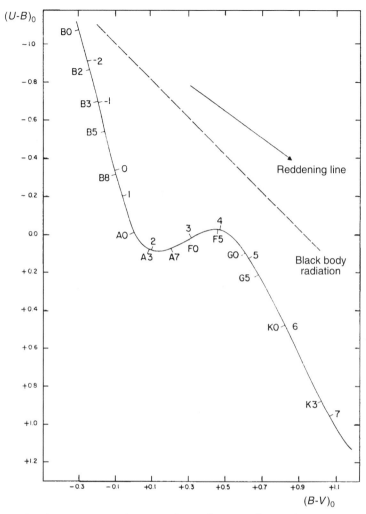

Fig. 8.4. The $(U - B, B - V)$ two-colour diagram for main-sequence stars. Note that $U - B$ is conventionally plotted as increasing downwards. The theoretical relationship for black bodies is also shown. The reddening line is explained in Section 12.2. (©1963 by The University of Chicago. All rights reserved.)

The Balmer continuum and line absorption arise in a thin surface layer of the star known as the photosphere. This layer is cooler than the interior and the atoms and ions tend to absorb the underlying continuum radiation; roughly speaking, the hot interior is a source emitting a black-body continuum spectrum and absorption in the photospheric layers modifies this black-body spectrum to produce the observed stellar spectrum. This process is shown schematically in Fig. 8.7. The atoms in the photospheric layers

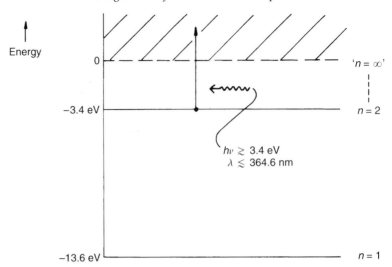

Fig. 8.5. Energy levels for the hydrogen atom. Photons with wavelengths less than the Balmer series limit at 364.6 nm have enough energy to ionize the hydrogen, causing continuous absorption shortward of this limit.

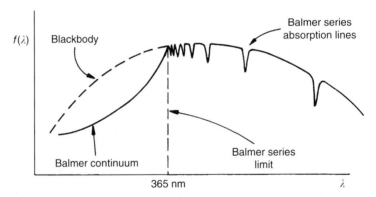

Fig. 8.6. Schematic spectrum of a star, showing hydrogen Balmer lines and continuum. There is a large departure from a black-body spectrum in the near ultraviolet.

absorb the radiation streaming out from the hot interior and re-radiate it in all directions, so that the radiation reaching the observer is reduced; another way of describing the absorption is to say that the continuum emission comes mainly from layers at the effective temperature, whereas the absorbing atoms are in cooler layers higher in the atmosphere where the emissivity is lower. Emission lines, by a similar argument, come from regions where the local temperature is higher than T_{eff}. Such regions will occur, for example, where energy is released by instabilities, such as the flares seen on the Sun in regions

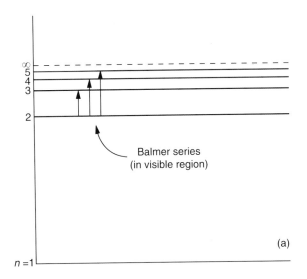

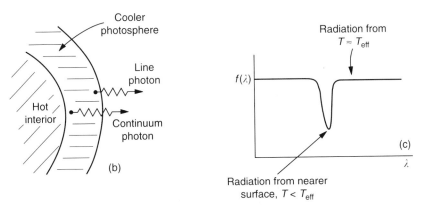

Fig. 8.7. (a) The line spectrum of atomic hydrogen. (b) Line photons are produced nearer to the surface, in cooler regions than the continuum photons. (c) Because of this, there is less radiation at line frequencies.

of strong magnetic field; similarly, emission lines in a stellar spectrum are an indication of possible surface activity in other stars.

Because every absorbing atom or ion has its own characteristic pattern of lines, the line spectrum can be used to determine which elements are present in the star's atmosphere. The relative strengths of the various lines can also be used to find the relative abundances of the various elements. It is clear that the strength of an absorption line must depend on the number of atoms doing the absorbing. However, a particular line arises from a transition upwards from a particular energy level, so the line strength only tells us how many atoms there are in that level. To relate that number to the total number of atoms of that species present we must study the distribution of atoms among the various possible excitation and ionization states. Suppose that the line arises from the ith level in the qth stage of ionization and let N_q be the number density of atoms in the qth ionization stage, of which $N_{q,i}$ are in the ith energy level. Then in thermal equilibrium the distribution of atoms among the different excitation levels of the ion follows a Boltzmann distribution:

$$\frac{N_{q,i}}{N_q} \propto \exp(-V_{q,i}/kT), \qquad (8.10)$$

where $V_{q,i}$ is the excitation potential of the ith level. Thus the strength of a particular line depends on temperature as well as on the number of ions present in the qth stage.

In ionization equilibrium, at a given temperature, the rate of ionization is exactly balanced by the rate of recombination. The rate of ionization must be proportional to the number of ions in the qth stage, while the rate of recombination is proportional both to the number of ions in the $(q+1)$th stage and to the number density of electrons available for recombination. The constant of proportionality depends on temperature and we have as the condition of ionization equilibrium:

$$\frac{N_{q+1}N_e}{N_q} = f_q(T) \propto \exp(-V_q/kT), \qquad (8.11)$$

where V_q is the ionization potential of the ion in the qth stage. This equation is known to astronomers as Saha's equation, after the Indian astronomer who first applied it to stellar atmospheres. By writing down Saha's equation for each ionization stage, we can relate the number of ions to the number of neutral atoms and finally to N, the total number density of the species. For both the Saha and the Boltzmann equations the 'constant' of proportionality in fact depends weakly on temperature, as well as on atomic and fundamental constants. However, no other physical variable is involved.

Thus the combination of excitation and ionization equilibrium tells us that:

$$N_{q,i} = f(N_e, T, N). \qquad (8.12)$$

This means that the strength of a spectral line depends not only on the abundance of the element concerned (N) but also on two other parameters, the electron density and temperature in the atmosphere, which we must know before we can deduce the chemical composition of a stellar atmosphere from the appearance of its line spectrum.

Remarkably, it is found that the chemical composition is about the same for the great majority of stars. By mass, hydrogen accounts for about 72%, helium for about 25% and the remaining elements† make up only about 3% of the mass. Within that 3%, the relative abundances of the various elements are also nearly the same from star to star. You would therefore expect that two parameters should be enough to describe the main features of the spectra of most stars.

This agrees with what is found empirically and is the basis of what is known as *spectral classification*. This is an ordering of stellar spectra into groups or classes according to the relative strengths of various absorption lines. Originally, it was purely empirical: a sort of 'botanizing', to try to produce order out of the great range of visual appearance. The first major catalogue of stellar spectra was produced over a period of 30 years at Harvard College Observatory and was published in about 1920 as the Henry Draper Catalogue; the corresponding spectral classes or types are known as HD spectral types. The MK system in current use was devised by W.W. Morgan and P.C. Keenan in about 1950 and is a two-parameter system.‡

The spectral type is the same as the HD type and the sequence of spectral types corresponds to a temperature sequence:

$$O \quad B \quad A \quad F \quad G \quad K \quad M$$

hot   cool

'early'   'late'

The curious ordering of the letters is a historical accident; originally, the classes were arranged in alphabetical order in order of decreasing hydrogen line-strength. Later, it was recognized that both hot stars and cool stars had weak hydrogen Balmer lines (the hydrogen is ionized in hot stars and in the ground state in cool stars) and the original HD classes were re-ordered into a temperature sequence; many classes were dropped. A useful mnemonic for the present order (introduced, it is said, by Henry Norris Russell's students in the 1930s) is: Oh Be A Fine Girl, Kiss Me! The terminology 'early' and 'late' refers to a long-dead theory of stellar evolution; there is no physical significance in the terms but the usage lingers on. Astronomy is full of such historical echoes.

---

† Because the most abundant of these other elements are metals, such as iron, astronomers usually refer to all elements other than hydrogen or helium as 'metals', even though they also comprise such obvious non-metals as carbon and oxygen!

‡ The 'metal-poor' stars, which are deficient in 'metals', and do not fit into the standard two-parameter classification scheme, are discussed in Section 11.4.

Table 8.2. *The temperatures and main lines that characterize the MK spectral types*

Type	T_{eff}/K	Dominant lines
O	>20 000	Ionized He
B	20–10 000	Neutral He, H
A	10–7000	Neutral H (Balmer lines)
F	7–6000	Ionized Ca (CaII), neutral H, metals
G	6–5000	CaII, neutral metals (FeI)
K	5000–3500	Neutral metals (Ca, Fe), molecular bands
M	3500–2000	TiO bands, neutral Ca

The second parameter is the luminosity class, which is related to the electron density in the atmosphere – stars with high luminosity at a given temperature are also large and have low electron densities in their tenuous extended envelopes. The luminosity classes are given Roman numerals:

I	II	III	IV	V
bright				faint
(low N_e)				(high N_e)
supergiant		giant		main
				sequence

The spectral types and luminosity classes are both defined by the ratios of strengths of various pairs of spectral lines. Because the great majority of stars are luminosity class V, I just indicate in Table 8.2 the way that line strengths vary along the spectral type sequence for main sequence (class V) stars. (Note that astronomers also use Roman numerals to denote the degree of ionization of a species, with I for the neutral atom, II for singly ionized, III for doubly ionized, and so on; do not confuse the examples in the table with luminosity classes!)

In the hotter stars, the metals are mostly ionized and their absorption lines are not in the visible part of the spectrum. Towards the cooler end of the sequence, neutral metals begin to dominate the appearance of the spectrum, with molecules making a significant contribution in the coolest stars. Each type is further sub-divided decimally, from 0 to 9 (not all the sub-types are used in practice); for example, the Sun is of type G2V, with effective temperature 5800 K. The zero of colour index is at A0, where $B - V = 0 = U - B$.

The concept of spectral type was devised as a quick way of estimating stellar properties from a single spectrum. Once the spectral types and luminosity classes were calibrated, it was possible, for example, to determine the rough temperature of a star from its spectral type, or the approximate surface gravity from its luminosity class (giant stars of a given mass being larger and therefore having a smaller gravity: $g = GM/R^2$). If the radius of the star can also be found, the estimate of gravity allows an estimate of the star's mass; since the mass depends on R^2, any error in R is magnified, and the estimate is very rough, but it may be the only possible estimate for a single star (cf. Section 9.2).

8.3 Summary

Stellar brightnesses are expressed on a logarithmic scale of magnitudes. There are many magnitude systems, depending on the spectral sensitivity of the receiver used to measure the star's brightness; one of the commonest is the broad-band *UBV* system. All practical systems are defined by two factors: the response function of the receiver and the standard stars used for calibration.

Most stellar spectra consist of a continuum plus absorption lines. The shape of the continuum, which is nearly that of a black body, allows an estimate of the surface (effective) temperature of the star, often expressed in terms of a colour index. The absorption lines are the signature of which chemical elements are present in the star's atmosphere, and enable the determination of the star's chemical composition. The relative strengths of the various lines also depend on two other factors: the temperature and electron density in the atmosphere.

Because most stars have similar chemical composition, the appearance of their spectra is mainly influenced by the other two factors, allowing a two-parameter classification of stellar spectra, by spectral type (roughly equivalent to temperature) and luminosity class (roughly equivalent to electron density). Stars of intermediate temperature show strong Balmer absorption lines due to hydrogen. Hotter stars mainly show lines of neutral and ionized helium, while cooler stars show lines from ionized and neutral metals, and from molecules.

Exercises

8.1   A star whose luminosity is $100\,L_\odot$ has an apparent bolometric magnitude $m_{bol} = 9.8$. Given that the Sun's absolute bolometric magnitude is $M_{bol}(\odot) = +4.8$, determine the distance of the star.

8.2 The constant in equation (8.6) is -21.58 when the visual flux is measured in $W\,m^{-2}$. Calculate the visual flux corresponding to a star of apparent V magnitude $+6$ (i.e. just on the verge of naked-eye visibility).

8.3 What is the radius (in solar units) of a star of the same effective temperature as the Sun but with a luminosity 10^4 times larger?

9

Properties of stars

9.1 Range of observed properties

Stars can be conveniently described by six fundamental properties. I discussed some of these in the preceding chapter and they are all summarized in Table 9.1.

The Sun is a fairly ordinary star, for which we can determine all these properties quite accurately. The chemical composition is summarized in Table 9.2. The age is determined both theoretically, by constructing models of the Sun, and by radioactive dating of the rocks of the Earth, which are presumably younger than the Sun but not by much. Geologists believe that the Earth is about 4.55×10^9 yr old and the age of the Sun seems to be about 5×10^9 yr. The other properties are given in Table 9.3.

An interesting feature is that the energy output of the Sun, measured per unit mass, is quite small – only 2×10^{-4} W kg^{-1}. Even the human body, at about 1 W kg^{-1}, is much more powerful! The reason the Sun is such an important source of power is simply that it has an enormous mass.

Because the Sun is an average star, it is useful to express the properties of other stars in solar values. The approximate ranges in L, T, R and M are given in Table 9.4.

However, the various properties do not range independently over these values. There are in fact two major relationships: luminosity with effective temperature and luminosity with mass. The first of these relations was discovered independently by E. Hertzsprung in Denmark in 1911, from a study of star clusters (Section 9.3), and by H.N. Russell in the United States in 1913, from a study of all nearby stars of known parallax (Section 14.2.2). The relation they found is shown schematically in Fig. 9.1 and is known as the Hertzsprung–Russell diagram (HR diagram for short) or as the colour–magnitude diagram. The main feature of the diagram is that about

Table 9.1. *Six fundamental properties of stars and how they are determined from observation*

Property	Determined from
Luminosity (L)	apparent magnitude and distance
	or spectrum (luminosity class)
Effective temperature	continuous spectrum
(T_{eff})	*or* spectral type
Chemical composition	line spectrum
Radius (R)	L, T_{eff} [$L = 4\pi R^2 \sigma T_{eff}^4$]
	or interferometry (angular diameter)
	and distance
Mass (M)	binary stars (Section 9.2)
	or g (spectrum) and R (Section 8.2)
Age (t)	star clusters and theory (Section 9.3)

Table 9.2. *The relative abundances in the Sun of the* 13 *elements which are present at the level of one part in a million or more relative to hydrogen, taken to have* $\log N = 12.00$ *for convenience. The value for helium is taken from stars other than the Sun. Although helium was first discovered in the Sun, the lines in the visible part of the spectrum are so blended with lines of other elements that it is impossible to obtain the abundance accurately.*

Element	$\log N$	Element	$\log N$
H	12.0	Mg	7.57
He	11.0	S	7.2
O	8.83	Al	6.4
C	8.53	Ca	6.36
N	7.91	Na	6.30
Si	7.65	Ni	6.25
Fe	7.6		

Table 9.3. *Values of solar properties*

$L_\odot$	$=$	3.83×10^{26} W
$T_{eff\odot}$	$=$	5780 K
$R_\odot$	$=$	6.96×10^8 m
$M_\odot$	$=$	1.99×10^{30} kg

Table 9.4. *Range of values of stellar properties*

10^{-4}	<	$L/L_\odot$	<	10^{+6}	– the most extreme range
$1/3$	<	$T_{\text{eff}}/T_{\text{eff}\odot}$	<	20	– the least extreme range
10^{-2}	<	$R/R_\odot$	<	10^{+3}	
10^{-1}	<	$M/M_\odot$	<	10^{+2}	

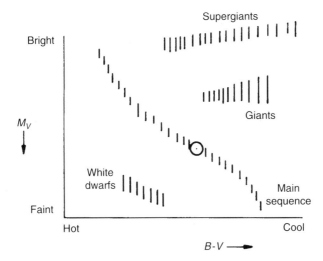

Fig. 9.1. Schematic Hertzsprung–Russell diagram: absolute visual magnitude against colour index. The Sun's position on the main sequence is shown by a circle. The other three hatched regions are giants and supergiants (top right) and white dwarfs (bottom left).

90% of all stars lie on a single narrow band running diagonally from hot, bright stars to faint, cool ones. This is known as the *main sequence* and corresponds to luminosity class V. *Giants* (class III) and *supergiants* (class I) are much more luminous at a given colour, at the cool end of the diagram; the three classes converge in luminosity at the hot end. Well below the main sequence are a few *white dwarf* stars – these are actually very numerous, but are so faint that they can only be seen if they are nearby. The position of a star in the HR diagram is mainly determined by two† parameters, mass and age; more precisely, by age I mean the stage of evolution – as you will see later, different stars evolve at different rates and therefore reach similar stages of evolution with very different ages in years. Main-sequence stars are believed to be unevolved and to differ from

† Chemical composition also affects the position, but to a lesser extent. I will discuss one important consequence in Section 14.4.3.

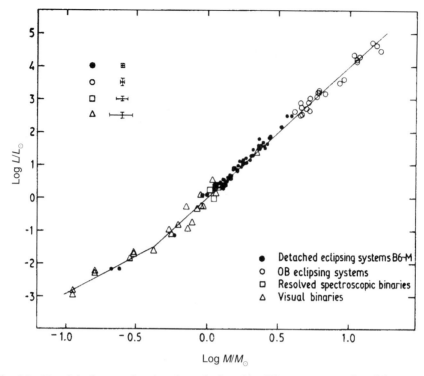

Fig. 9.2. Empirical mass–luminosity relationship. There are very few faint stars with well-determined masses, and few massive stars of any kind, so the relationship is uncertain at low masses and high masses. This plot contains about 125 stars. (*The Observatory* Magazine.)

one another only in mass, the hot, luminous stars being also the most massive.

The narrowness of the main sequence reflects the fact that the other major correlation is between mass and luminosity for main-sequence stars. This mass–luminosity relation is shown in Fig. 9.2. It is only well determined for stars with masses between about one and ten solar masses, where the slope in a $\log L - \log M$ plot is about 4. The slope seems to be somewhat less steep for both more massive and less massive stars. The mass–luminosity relation is less well determined than the colour-magnitude relation because mass is a difficult quantity to measure accurately. It can only be directly determined if the star has a companion, so that we can observe the mutual gravitational interaction. Although the majority of stars are probably in binary or multiple systems, only a few systems have sufficiently well-determined properties to yield accurate masses, as you will see in the next section.

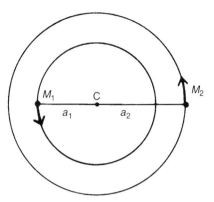

Fig. 9.3. Two stars, of masses M_1 and M_2, are in circular orbits of radii a_1 and a_2 about their common centre of mass, C.

9.2 Binary stars

Binary stars are two stars that are in orbit around one another under their mutual gravitational attraction. Although close pairs of stars had been recognized since antiquity, they were thought by most people just to be chance superpositions along the line of sight of two unrelated stars. It was not until the eighteenth century that it was realized by the English clergyman John Michell that there were too many doubles to be explained by chance and that in many cases the two stars must actually be moving together through space. The first major catalogue of binary stars was made by his more famous compatriot, Sir William Herschel, in about 1800, and today many thousands of binary stars are known. Some binaries are members of larger groupings, triple or even quadruple star systems, but many of these are not stable and are expected to dissolve within a few rotation periods of the Galaxy.

Historically, the discovery of binary stars was extremely important because they provided the first proof that Newton's law of gravitation held outside the solar system and gave astronomers confidence in applying other physical laws elsewhere in the Universe. Today we can use Newton's law to determine the masses of the two stars in a binary system, as I will now demonstrate for the simple case of circular orbits. Consider two stars, masses M_1 and M_2, moving in circular orbits of radii a_1 and a_2 round a common centre of mass C (Fig. 9.3) defined by:

$$M_1 a_1 = M_2 a_2. \tag{9.1}$$

Then if we can locate the centre of mass of the system and measure the ratio of a_1 to a_2 we can obtain the ratio of the masses:

$$M_1/M_2 = a_2/a_1. \tag{9.2}$$

This ratio can be found without knowing the distance of the binary.

The two stars must clearly have the same angular velocity, ω say, about the centre of mass. Then, using Newton's law of gravitation for the force between the two stars and equating it to the centrifugal acceleration, we find:

$$\frac{GM_1M_2}{a^2} = M_1\omega^2 a_1 = M_2\omega^2 a_2, \tag{9.3}$$

where, using equation (9.2),

$$a = a_1 + a_2 = a_1\left(1 + \frac{M_1}{M_2}\right). \tag{9.4}$$

Hence

$$GM_1M_2/a^2 = M_1\omega^2 a M_2/(M_1 + M_2)$$

or

$$\omega^2 a^3 = G(M_1 + M_2). \tag{9.5}$$

Since

$$\omega = 2\pi/T, \tag{9.6}$$

where T is the period of the orbit, we find

$$\frac{4\pi^2 a^3}{T^2} = G(M_1 + M_2). \tag{9.7}$$

This is Kepler's third law of planetary motion, generalized to any two bodies. Although I have derived it for circular orbits, it is also true for elliptical orbits, in which case a is the semi-major axis of the orbit of one star relative to the other, which is at the focus of the relative orbit. (The centre of mass is the common focus of the two individual elliptical orbits.)

Thus if we can measure the period and the separation of the stars (which *does* depend on knowing the distance) we can obtain the sum of the masses. This measurement is independent of knowing the position of the centre of mass.

If we know both the centre of mass and the distance, then we can in principle obtain both the ratio and the sum of the masses and hence find the individual masses. However, it is not always possible to achieve this, and astronomers must often make do with partial information. How partial it is depends on how the binary system is observed. There are three ways in

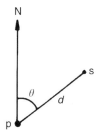

Fig. 9.4. The position angle θ of a visual binary of (angular) separation d is defined at the primary, p, by the angle between north and the secondary, s.

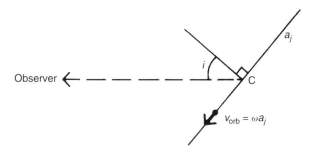

Fig. 9.5. The angle of inclination of a binary orbit is defined as the angle i between the normal to the orbit and the direction to the observer. For a circular orbit, the observed component of the orbital velocity is then $v_r = v_{orb} \sin i$.

which we can detect binary stars and three corresponding classes of binary, which I will now describe.

9.2.1 Visual binaries

Double stars are called visual binaries if the two stars can be resolved using a telescope. For visual observation, the best that can be done is determined by the resolving power of the telescope and is, at least for small telescopes, given by Rayleigh's criterion, $\alpha = 0.14\,\text{arcsec}/(D/1\,\text{m})$ (equation (2.10)), where α is the smallest angular separation that can be resolved with a telescope of aperture D. Direct visual observation was until recently preferred for observing visual binaries because photographic resolution is limited by the smearing effect of seeing, while visual observations can take advantage of moments of good seeing. Recently, higher resolutions have become possible using the technique of speckle interferometry (Section 3.4.5).

For visual binary stars, the (angular) separation d and the position angle θ

(Fig. 9.4) are measured as functions of time. This information gives the relative orbit of one star around the other, as projected on the plane of the sky: the apparent orbit. If the distance is known, d can be converted to an actual separation in metres. To find the sum of the masses, it is necessary to convert the apparent separation to the real separation in the plane of the orbit, which will in general be inclined to the plane of the sky. The inclination i (Fig. 9.5) can be found by requiring the primary star to be at the focus of the 'true' relative orbit (the real relative orbit in space). This is not true for the projected orbit on the plane of the sky, as can easily be seen by imagining a true circular orbit: when the circle is projected onto the plane of the sky it gives an elliptical apparent orbit, but with the primary star at the centre of the ellipse.

This technique of measurement only gives the relative orbit and so can only give the sum of the masses. To find the centre of mass, it is necessary to observe the absolute motions of the two stars, for example with respect to faint background stars that are supposed to be so far away that their own motions can be ignored. Because the binary is an isolated system, with no external forces on it, the centre of mass must move in a straight line and this constraint enables us to deduce the absolute orbits of the two stars about the centre of mass (again, it is also necessary to correct for the inclination) and hence to obtain the mass ratio.

The most interesting example for which the absolute motion has been used to find the orbital properties is the brightest star in the sky, Sirius, which lies just to the southeast of Orion. It has a period of just under 50 years and by 1844 it had been sufficiently well observed for the motion about the centre of mass to be detected. Since there was no sign of any other star, Bessel suggested that Sirius had a dark companion. In 1862, the great American optician Alvan Clark succeeded in detecting a faint companion, Sirius B, and finally, in 1915, Adams managed to obtain a spectrum which demonstrated that Sirius B was a white dwarf, the first one to be discovered. The significance of its being white was that it had a fairly high temperature, despite being faint: all previously known faint stars had been cool and red. The low luminosity and high temperature combined to imply a very small radius ($R \propto L^{1/2} T_{\text{eff}}^{-2}$ – equation (8.7)). The tiny size of Sirius B, only one-fiftieth of the Sun's radius, coupled with a mass nearly equal to the Sun's, gives it a density some half a million times that of the Sun. At the time of Adams' discovery, this was incomprehensible, and the nature of Sirius B was not understood until R.H. Fowler argued in 1927 that white dwarfs were made of degenerate matter (Sections 10.3.3 and 10.4.1). However, it was at least clear that the surface gravity was very large, and Adams took another spectrum in 1925, which was used to verify the gravitational redshift predicted by general relativity.

To summarize:

apparent relative orbit + distance + i → sum of masses;
apparent absolute orbit + i → ratio of masses.

Really good orbits are available only for about 25 visual binaries and it is these that are the basis for the mass–luminosity relation (Section 9.1); some 600–700 orbits are known to some degree of accuracy, out of a total of some 50 000 known systems.

9.2.2 Spectroscopic binaries

Visually unresolved double stars may still have their binary nature revealed through their orbital motion, if the two stars are close enough to one another for the orbital velocity to be more than about $1\,\mathrm{km\,s^{-1}}$. If a spectrum is taken, then binary stars will show periodic shifts in the positions of the spectral lines, and possibly line doubling at certain phases of the orbit. The line shifts are caused by the Doppler effect, as the stars alternately approach us and recede from us in their orbital motion relative to their centre of mass. If both spectra are visible, then line doubling will occur when one star is approaching and the other receding, so that their Doppler shifts are of opposite sign. When the stars are moving across the line of sight, the lines will coincide and any Doppler shift with respect to the laboratory wavelength will be that of the centre of mass.

Measurement of the Doppler shift as a function of time gives the line-of-sight velocity, usually misleadingly called the radial velocity, and typical radial velocity variations deduced from the Doppler shifts are shown in Fig. 9.6; the variations are sine curves. The amplitude of the radial velocity curve for star j, v_j ($j = 1, 2$), is related to the radius of the orbit (assumed circular for illustration) by (Fig. 9.5):

$$v_j = \omega a_j \sin i. \tag{9.8}$$

Here a_j is the radius of the orbit of star j relative to the centre of mass; if both spectra are measured, we have:

$$v_1/v_2 = a_1/a_2 = M_2/M_1. \tag{9.9}$$

That is, the $\sin i$ factor cancels and we can at once obtain the ratio of the masses, independent both of the distance and of the inclination. Unfortunately, there is no way of finding the inclination for spectroscopic binaries,

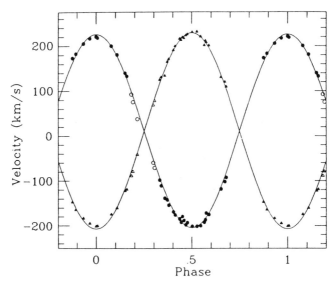

Fig. 9.6. Radial velocity variations observed for a spectroscopic binary in which the spectra of both stars are visible. The velocities of the two stars vary in anti-phase, with one star moving towards the observer while the other is moving away. (*The Observatory* Magazine.)

so we cannot find the sum of the masses. The best we can do is to measure

$$v_1 + v_2 = \omega(a_1 + a_2)\sin i = 2\pi a \sin i / T$$

and use equation (9.7) to find

$$(M_1 + M_2)\sin^3 i = T(v_1 + v_2)^3 / 2\pi G. \tag{9.10}$$

For elliptical orbits, the radial velocity curves are no longer pure sine curves, and the eccentricity e can be deduced from the degree of distortion of the curves.

Even this information is only available if both spectra are visible – the shorthand for such systems is SB2, standing for double-lined spectroscopic binary. If one star is too faint for its spectrum to be detected, which will be the case if it is more than about one magnitude fainter than its companion, it is known as an SB1 – a single-lined spectroscopic binary. Then we can only measure one radial velocity curve and determine one velocity amplitude, v_1, say. We can then find just a single combination of the masses, called the *mass function*:

$$\frac{M_2^3 \sin^3 i}{(M_1 + M_2)^2} = \frac{Tv_1^3}{2\pi G}. \tag{9.11}$$

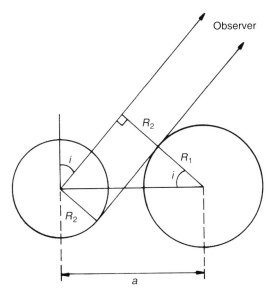

Fig. 9.7. Eclipse limits for a binary system whose stars have radii R_1 and R_2 and separation a. For the configuration shown there will just be a grazing eclipse, and $\cos i = (R_1 + R_2)/a$.

We can therefore only obtain statistical information about the masses of most spectroscopic binaries, for example by assuming that the inclinations to the line of sight are randomly distributed. About 2000 systems are known, of which about half have their orbits known to some degree of accuracy, so quite useful statistical results can be obtained.

9.2.3 Eclipsing binaries

There is one class of spectroscopic binaries for which the inclination *can* be determined: the eclipsing binaries, which show regular light variations caused by one star passing directly in front of its companion, as viewed from the Earth. For these systems, $i \simeq 90°$; more precisely, $\cos i < R/a$, where R is the sum of the stellar radii and a is the separation of the stars (see Fig. 9.7). Because of the $1/a$ dependence in this condition, eclipses are extremely unlikely for wide binaries and are essentially never seen in visual binaries. Typically, $a \simeq 20R$ and $i > 87°$ for eclipses to occur. For simplicity, I will consider only the case when the eclipses are central ($i = 90°$), although in many real systems the eclipses may in fact be only partial. The inclination can be estimated by looking at the detailed light variations through eclipse.

A schematic eclipse light curve is shown in Fig. 9.8. Unless the stars are of very different luminosities, there are two eclipses per orbital period.

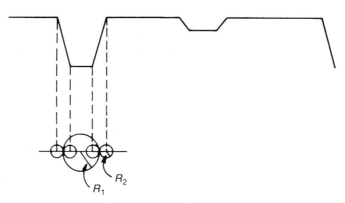

Fig. 9.8. A light curve is shown for an ideal binary system consisting of two spherical stars of uniform surface brightness whose orbit is exactly in the line of sight. Below the primary minimum (brighter star eclipsed) is shown the corresponding geometrical configuration. The light drops smoothly between the first and second contacts, remains constant while the smaller star moves across the diameter of the larger, then rises again between the third and last contacts. Note that at primary eclipse it is always the star with the larger surface brightness (i.e. the hotter star) that is eclipsed. If the smaller star is the hotter one, it will be completely hidden and during the eclipse only the large cool star will be visible; this is a total eclipse. However, if the smaller star is the cooler star it will pass in front of the large hot star, blocking out only part of its light; this is a partial eclipse.

The same surface area is hidden at each eclipse, so the primary, or deeper, eclipse occurs when the star with the higher surface brightness is hidden, that is, when the hotter star is eclipsed. If the two stars are of different radii, then there is a period of total eclipse when the light is constant; as can be seen from the diagram, the duration of totality, t_{tot}, is proportional to the difference in the radii. Since the two stars are moving across the line of sight during the eclipse, the speed with which one moves past the other is approximately given by their relative orbital motion; for circular orbits, this gives:

$$t_{tot} = \frac{2(R_1 - R_2)}{(v_1 + v_2)} = \frac{(R_1 - R_2)T}{\pi a} \, , \qquad (9.12)$$

where I have used equations (9.4), (9.6) and (9.8). Similarly, the time for the drop from maximum to minimum, t_{drop}, is the time between first and second contact (Fig. 9.8) and is proportional to $2R_2$, the diameter of the smaller star. Thus:

$$t_{drop} = \frac{R_2 T}{\pi a}. \qquad (9.13)$$

This simple case can be generalized to more realistic geometries, so we can obtain information about the radii of the two stars, in terms of the orbital separation, just from eclipse timings. If we also have radial velocities, then we can find a from $v_j = 2\pi a_j / T$ $(j = 1, 2)$ and $a = a_1 + a_2$ and hence find the actual radii.

Because the light curve repeats regularly once every orbital period, the position of features in the light curve can be expressed in terms of the fraction of an orbital period since some arbitrary zero point. This fraction is known as the *orbital phase*. For circular orbits, the two eclipses will be at orbital phases 0.0 and 0.5, where zero phase corresponds to zero radial velocity relative to the centre of mass of the system; that is, the time when the two stars are moving exactly at right angles to the line of sight. For elliptical orbits, the orbital phases of the eclipses depend on both the eccentricity and the orientation of the major axis to the line of sight. Since the shape of the radial velocity curve also depends on these two properties, it is possible to determine them both by combining eclipse and radial velocity information; the details are too complicated to give here.

At one of the minima, the smaller star is entirely hidden (at least in a central eclipse) and the luminosity of the system is just that of the larger star. If the larger star is also known to be the hotter star (for example, from the spectral types and luminosity classes of the two stars), then its companion is completely hidden at secondary minimum, and by comparing the magnitudes of the system at secondary minimum and at maximum, when we see both stars, we can find the relative luminosities of the two stars. With the mass ratio determined from the radial velocity curve, this gives a useful check on the mass–luminosity relation, a check which is independent of knowing the distance of the system. The distance itself can be found if we can measure the effective temperatures of the stars and find the absolute luminosities by using $L = 4\pi R^2 \sigma T_{\text{eff}}^4$ (equation (8.7)).

To summarize:

SB orbits	$\rightarrow$	$M_1 \sin^3 i, M_2 \sin^3 i, a \sin i, e$
light curve	$\rightarrow$	$i(\simeq 90°), R_1/a, R_2/a, L_1/L_2$
combination	$\rightarrow$	M, R for each component;
		also L (if T_{eff} known) and hence distance.

Thus eclipsing binaries are extremely useful systems, which enable us to measure almost every interesting property of the stars in them. Unfortunately, although some 4000 eclipsing systems are known, less than 10% of them have had their orbits determined and only a much smaller number have orbits good enough to determine all properties of interest; in particular, many eclipsing systems do not have measured radial velocity curves. This is because it is easier to measure the total brightness of an object than its spectrum; for faint objects, it may be possible to detect a drop in brightness even though the spectrum is too faint for any radial velocity measurements to be made. There are also various complications in interpreting the light curves of very close binaries, in which the two stars may be distorted in shape and/or the radiation from one star may be heating the face of the other. Since these close systems are the ones that are most likely to eclipse, there are relatively few light curves which are straightforward to interpret.

9.2.4 Close binaries and Roche surfaces

Binary systems are important not only because they enable us to determine stellar masses but because they are very common and thus may tell us something about how stars are born. About 85% of all stars are in double or multiple systems, although only a tiny fraction of these are sufficiently well observed for masses and radii to be obtained. Very close binaries are particularly interesting because the components interact with each other in complex ways – tidal forces may cause gas streaming or even mass transfer between the two stars, drastically affecting the stars' evolution. I will discuss some examples of these systems in the chapter on stellar evolution (Section 10.5). For the moment, I will just describe a useful classification system based on the degree of interaction.

When two binary components are very close together, they distort each other's shapes by tidal forces. An estimate of this effect can be made by treating each star as a point mass when calculating the gravitational forces. The orbital motion is taken into account by including centrifugal force. The surfaces of constant gravitational potential, which are spheres for single stars, are then distorted into tear-drop shapes, with the 'pointed ends' facing the other star. There is a critical point on the line joining the centres of the two stars at which the total force felt by a test particle is zero – this is known as the *inner Lagrangian point*, after the French mathematician Lagrange, or sometimes the L_1 point. The potential surface through that point is common to both stars and is generally known as the *critical* or *Roche* surface, after another French mathematician, who introduced the point-mass approximation. The material within each star's Roche 'lobe' is confined within that star's potential well and clearly belongs to a single star. If a star extends beyond its own Roche lobe, its outer layers, at least on the line of centres, are attracted to its companion more strongly than to itself and the star starts to transfer mass into the potential well of its companion.

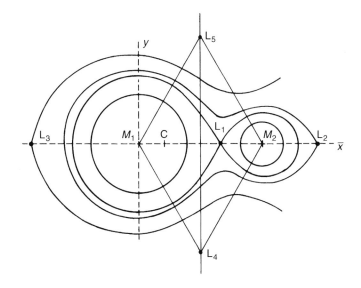

Schematic cross-sections through the Lagrangian points:

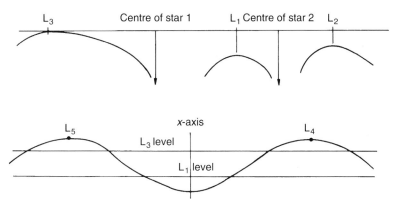

Fig. 9.9. The critical Roche surface has a figure-of-eight contour, passing through the L_1 point; the three-dimensional shape resembles an egg-timer. The contours in the upper half of the diagram show the potential in the orbital plane, the x-axis being the line of centres. C is the centre of mass of the system. Cross-sections of the potential along the line of centres and perpendicular to it are shown below: the Lagrangian points L_1 to L_5 are local maxima in the potential, where the total force on a test particle is zero (in a frame rotating with the orbital angular velocity). The inner Lagrangian point is in fact a saddle point: a cross-section at right angles would show a local minimum at L_1, similar to the one shown in the lowest sketch. When one star swells to fill its Roche lobe, the gas expands above the level of the L_1 point and flows over the 'mountain pass' into the neighbouring valley: the potential well of the other star.

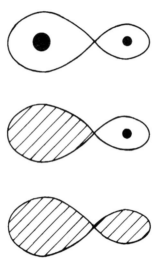

Fig. 9.10. Theoreticians classify close binary stars as detached, semi-detached or contact according as neither star, one star or each star fills its Roche lobe. A contact system overfills both Roche lobes and has a shallow common envelope.

The process is analogous to filling up a jug until it overflows, as can be seen from the cross-section of a potential surface shown in Fig. 9.9.

These theoretical ideas give rise to the classification scheme illustrated in Fig. 9.10. Stars that are both well within their Roche lobes are said to be *detached*, while systems with one star filling its Roche lobe are *semi-detached*. If both stars slightly overflow their Roche lobes, we have a *contact binary*, with a common envelope of material which is not attached uniquely to either star. The concept of Roche lobes is important for understanding the evolution of stars in binaries, as you will see in Section 10.5.

The three types of binary give rise to distinctly different light curves. Detached eclipsing systems with a wide separation have light curves like the schematic one shown in Fig. 9.8 – the light remains constant between the eclipses. These are known as EA light curves; Algol is a well-known example. As the separation decreases, the distortion of the stars increases and the radiation from one star heats the facing side of the other star. This means that even outside eclipse the light varies, both because the observer is seeing regions of different temperature and because the projected area of the distorted star is varying; light curves with continuous variation between eclipses, and with eclipses of unequal depth, are known as EB curves; an example is Beta Lyrae. Contact systems are an extreme example of this, which show minima of nearly equal depth and a smooth variation around the orbit; their light curves are known as EW, after their prototype system, W Ursae Majoris.

9.3 Star clusters

Stars seem to be naturally gregarious and are often found moving through space together in small groups or larger clusters, probably because they were all formed out of the same initial gas cloud. There are two major types of cluster:

1. *Galactic, or open, clusters.* These are loose structures, containing some 10^2–10^3 stars, the brightest of which are mostly blue in colour. They are irregular in shape and there is a great range in size (1–20 pc) and in number of members. Some 700–800 are known in our Galaxy, and their distribution is concentrated to the galactic plane. Many of them contain gas and dust, as in the well-known Pleiades cluster. Others, such as the Praesepe cluster shown in Fig. 9.11, are very sparse and difficult to detect against the normal stellar background. The chemical composition of the stars is similar to that of the Sun (Table 9.2).

2. *Globular clusters.* These clusters are different in almost every possible way. They are centrally condensed and very regular in structure, most of them being almost perfectly spherical (Fig. 9.12); an exception is the slightly flattened southern cluster ω Centauri. They contain many more stars, probably between 10^5 and 10^6, the brightest of which are red. A typical linear diameter is about 40 pc. About 150 are known in our Galaxy and the total galactic population is probably about 500. They are not confined to the galactic plane but form a roughly spherical distribution about the centre of the Galaxy. None shows any significant amounts of dust or gas. The chemical composition differs from that of the Sun: globular cluster stars are 'metal-poor', that is (see Section 8.2), the abundances of elements other than hydrogen and helium are much less than in the Sun, although the relative proportions of these heavier elements to one another are about the same as in the Sun. The overall fraction of heavy elements is less than in the Sun by a factor of between 10 and 1000.

Other groups of stars include *associations*, loose groupings of some 100 stars with a common space motion, and *moving groups*, stars whose only identifying feature is their common space motion. These extremely diffuse groups may contain a few associations and open clusters as tighter condensations within them, and may consist of stars which originally belonged to

Fig. 9.11. The Praesepe star cluster is an open or galactic cluster, a loosely bound collection of stars moving together through the galactic disc. (Royal Astronomical Society.)

Fig. 9.12. The star cluster M3 in the southern hemisphere is a typical globular cluster. It is tightly bound and almost spherical. (Photograph from the Hale Observatories.)

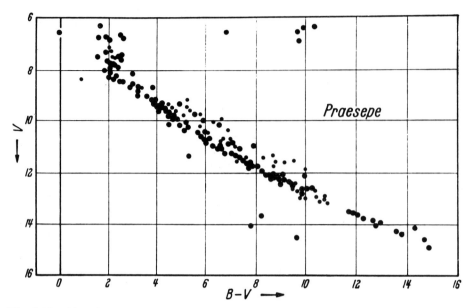

Fig. 9.13. The observed colour–magnitude diagram for the galactic cluster Praesepe. Most stars lie on a well-defined main sequence; there are a few giants. (H.C. Arp.)

the tighter groupings but have gradually become detached under the tidal influence of the rest of the Galaxy. Even open clusters can only survive disruption by tidal forces for some 2×10^9 yr, about 10–20% of the age of the Galaxy, and the looser associations must be younger than about 10^7 yr to have survived. Only globular clusters, and a few of the more concentrated galactic clusters, are sufficiently tightly bound to have survived since the Galaxy was formed.

The study of star clusters is very useful for many purposes but especially for obtaining observational information about the evolution of stars. It is impossible to follow the evolution of *individual* stars observationally – the timescales are far too long, being thousands of years even at the fastest evolutionary stages, and more typically hundreds of millions of years, and we simply observe stars with a range of properties, one of these being age. However, if we look at clusters we can disentangle the effects of age from the other variables, since all the stars in a particular cluster were presumably formed together at the same time, out of the same material. They are therefore the same age and have the same chemical composition, varying from one another only in mass. Because the size of a cluster is typically small compared to its distance from us, all the stars in a given cluster are also effectively at the same distance.

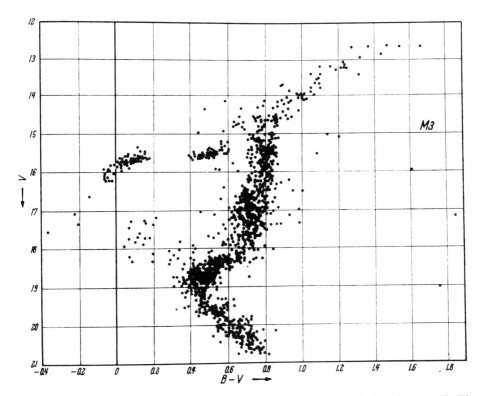

Fig. 9.14. The observed colour–magnitude diagram for the globular cluster M3. The main sequence is short, and all the bright stars are on the well-developed giant branch or on the horizontal branch. (H.C. Arp.)

This last fact is crucial, for it means that the *relative* brightnesses of the stars are well-known, even if the distance itself is poorly determined and the absolute luminosities are not known. We can therefore find the *shape* of the colour–magnitude diagram for a cluster, without having to know the zero point of the magnitude scale. The shapes of the colour–magnitude diagrams turn out to be very different for galactic and globular clusters, as can be seen in Figs. 9.13 and 9.14.

Most of the stars in a galactic cluster lie in the narrow main-sequence band, with only a few stars in the giant region to the upper right of the diagram. The most prominent stars are the blue stars at the upper end of the main sequence. The main sequence in a globular cluster is much less prominent and the main features are a well-populated giant branch and a horizontal branch which stretches out to the blue about halfway up the giant branch. The most prominent stars are the luminous cool supergiants at the tip of the red giant branch. Different globular clusters show very similar

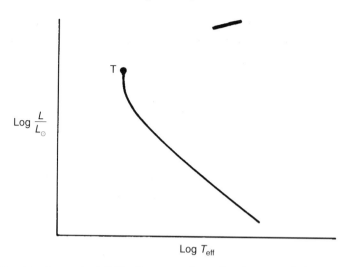

Fig. 9.15. In this theoretical HR diagram, T marks the top of the main sequence: the star which has just reached the end of its main-sequence lifetime. The age of the cluster, assuming all the stars were born at the same time, is the same as the main-sequence lifetime of this star.

colour–magnitude diagrams to one another, whereas the colour–magnitude diagrams for galactic clusters show a considerable variation in the length of the main sequence and the proportion of giants. For example, the galactic cluster M67 has a colour–magnitude diagram that at first sight looks very similar to Fig. 9.14, although the main sequence is rather more densely populated.

These differences are interpreted as *age differences*, with the galactic clusters being younger but having a larger range of age. Globular clusters are believed to have ages which are all within a factor of 2 of 10^{10} yr, while galactic cluster ages range from less than 10^6 yr to about 10^{10} yr. To see why the different colour–magnitude diagram shapes arise, we need to understand something about how stars are born, evolve and die. I will discuss this in detail in the next chapter but for the moment I will just concentrate on one point: how we can use the diagrams to estimate the ages of clusters.

In the youngest clusters, virtually all the stars are on the main sequence. Astronomers believe that main-sequence stars are in equilibrium, with a balance between the loss of energy by radiation from their surfaces and the production of energy by nuclear reactions in their centres. After all its nuclear fuel has been used up, a star leaves the main sequence and moves into the giant region, for reasons which I will discuss in Section 10.3. The time that a star spends on the main sequence will depend on how much nuclear fuel it possesses, which depends on the mass of the star, M, and on the rate at which it uses its nuclear fuel, which depends on the luminosity of the star, L. We can therefore define a *main-sequence lifetime*, t_{ms}, for a star by:

$$t_{ms} \propto \frac{M}{L} = 10^{10} \text{ yr} \frac{M/M_\odot}{L/L_\odot} \qquad (9.14)$$

where 10^{10} yr is the theoretical main-sequence lifetime for the Sun. Since we know from the main-sequence mass–luminosity relation that $L \propto M^\alpha$, where $\alpha \simeq 4$, this shows that more massive stars spend a shorter time on the main sequence before becoming giants. Thus, as a cluster becomes older, the more massive blue stars at the upper end of the main sequence evolve faster and leave the main sequence before the less massive redder stars; for a particular cluster, the stars now at the very top of the main sequence are those which have just reached the end of their main-sequence lifetimes. The age of a cluster is therefore equal to the main-sequence lifetime of the most massive star still on the main sequence. If we measure the luminosity of the turn-off point on the main sequence (Fig. 9.15), and use the mass–luminosity relation to find the corresponding mass, then equation (9.14) gives the age of the cluster. The ages found in this way are consistent with the limits quoted above from estimates of how long clusters could survive disruption by galactic tidal forces.

9.4 Variable stars

It has been known for centuries that some stars vary in brightness, the most dramatic examples being supernovae – brilliant new stars appearing where none was known before. In the eighteenth century, John Goodricke recognized, correctly, that the periodic decreases in brightness of the well-known variable star Algol were due to eclipses of one star by another orbiting around it. With the development of more precise measurement of stellar brightness, ever more subtle variations in brightness became detectable, until now it is believed that every star varies in brightness to some extent, although the amplitude is extremely small in most cases.

Three broad classes of variable star are recognized, with a large range of amplitude and timescale. I have already discussed the *eclipsing variables;* the other two classes are the *pulsating variables* and the *eruptive variables*. There are many subclasses, forming a veritable zoo of different kinds of star, but perhaps the most important are:

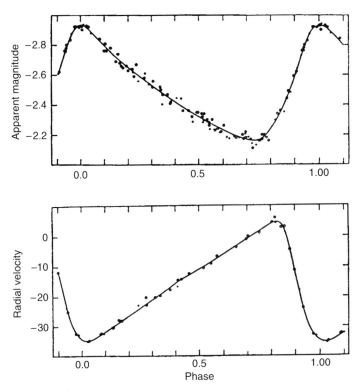

Fig. 9.16. The light and radial velocity variations for δ Cephei, the prototype Cepheid variable. (L. Goldberg and L.H. Aller.)

pulsating variables   –   Cepheids and RR Lyrae stars
eruptive variables   –   supernovae and novae.

Pulsating variables are particularly important, because they not only act as important tests of stellar evolution but they serve as 'standard candles' that can be used to estimate distances. Supernovae have also recently become important distance indicators, as you will see in Chapter 14.

9.4.1 Pulsating variables

9.4.1.1 Cepheids

The Cepheid variables are F–K supergiants, named after their prototype δ Cephei. They show asymmetric variations in light and radial velocity, as shown for δ Cephei in Fig. 9.16, which are attributed to a radial pulsation of the whole star; the star remains spherical, but varies in radius, breathing in and out with a regular period which is in the range 1–80 days. The pulsation

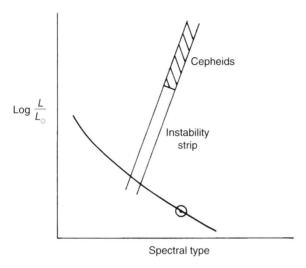

Fig. 9.17. Cepheids occupy the upper part of a narrow instability strip in the HR diagram.

is the star's response to an instability that occurs in the surface layers of the star when it is in a particular temperature range, so Cepheids are restricted to a fairly narrow band in the HR diagram, the so-called 'instability strip' (Fig. 9.17), which is only a few hundred degrees wide. The instability can be understood by considering the structure of the outer layers of a star. The gas in the deep interior is so hot that it is fully ionized, but towards the surface the temperature drops, and in cool stars, with spectral types later than A, the surface layers are largely composed of neutral gas. This means that there must be regions below the surface – the *ionization zones* – where the gas is partially ionized and ionization and recombination processes dominate the atomic physics. This affects the way in which radiation is transmitted through these layers; in particular, if a star is compressed, the ionization layers become much less able to transmit radiation and the energy coming up from the interior is dammed up. This stored-up energy causes the star to re-expand to a larger radius than the one from which it started, and produces an oscillation of increasing amplitude, which grows until the energy dissipated in the pulsation balances the energy supplied by the instability. It is this unstable growth of any perturbation in the radius that accounts for the Cepheid pulsations; the growth is driven mainly in the layers where singly ionized helium is being further ionized. In stars hotter than about spectral type F these layers are too close to the surface for the driving to be effective; in cool stars, convection is the main method of energy transport

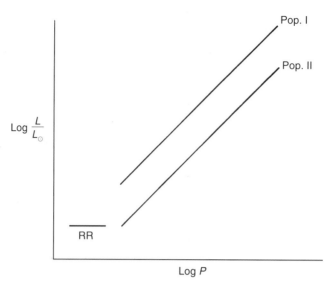

Fig. 9.18. Cepheid variables define a line $L \propto P$ in the period–luminosity diagram, with some intrinsic scatter in practice because of a secondary dependence on colour (or temperature). Population I Cepheids are about four times brighter than Population II ones, but the slope of the relationship is the same. RR Lyrae variables, which are also Population II, have a luminosity that is essentially independent of period and fits fairly smoothly onto the bottom end of the Population II relationship.

and the damming-up of the radiation is unimportant. These two effects confine stellar pulsation to the instability strip.

9.4.1.2 RR Lyrae stars

The RR Lyrae stars, also named after their prototype, are giants that show similar pulsation properties to the Cepheids, with smaller amplitudes for the radial velocity and brightness variations. They are also much less luminous and have considerably shorter pulsation periods – typically half a day. RR Lyraes are found in the instability strip in the HR diagram, at luminosities corresponding to the horizontal branch in globular clusters. They are never found in open clusters, but are common in globular clusters and in the solar neighbourhood.

9.4.1.3 The period–luminosity relationship

Perhaps the most important property of these two classes of variable star is the famous period–luminosity relationship, illustrated in Fig. 9.18. This was first discovered for Cepheids in the Small Magellanic Cloud, where, in 1912, Miss Henrietta Leavitt of Harvard noticed that there was a relationship between the period of pulsation and the apparent magnitude, and deduced a relationship with the luminosity because all the stars were at the same distance. The relationship between magnitude and $\log P$ has a slope of 2.5, which means, from the definition of magnitude (equation (8.2)), that $L \propto P$.

The relationship is complicated by the fact that there are two main subgroups of Cepheids, found in galactic and globular clusters. The main differences between them are in their chemical composition and their luminosity. The Cepheids in globular clusters, known as Population II Cepheids, are fainter at a given period by about 1.5 mag, or a factor of four, than are the galactic cluster, or Population I, Cepheids. (The significance of the terms Population I and Population II will be explained more fully when I discuss galaxies in Chapter 11. For the moment it is enough to know that Population I stars have a similar chemical composition to the Sun while Population II stars have a markedly smaller fraction of elements heavier than helium.) The relationship for the globular cluster Cepheids fits smoothly at its lower end onto the relationship for RR Lyrae stars, which is simply that the luminosity is constant: all the RR Lyrae stars lie at a fixed luminosity on the horizontal branches of globular clusters. The best current estimate of the absolute magnitude is $M_V = 0.6$, but there is still considerable uncertainty in this figure.

Pulsating variables are easy to recognize from the shapes of their light variations, and their periods can usually be measured with high accuracy. In principle, the period–luminosity relationship is therefore a very useful way of finding stellar distances, and RR Lyrae and Cepheid stars are amongst the commonest objects used as distance indicators. Since the long-period Population I Cepheids in particular are very luminous and can easily be seen in many external galaxies, they form a very important link in the long chain of argument that is used to establish the distances of the furthest galaxies.

The main practical difficulty in using the $P–L$ relationship is in finding its zero point – data from the Magellanic Clouds and globular clusters tell us that $L \propto P$ for Cepheids and that L is a constant for RR Lyrae stars, but they cannot give the constant of proportionality unless we already know the distances of the Clouds or the globular clusters. I will discuss this problem of the calibration of the $P–L$ relationship in Chapter 14.

9.4.1.4 The Baade–Wesselink method of determining radii

One of the more difficult properties to measure accurately for a single star is its radius, both because the angular diameter of most stars is extremely small and can be measured only by delicate interferometric techniques (Sections 3.4.4 and 3.4.5) and because the precise distance of an isolated star

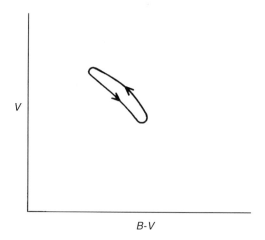

Fig. 9.19. Because of dissipation during its pulsation cycle, a Cepheid traces out a hysteresis loop in the colour–magnitude diagram. The anti-clockwise motion shown corresponds to a pulsation of growing amplitude.

is difficult to determine. Variable stars offer the unique opportunity of determining the radius independently of the distance, using a method first suggested by W. Baade in 1926. At that time, the observations were not accurate enough, and the method was first used successfully some 20 years later by A.J. Wesselink.

Part of the argument simply makes use of the fact that the radius is varying with time and that its rate of variation can be measured *directly* by using the Doppler effect. That is, assuming the star to be spherically symmetric, the rate of change of radius at any time t is given by

$$\frac{\mathrm{d}R}{\mathrm{d}t} = v(t), \tag{9.15}$$

where $v(t)$ is the line-of-sight pulsation velocity (corrected for the relative motion of the Earth and the star, which can be determined by looking at a complete cycle of the radial velocity curve). Integrating this between any two times gives the difference between the radii of the star at these times:

$$R(t_1) - R(t_2) = \int_{t_1}^{t_2} v(t)\,\mathrm{d}t. \tag{9.16}$$

Further progress is made possible by the fact that the pulsation is a hysteresis phenomenon and the star performs a small loop in the HR diagram during a pulsation cycle (Fig. 9.19). If we then choose the times t_1 and t_2 such that the colour of the star is the same at these times, but the magnitude is not, the difference in magnitude is simply a result of the star's having a different radius at the two times. This follows from the general result (equation (8.7))

$$L \propto R^2 T_{\text{eff}}^4,$$

so long as there is a unique relationship between colour and effective temperature and between luminosity and visual magnitude. For then, equal colours imply equal effective temperatures and equal corrections from visual to bolometric magnitudes and we can write

$$\frac{L(t_1)}{L(t_2)} = \frac{R^2(t_1)}{R^2(t_2)} \qquad (9.17)$$

and also

$$\Delta V = V(t_1) - V(t_2) = -2.5 \log_{10} \frac{L(t_1)}{L(t_2)} = -5 \log_{10} \frac{R(t_1)}{R(t_2)},$$

which yields

$$\frac{R(t_1)}{R(t_2)} = 10^{-0.2\Delta V}. \qquad (9.18)$$

From equations (9.16) and (9.18), the radii at times t_1 and t_2 can be found, and by extending this argument to other times we can construct $R(t)$ and the mean radius.

There are various practical difficulties, but it is possible to obtain radii with an uncertainty of 15–20%, which is considerably more precise than most other radius estimates. Where other estimates are possible, for example for Cepheids in eclipsing binary systems or if angular diameter measurements have been made, the agreement between the different methods is satisfactory.

An important byproduct of this method is that the distance of the star can also be found. If a good estimate of the effective temperature can be made from the shape of the continuum, then equation (8.7) ($L \propto R^2 T_{\text{eff}}^4$) gives the luminosity directly. The observed visual magnitude, together with a bolometric correction, then gives the distance from equation (8.4). This is the basis for a recently developed method of distance determination using observations of the expansion of supernova shells, as you will see in Chapter 14.

Further, it is possible to estimate the mass of a pulsating star if its radius is known. The theory of stellar structure tells us that the natural period of vibration of a pulsating star is

$$P \approx (G\bar{\rho})^{-\frac{1}{2}} \propto R^{\frac{3}{2}}/M^{\frac{1}{2}}$$

(equation (10.2)). We can measure P easily, so if we can estimate R from the Baade–Wesselink method we can find M. Unfortunately, because

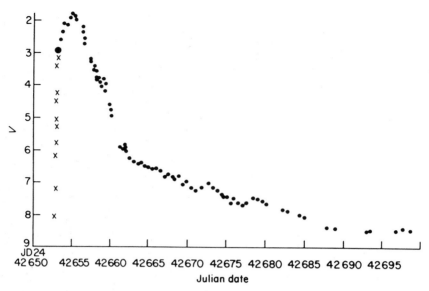

Fig. 9.20. A nova shows a very rapid increase in brightness followed by a more gentle decline. The timescale for the decay varies from a few days to many years. This example is Nova Cygni 1975, which showed an exceptionally large outburst. (M. Petit.)

$M \propto R^3/P^2$, the uncertainty in M is three times that in R, so this is not a very reliable way of determining masses.

9.4.2 Eruptive variables

9.4.2.1 Novae and dwarf novae

Novae are defined as stars that show a sudden large outburst of energy, causing their visual brightness to increase by 11 magnitudes or more in a few days. After a short time at maximum light, the visual brightness declines over a timescale of tens or hundreds of days, depending on the individual nova. The lightcurve of Nova Cygni 1975, shown in Fig. 9.20, is typical of many fast novae. Measurements in the infrared and ultraviolet for some novae show a slower rise and fall and suggest that the total, or bolometric, luminosity remains constant for several hundred days after the outburst, although the spectral distribution changes. The total energy released in the explosion seems to be in the range 10^{38}–10^{39} J.

The main characteristic of the outburst spectrum is that the lines show a large Doppler shift, indicating that a shell of material has been ejected at speeds of about $1000 \, \mathrm{km \, s^{-1}}$. In some novae, this shell can be detected

Fig. 9.21. A 1949 photograph of Nova Persei 1901, showing the expanding shell. (Photograph from the Hale Observatories.)

visually some time after the outburst as an expanding nebula surrounding the fading central star (Fig. 9.21). The density of the shell is quite low, and the total mass lost in this shell ejection is only about $10^{-5}\,M_\odot$.

Novae are comparatively common. Observations of external galaxies show explosions at a rate of about 30 per year, although in our own Galaxy most are too faint to be observed easily through the interstellar dust and gas. This is such a large rate compared to the possible pre-nova objects that it seems very likely that most novae are not exploding for the first time but that explosions recur at intervals of some 10 000 years. This is consistent with ideas about the mechanism for the explosion, which I will discuss shortly. First I will introduce some related objects.

Novae are now recognized as being members of a more general class of object known as *cataclysmic variables*. The other main members of this class are the dwarf novae, whose outbursts at first sight resemble miniature nova explosions. However, they have much lower amplitude (from 2 to 6 mag), recur at much shorter intervals (from 10 d to 500 d) and are believed to have

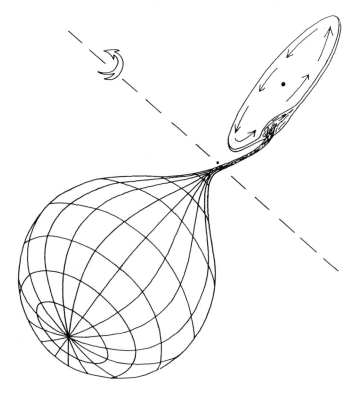

Fig. 9.22. The canonical model for a cataclysmic variable is a binary system consisting of a faint red dwarf filling its Roche lobe and transferring mass via an accretion disc to a white dwarf. (M.T. Friend.)

a different cause. There are also a few intermediate objects, the recurrent novae, that show outbursts of about 8 mag and recur at intervals of tens of years, and there are various nova-like variables that show no outbursts but possess properties that are similar to novae and dwarf novae between outbursts. Some of the nova-like variables show signs of strong magnetic fields.

An important property that is shared by all these objects is that they are all extremely close binary systems, in most cases with orbital periods of less than half a day. The canonical model, illustrated in Fig. 9.22, consists of two very different stars. One is a more or less normal cool dwarf star, usually of spectral type K or M, while the other is a hot white dwarf. I will discuss the properties of white dwarfs in more detail in Section 10.4; they are stars near the ends of their lives, which are very compact and dense. Their masses are typically a solar mass or a little less, but their radii are only about 1/100 of the Sun's radius, so the white dwarf component of a cataclysmic binary is much smaller than its red companion, although it is usually more massive.

The typical separation of the red and white dwarfs is less than a solar radius. It is therefore not surprising that perhaps the most important common feature of all these systems is that they are *semi-detached* binaries (Section 9.2.4), in which the red star is overflowing its Roche lobe and transferring material to the white dwarf. Because the white dwarf is so compact, it is much smaller than its Roche lobe and the material that falls into its potential well does not collide directly with its surface. Instead, the orbital motion gives angular momentum to the falling material, which tries to go into orbit around the white dwarf. When mass transfer starts, an initially elliptical orbit is rapidly circularized by collisions with further infalling material and a ring is formed, fed by a continuous stream of material from the red star. If there is friction within this ring, as seems to be the case, angular momentum is exchanged between neighbouring orbits and the ring gradually spreads out into a disc, with high angular momentum material moving outwards and low angular momentum material drifting slowly inwards towards the white dwarf. Eventually, a quasi-steady state is reached, with the radius of the outer edge of the disc being determined by gravitational interactions with the red star, which raises huge tides on the edge of the disc. At the inner boundary of the disc, material is accreted onto the surface of the white dwarf.

This basic model of a cataclysmic binary has many free parameters (orbital period, masses of the two stars, mass transfer rate, magnetic field, etc.), so it is not surprising that it can be invoked to explain almost every known variety of cataclysmic variable. Novae and dwarf novae differ in their mass transfer rates, with old novae showing a considerably higher rate. This difference is currently believed to be related to the mechanisms for the outburst. In dwarf novae, it is thought that the rate at which mass can be transported *through* the quasi-steady disc is lower than the rate at which mass is being transferred from the red star. Mass therefore gradually piles up in the disc until a critical density is reached above which there is no neighbouring quasi-steady solution. A thermal instability (one which occurs on a timescale determined by the heating and cooling processes in the disc) then causes an abrupt transition to a state in which the mass transport rate through the disc is much higher, and the disc suddenly empties, releasing gravitational potential energy as the mass from the disc falls down the deep potential well onto the surface of the white dwarf. Once the disc has emptied, it reverts to its previous state and the cycle starts again.

It is believed that the thermal instability only occurs in a fairly narrow range of mass transfer rates. The observed rates in old novae are considerably higher, and accretion onto the white dwarf proceeds in a steady manner, with material transported through the accretion disc at the same rate as it is being transferred from the red star. Gravitational energy is thus released continuously and novae between their outbursts have a similar brightness to dwarf novae in outburst. The steady accretion leads to a gradual build-up on the surface of the white dwarf of a layer of hydrogen-rich material from the unprocessed envelope of the red dwarf. As you will see in the next chapter, white dwarfs are old stars that have either consumed or lost all their hydrogen. The addition of new hydrogen rejuvenates the stars, at least in their surface layers: the weight of the accreted gas, and the release of gravitational energy, both act to heat up the base of the hydrogen until it becomes hot enough for nuclear fusion reactions to start. Because the density is very high, the electrons in the ionized hydrogen gas are degenerate and hydrogen ignites explosively, as in the helium flash (Section 10.3.4). This nuclear explosion, like a gigantic hydrogen bomb, blows off essentially the whole layer of hydrogen. The expanding shells of several novae have shown excesses of carbon and nitrogen which suggest that a tiny fraction of the underlying white dwarf may also be ejected. The system is therefore restored more or less to the state it was in before any mass transfer started, and it seems very likely that the whole cycle will be repeated. Since it takes some 10 000 years to build up a thick enough layer of hydrogen for an explosion, we cannot test this directly but must rely on numerical simulations.

9.4.2.2 Supernovae

The first supernova to be observed using modern techniques was the variable star S Andromedae in M31, the galaxy in Andromeda that is the nearest large spiral galaxy (Chapter 11). When it was first observed in 1885, it was believed to be a normal nova, with an absolute magnitude at maximum of about -8. This led to estimates of the distance of M31 that misled people's ideas about the sizes of galaxies until the 1920s. In 1924, Hubble discovered Cepheids in M31 and was able to make an independent estimate of the distance, using the period–luminosity relationship described above. It then became clear that S And was more than ten thousand times brighter than an ordinary nova; according to a modern estimate, its absolute magnitude at maximum was about -19. The term 'supernova' was coined for this new type of violent event.

Supernovae are rare events, with a typical external galaxy showing no more than about one every 30 years; the rate depends somewhat on galaxy

 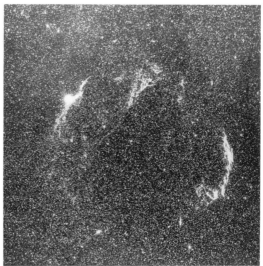

Fig. 9.23. Two examples of expanding shells around supernovae. The Crab Nebula (*left*; photograph from the Hale Observatories) is much younger than the Cygnus Loop supernova remnant (*right*; Royal Astronomical Society), and is much smaller, but both clearly show the violence of the explosion that has created them.

type. All the known pre-telescopic supernovae were in our own Galaxy and the well-attested ones in our Galaxy number only seven over the last 2000 years. The best-known ones are the explosion in 1054 that gave rise to the Crab nebula and the supernovae of 1572 and 1604 associated with the names of Tycho Brahe and Johannes Kepler.

At first sight, a supernova outburst looks rather similar to that of a nova, but it is much more violent. The absolute magnitude at maximum is in the range −15 to −21: comparable to the light from an entire galaxy. The energy output is 10^{42}–10^{44} J, which is comparable to the gravitational potential energy of the star. This is no minor shell-ejection episode but a catastrophic explosion that almost certainly disrupts the whole star, leaving behind at best a highly compact remnant in the form of a neutron star or even a black hole. The bulk of the star is flung outwards at speeds of up to $20\,000$ km s^{-1}, in a shell whose ragged structure displays the violence of the event (Fig. 9.23).

The spectrum of a supernova is difficult to interpret, because the lines are extremely broad and so blended with one another that it is difficult to decide where the continuum lies – to begin with it was not even clear whether most of the lines were in emission or absorption and it took a long time to identify the lines. It is now recognized that there are two principal types

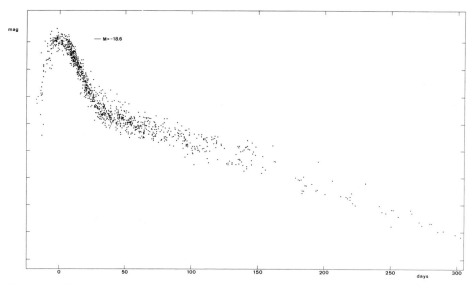

Fig. 9.24. The average blue light curve of 38 Type I supernovae, scaled so that their maxima coincide; one magnitude intervals are marked on the axis (from R. Barbon, F. Ciatti and L. Rosino, *Astronomy and Astrophysics*, **25**, 241, 1973). The light curve of a supernova is similar to that of a nova but has a much greater amplitude. The final decline is exponential, and is powered by radioactive decay. At maximum, Type I supernovae are about three magnitudes brighter than Type II supernovae.

of supernova, distinguished by their spectra. Type I supernovae show no Balmer lines of hydrogen in their spectra; this is believed to reflect a genuine lack of hydrogen in the star that explodes. Supernovae that do show Balmer lines in the spectrum of the expanding shell are known as Type II and their progenitor stars must be of more normal composition.

Type I and Type II supernovae also differ in their light curves. To a first approximation, all Type I supernovae show the same light curve, with an absolute magnitude at maximum of about −20 (a recent review of many examples found an average of −19.6 ± 0.2 for the absolute blue magnitude, about a magnitude brighter than is shown in the 1973 diagram of Fig. 9.24). The rise to maximum is so fast that it is often missed; the last three magnitudes of the rise take about two weeks. After a somewhat less rapid decline from the peak, taking a month or two for a drop of three magnitudes, there is a fairly abrupt change to a more gradual decline, the magnitude decreasing linearly with time. This corresponds to an exponential decay in the luminosity, with an e-folding time of 50–100 d.

The light curves of Type II explosions are much more variable, but the majority show a fainter and broader maximum, followed by a plateau before

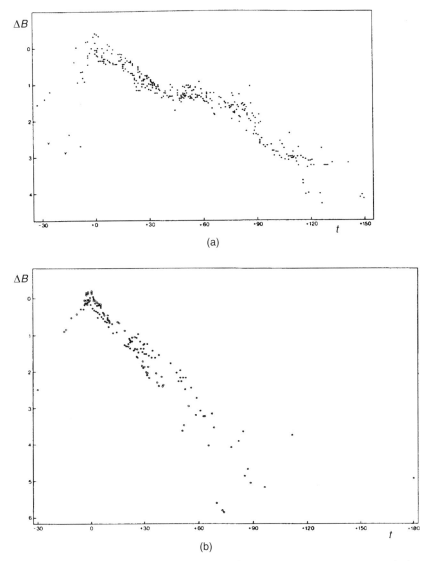

Fig. 9.25. (a) The average blue light curve of 15 Type II supernovae, scaled so that their maxima coincide. The initial decline turns into a plateau at nearly constant luminosity. (b) The average blue light curve of 6 Type II supernovae which show an almost linear decline. (Both diagrams from R. Barbon, F. Ciatti and L. Rosino, *Astronomy and Astrophysics*, **72**, 287, 1979.)

the final fading; the maximum is about three magnitudes fainter than for Type I supernovae. A smaller group show an almost linear decline, with at most a faint hint of a plateau (Fig. 9.25). The final stages of Type II lightcurves are less well documented, because they are fainter and thus more difficult to observe in distant galaxies, but they are believed to be similar to those of Type I.

In 1987, supernova research received a tremendous boost from the discovery of a Type II supernova in the Large Magellanic Cloud. Although the initial behaviour was unusual, with the broad peak being replaced by a slow climb to the plateau, this unique opportunity to observe a nearby supernova has led to a much clearer understanding of what goes on in a supernova explosion. Observations of the decline have shown the same exponential decay as can be seen in Type I systems, but the most important observation was the first-ever detection of a burst of neutrinos from an object outside the solar system. As you will see in Section 10.4, the detection of these neutrinos confirms that our ideas about stellar evolution are correct to a better accuracy than we had any right to expect and shows that the brilliant fireworks that we see as the optical explosion of a supernova represent only 1% of the energy of the explosion – 99% emerges in the form of neutrinos.

It seems clear that the causes of Type I and Type II explosions are quite different. Type II progenitors are thought to be massive supergiant stars while Type I progenitors are probably hydrogen-poor white dwarfs in binary systems. I will discuss supernova explosions in more detail in Section 10.4.5 after I have given an outline of stellar evolution.

9.5 Summary

Stars have six fundamental properties (L, T_{eff}, composition, R, M, t) and I related their range to solar values. The Hertzsprung–Russell, or colour–magnitude, diagram shows that there is a relationship between luminosity and effective temperature, with most stars lying on the main sequence. For main-sequence stars, there is also a strong correlation between luminosity and mass.

Masses can be found only from binary stars. In principle, given enough information, both the mass ratio and the sum of the masses can be found, but in practice we must often be content with partial information. For well-observed visual binaries, in which the individual stars can be resolved, full orbital information can be obtained if the absolute motions of the two stars can be measured. For spectroscopic binaries, detectable only by observing the Doppler shifts of lines in the stars' spectra, the inclination i of the orbit is not usually available; the mass ratio can be found if both stars are of similar brightness, so that both spectra are measurable, but only a lower limit to the sum of the masses can be found; if only one spectrum is measurable, only a single combination of the masses and inclination, the mass function, can be determined. For eclipsing binaries, the inclination can be deduced from the eclipse data, which also enable the determination of the relative brightnesses of the two stars and their sizes in terms of the orbital separation; the combination of spectral and photometric information allows the measurement of not only the individual masses of the stars, but also their sizes and luminosities, and the distance of the system.

Very close binaries interact in interesting ways, and in particular may exchange mass. They may be classified in terms of how fully they fill their Roche lobes (the region within which the gravitational influence of one star dominates over that of the other); most systems are detached, but if they are semi-detached then mass transfer will occur, while contact systems share a shallow common envelope.

Stars, whether single, double or multiple, also occur in larger groupings, known as clusters, associations and moving groups. Galactic (open) clusters contain some hundreds of stars and are irregular in shape, while globular clusters are nearly spherical and contain up to a million stars. Globular cluster stars also differ in composition and age: they are older and more 'metal-poor' than typical galactic cluster stars. Star clusters are of vital importance for the observational study of stellar evolution, because the stars in a particular cluster are believed to have been formed from a single initial gas cloud and therefore provide a sample of stars with the same composition, age and distance.

The shape of cluster colour–magnitude diagrams is characteristically different for galactic and globular clusters, with galactic clusters showing most stars on the main sequence while many globular cluster stars are giants. This is interpreted as mainly arising from a difference in the ages of the clusters, with globular cluster stars being older and more evolved. Galactic clusters have a much greater range of age, and the brightness of the 'turn-off' point at the top of the main sequence can be used, together with the mass–luminosity relationship, to estimate ages of galactic clusters.

Many stars show variations in brightness, with the main classes being the eclipsing, pulsating and eruptive variables. The most important pulsating variables are the Cepheids and RR Lyraes, which are supergiants and giants that show regular asymmetric variations in light and radial velocity with a period in the range of half a day to perhaps 80 days. The stars are undergoing regular radial pulsation, driven by an instability in their ionization zones.

Perhaps their most important property is the period–luminosity relationship: RR Lyraes all have the same luminosity ($M_V \simeq 0.6$), while longer-period Cepheids have higher luminosities. This makes them very valuable as distance indicators, since periods are easy to measure accurately, while long-period Cepheids are very luminous and can be seen to great distances.

Pulsating variables can also be used to measure stellar radii: in the Baade–Wesselink method, differences in radii are deduced from the radial velocity variations, while ratios of radii are found from ratios of luminosities at times in the pulsation cycle when the temperatures are the same.

The other main class of variable star is the eruptive variable, which reveals itself by a sudden large outburst. Novae and dwarf novae are common examples, which form part of the larger group of cataclysmic variables, all of which are extremely short-period binary systems. The canonical model consists of a red star transferring mass to its companion white dwarf via an accretion disc. Novae seem to explode at intervals of some 10 000 years, as a result of hydrogen from the red star accumulating on the surface of the white dwarf until it is hot enough for a thermonuclear explosion to occur. Although the outburst appears to be very violent, very little mass is ejected and the system settles down to more or less the same state that it was in before the explosion. Dwarf novae show smaller outbursts much more frequently, releasing purely gravitational energy as a result of an instability in the accretion disc.

By far the most violent eruptive variable is the supernova, which can outshine an entire galaxy when at the peak of its outburst. Supernovae are much rarer, occurring at most two or three times per century in a typical galaxy. The energy released is enough to blow the entire star to pieces, perhaps leaving behind a compact remnant in the form of a neutron star or black hole. The disrupted star is seen much later as a ragged shell of gas, expanding into the interstellar medium and gradually mixing with it.

Type I supernovae show no hydrogen in their spectra and are thought to result from the disruption of a hydrogen-poor white dwarf in a binary system. Type II supernovae do show hydrogen lines in their spectra, and are believed to represent the final state of a highly evolved massive star that has burned all its nuclear fuels at its centre while retaining an unburned hydrogen envelope. The observation of a Type II supernova in the Large Magellanic Cloud in 1987 enabled this model to be tested: the most dramatic success was the detection of a neutrino burst from deep within the exploding star. The energy emerging in the form of neutrinos amounts to 99% of the total energy released in the explosion – the optical outburst is just a tiny hint of the real power behind the event.

Exercises

9.1   Derive equation (9.11) for the mass function for the case of a circular orbit, assuming that star 1 can have its radial velocity measured.

9.2 A double-lined spectroscopic binary has a period of 125 days. It is deduced from the velocity curve that the orbit is circular and that the velocities of the components about the centre of gravity are $51 \, \text{km s}^{-1}$ and $36 \, \text{km s}^{-1}$. Find the radius of the relative orbit in astronomical units and the masses of both components in solar units. (Note that Kepler's third law can be written as

$$a^3/T^2 = M_1 + M_2,$$

where a is measured in astronomical units, T in years and M_1, M_2 in solar masses. You may take $1 \, \text{AU/day} = 1730 \, \text{km s}^{-1}$.)

9.3 The binary system of exercise 9.2 shows eclipses. The drop to minimum takes 11 h and the minimum lasts for 99 h. Assuming that the eclipse is central (one star passes along the diameter of the other), find the radii of the two stars in solar units. (You may assume that $1 \, \text{AU} = 215 \, \text{R}_\odot$.)

9.4 The bright supernova in M31, S Andromedae, reached an apparent visual magnitude of 5.5 at its maximum in August 1885. Calculate the distance that it would have had if it had been an ordinary nova, with an absolute magnitude at maximum of -8 (your answer should be less than the distance to the Galactic Centre). Given that it was a supernova, with an absolute magnitude at maximum of -18.8, find the actual distance of M31 in kpc.

9.5 A galactic cluster has a well-defined main sequence with a turn-off point at $L/L_\odot = 81$. Assuming that the mass–luminosity relation for stars near the turn-off point is $L/L_\odot = (M/M_\odot)^4$, and that the main-sequence lifetime of the Sun is 10^{10} yr, estimate the age of the cluster.

9.6 The semi-amplitudes of the radial velocity curves in a spectroscopic binary are found to be $18 \, \text{km s}^{-1}$ and $36 \, \text{km s}^{-1}$. If the star with the larger semi-amplitude has a spectral type of G2V, *estimate* the masses of the two stars.

9.7 A Cepheid variable star is observed at two times in its cycle, t_1 and t_2, at which the effective temperature is the same. At these times, its apparent visual magnitudes are 4.27 and 4.09 respectively. Between t_1 and t_2 the radial velocity is observed to be approximately constant, with a value of about $13 \, \text{km s}^{-1}$ towards the observer. If this velocity is measured relative to the motion of the centre of the star, estimate the radii of the star at the two times if $t_2 - t_1 = 1.34 \, \text{days}$.

10

Stellar structure and evolution

Studies of the Hertzsprung–Russell diagrams of star clusters (Section 9.3) began at about the same time as Sir Arthur Eddington began to lay the foundations of our modern ideas of how stars are born, live and die. In his classic book *The Internal Constitution of the Stars*, published in 1926, he was able to explain the basis of stellar structure and to give a quantitative explanation for the observed mass–luminosity relation, despite the fact that nuclear reactions were not yet known, so he had no satisfactory mechanism for the internal energy source of stars. He succeeded because stars are, in outline at least, remarkably simple objects. In detail, there are still many problems, but at the level of this book we can say that the structure and evolution of stars are very well understood. Certainly they are much better understood than the evolution of galaxies.

One of the reasons for this better understanding is that we can use the HR diagrams of star clusters as testing grounds for our theories, trying to explain the observed shapes in terms of the evolutionary paths of stars and the observed distributions of stars within the diagrams in terms of the time taken at various stages of evolution. In this chapter I will describe current theoretical ideas about stellar structure and evolution and relate them to the observed HR diagrams of clusters.

10.1 The formation of main-sequence stars

10.1.1 From gas cloud to protostar

We must first ask how we think stars are formed in the first place. This is actually a much more difficult question than the question of the present structure of a main-sequence star, and there is still no complete answer. However, there is general agreement on some basic ideas. There seems little doubt that what we now see as a star was once part of a very tenuous gas cloud of the sort that we see in large numbers throughout the interstellar medium. The effect of gravity on a low-density gas cloud is to make it contract, so you will see an immediate problem – why are there any gas clouds at all? To see how great the problem is, I will estimate the time for a gas cloud to contract under gravity. In the absence of any other forces, a spherical gas cloud of radius r and mass M changes its radius according to Newton's second law:

$$\ddot{r} = -\frac{GM}{r^2}. \tag{10.1}$$

If I approximate $\ddot{r}$ by r/t_{ff}^2, where t_{ff} is the time for free fall under gravity (the *dynamical timescale* of the cloud), then:

$$t_{\mathrm{ff}}^2 \simeq r^3/GM$$

or

$$t_{\mathrm{ff}} \simeq (G\bar{\rho})^{-\frac{1}{2}} \tag{10.2}$$

where $\bar{\rho}$ is the mean density of the interstellar cloud. For a typical cool cloud density of $10^{-17}\,\mathrm{kg\,m^{-3}}$ this gives a free-fall time of only about a million years: much shorter than the lifetime of our Galaxy. You might therefore expect that all the gas in interstellar clouds would long since have collapsed into stars. The reason that it has not is that there are other forces opposing gravity, as I will explain shortly.

However, the short free-fall timescale does suggest that stars form very rapidly and that there is rather a small chance of seeing one actually in the process of formation. This conclusion is largely borne out by observation. It is therefore difficult to test theories of star formation. On the other hand, main-sequence stars are abundant, so it is easy to test our ideas about their structure. For this reason, newly forming stars are far less well understood than main-sequence stars and I will simply give the barest outline of how astronomers think stars form, before going on to discuss main-sequence stars in more detail.

In equilibrium, a self-gravitating gas cloud is supported against collapse mainly by its internal pressure. The source of the pressure inside a main-sequence star is usually thermal, arising from random motions of the individual atoms and molecules. A molecular cloud, with a temperature of some 10 K, is far too cold for thermal pressure to play an important role. However, observations of cold molecular clouds show that the gas is in rapid turbulent motion, and the effect is to provide a form of 'turbulent pressure' that supports the cloud against collapse. In molecular clouds, which are probably the main sites of star formation, turbulent pressure is the most important form of support, but if the cloud is rotating and even slightly ionized then centrifugal and magnetic forces also play a role. Thermal pressure may be important in hotter, more diffuse clouds.

This internal support may be overcome either by a sudden increase in external pressure, perhaps caused by a shock wave from a supernova explosion hitting the cloud, or, if the internal pressure is partly thermal, by a decrease in the gas particle number density, perhaps as a result of molecule formation or the condensation of some of the gas into dust. In either case, once the balance between pressure and gravity has been disturbed, collapse under gravity starts at about the free-fall rate. If the initial gas cloud is very massive, it will also begin to fragment into smaller clouds, which will fragment in their turn, ultimately producing clouds of stellar mass.

As the gas in a stellar-mass fragment falls down its own gravitational potential well, a large amount of energy is released. Some of this goes into kinetic energy of collapse, and some into random motions of the gas particles, which has the effect of heating the gas. Initially, the gas cloud is of very low density and radiation can escape freely, keeping the gas cool. However, as the collapse proceeds the density increases until the cloud finally becomes opaque and most of the radiation emitted by the atoms and molecules in the gas is trapped within the cloud. The cloud then begins to heat up in the interior, although the surface will remain fairly cool because radiation can still escape freely from the surface of the cloud. A temperature gradient is built up, with energy flowing down the temperature gradient from the centre to the surface, where it is radiated away from the cloud. Associated with the temperature gradient is a thermal pressure gradient, which begins to resist the collapse. Eventually, after about a free-fall time (equation (10.2)), the cloud reaches a state of quasi-equilibrium, in which the pressure gradient nearly balances gravity. A slow contraction continues, at a rate that is just enough to supply the energy needed to balance the rate of energy loss from the surface by radiation.

There is one complication in this picture which has an important qualitative effect. Not only is the centre of the cloud hotter than the surface, it also has a higher density. From equation (10.2), this means that the free-fall time at the centre of the cloud is shorter than at the outside, and a runaway develops, with the centre contracting faster than the outside, building up an even larger density contrast and contracting even faster. Thus a pressure gradient sufficient to halt the rapid collapse is produced in the centre long before the outside of the cloud is even opaque, and the cloud develops a core–halo structure, with a slowly contracting central core surrounded by an infalling envelope. This accreting material is still cool, and may contain large amounts of dust as well as gas. The central core is therefore buried in a cocoon of dusty gas and cannot be seen directly. However, the radiation from the core heats up the surrounding envelope to temperatures of a few hundred degrees, and the cocoon is detectable as a strong infrared source. With the development of infrared astronomy since the late 1960s, more and more strong infrared sources have been found in regions where there are also gas and dust clouds and young stars, and astronomers believe that they are seeing the accreting shells around still-forming stars. Eventually, all the material in the surrounding shell is either accreted onto the surface of the core, adding considerably to its total mass, or is driven off by the pressure of the radiation from the core. Observations of the infrared sources show evidence for both infall and outflow, with some of the outflow in the form of strong, rapidly moving jets of material. The more massive the core becomes, the more powerful is the radiation from its surface, and it may be that the maximum possible mass for a star is determined by a self-limiting process: once the star reaches a certain mass, around 100 solar masses, it is so luminous that no more material can be accreted, even if the original gas cloud was more massive than 100 solar masses. This seems to be consistent with the largest stellar masses observed.

This simple description needs to be modified if the initial gas cloud is rotating or has a magnetic field. Conservation of angular momentum and magnetic flux causes centrifugal forces and magnetic forces to increase as the cloud contracts, and both will tend to oppose the contraction. Since stars do form, there must be loss of both angular momentum and magnetic flux during the collapse, but there is as yet no consensus on how this happens. In some models, the slowing down of the cloud contraction is a necessary part of the star formation process, allowing time for cooling. In others, angular momentum is associated with the formation of binary stars. In this book I will not discuss such models further and will simply assume that, somehow, the forming star has managed to lose its angular momentum and magnetic flux and has become a spherically contracting cloud.

The slowly contracting core is in approximate equilibrium and is self-luminous; it can now be called a protostar rather than a gas cloud.

10.1.2 Energy sources

At one time it was thought that all stars must be contracting slowly, because there seemed to be no source of energy other than gravity that was sufficient to keep a star shining for an astronomical time. The total gravitational energy released by a mass M contracting from a large (essentially infinite) radius down to radius R is

$$E_{\text{grav}} \simeq \frac{GM^2}{R}, \tag{10.3}$$

as can be seen by calculating the work done to move an element of mass dM from R to infinity against the gravitational pull of a mass M and then integrating over mass from 0 to M. Most of this energy is of course released at radii not much larger than R, and so E_{grav} is a measure of the energy source that has been available to keep the star shining at about its present radius. More precisely, the total thermal, or heat, energy of the protostar is half of E_{grav} (by the virial theorem – equation (11.1)). We can then calculate the maximum time for which the star could have been shining at its present rate, L say, if gravitational contraction were the only energy source. This gravitational timescale is

$$t_{\text{grav}} \simeq \frac{GM^2}{2RL}. \tag{10.4}$$

Because a calculation of this kind was first made for the Sun in the middle of the nineteenth century by Lord Kelvin in Britain and by H. von Helmholtz in Germany, the timescale is commonly called the Kelvin–Helmholtz timescale, t_{KH}. It is also known as the thermal timescale, because it measures the rate at which the heat content of the star is being used up. The value for the Sun is about 17 million years.

Not long after this first calculation, it became apparent from geological evidence that the Earth was much older than 30 million years, so that a longer-lived energy source was needed. This source remained a mystery for more than 50 years until the birth of nuclear physics and the realization that fusion of light nuclei could supply enough energy to keep the Sun shining at its present rate for about the age of the Universe. The ideas of quantum mechanics were also needed, to explain why fusion could occur at relatively low temperatures, at which the nuclei had apparently too little energy to overcome the Coulomb repulsion arising from their positive charges. The idea that particles could tunnel through the Coulomb barrier, instead of having to climb over it, was fundamental to the understanding of nuclear energy sources in stars.

Our slowly contracting core continues to heat up slowly in the centre and eventually the central temperature becomes high enough for nuclear fusion reactions to start. The composition of the cloud is 90% hydrogen (by number of particles), so the principal reaction is the conversion of hydrogen to helium. This reaction becomes significant at temperatures of about 10^7 K, when the hydrogen nuclei have enough kinetic energy to overcome their

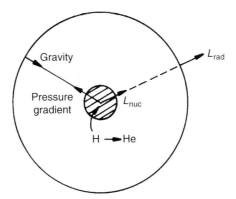

Fig. 10.1. The equilibrium of a main-sequence star is determined by two factors. Hydrostatic equilibrium is maintained by a balance of forces: gravity is balanced by an internal (thermal) pressure gradient. Thermal equilibrium depends on energy balance: the rate of production of nuclear energy at the centre (L_{nuc}) is exactly balanced by the rate of loss of energy at the surface by radiation (L_{rad}). This balance requires energy *transport* to occur; in most stars, the mechanisms are radiation and convection.

mutual Coulomb repulsion by quantum tunnelling through the Coulomb barrier.

The exact reactions that occur depend on temperature and involve a number of intermediate stages which I will not discuss in this book; the net result in all cases can be expressed as

$$4H^1 \rightarrow He^4 + 2e^+ + 2\nu + 6.3 \times 10^{14}\,\mathrm{J\,kg^{-1}}. \tag{10.5}$$

Since the energy released in the fusion is known to be about 0.7% of the rest-mass energy of the hydrogen nuclei, we can estimate the maximum time for which nuclear reactions can sustain a given luminosity. If the whole star, of mass M, were to be converted from hydrogen into helium, then it could remain shining with luminosity L for a nuclear timescale

$$t_{nuc} \simeq 0.007 Mc^2/L \simeq 10^{11}\frac{M/\mathrm{M_\odot}}{L/\mathrm{L_\odot}}\ \mathrm{yr}. \tag{10.6}$$

In practice, only the inner 10% or so of a star is hot enough for hydrogen fusion reactions, so we expect the the Sun to be able to live on hydrogen fusion for about 10^{10} years. Since geological and other evidence suggests that it is currently about 5×10^9 years old, the Sun is only about halfway through its supply of hydrogen, verifying that nuclear reactions are indeed a more than adequate energy source.

Once nuclear reactions start inside a slowly contracting protostar, the

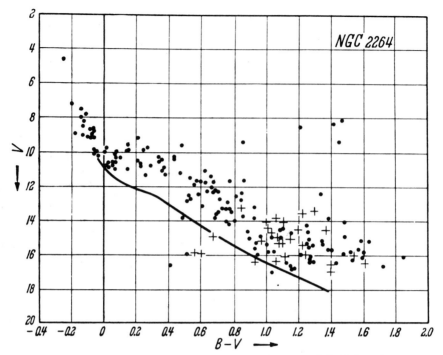

Fig. 10.2. HR diagram for the very young star cluster NGC 2264. The stars on the upper left of the diagram have reached the main sequence, but the stars at the lower right are still contracting onto it. (H.C. Arp.)

main energy balance is between nuclear energy production at the centre and energy loss by radiation from the surface. The slow contraction ceases, because there is no longer any need to draw on gravitational energy sources, and the protostar has finally become a star, in equilibrium. Such a star is on the main sequence in the HR diagram and is shown schematically in Fig. 10.1. Its route to the main sequence on the HR diagram is rather uncertain in detail, but it must start originally very large, very cool and very faint, somewhere in the lower right-hand corner of the diagram. In very young open star clusters and associations, some of the cool stars are above and to the right of the main sequence, showing that they are still in the final stages of contraction onto the main sequence (Fig. 10.2).

10.2 The structure of main-sequence stars

Although a main-sequence star is, by definition, in hydrostatic and thermal equilibrium, it is not quite unchanging, because of the nuclear reactions that are going on in the interior. Eventually, a main-sequence star will evolve away from the main sequence again and become a giant star; I will discuss that process in the next section. However, the nuclear evolution timescale is so long compared with the timescales for thermal and dynamical change that the structure of a main-sequence star can be studied by assuming that it is in exact equilibrium. In this section I will outline the equations that determine this structure.

Astronomers observe directly that the Sun is almost exactly spherical, with just a slight flattening due to rotation. Although we know that some stars rotate much faster than the Sun, that others possess strong magnetic fields and that the presence of a companion must also distort the shape of a star, I will ignore these effects here and consider perfectly spherical stars. This approximation is justified by the good qualitative agreement between the predictions of spherical stellar models and the properties of real stars. I will also assume that our stars are the same chemical composition throughout. This will be the case if, as is generally believed, stars are well-mixed by convection before they reach the main sequence.

10.2.1 Force and energy balance

Our model of a zero-age, main-sequence star is, then, a spherical, chemically homogeneous ball of gas in thermal and hydrostatic equilibrium, with nuclear reactions just starting to alter the composition at the centre. For this simple model, hydrostatic equilibrium corresponds to a balance between gravity and internal pressure. The inward gravitational force on an infinitesimal cube of height δr and surface area A at distance r from the centre is

$$F_g = \frac{GM(r)}{r^2} \rho(r) A \delta r, \tag{10.7}$$

where $M(r)$ is the mass within radius r and $\rho(r)$ is the density at r. Assuming spherical symmetry, the horizontal pressure forces on the cubical element balance each other and the net pressure force inward is

$$F_p = (P(r + \delta r) - P(r)) A. \tag{10.8}$$

In equilibrium, the total inward force must be zero and so

$$F_p + F_g = 0 \tag{10.9}$$

or, taking the limit $\delta r \to 0$,

$$\frac{dP}{dr} = -\frac{GM\rho}{r^2}. \tag{10.10}$$

This is the equation of hydrostatic equilibrium. In this equation, M is the total mass contained within a sphere of radius r and is given by

$$M(r) = \int_0^r \rho(r') 4\pi r'^2 \, dr' \qquad (10.11)$$

or, in differential form,

$$\frac{dM}{dr} = 4\pi r^2 \rho. \qquad (10.12)$$

This equation expresses the conservation of mass, while the hydrostatic equation expresses the conservation of momentum (since, by Newton's second law of motion, there is no change of momentum if the total force is zero). The conservation of energy leads in general to a rather more complicated equation. However, if there are no motions, and no energy sources other than nuclear reactions, energy balance simply requires that any net outflow of energy from a small volume is exactly balanced by the production of nuclear energy within that volume. If we consider a spherical shell, with inner and outer radii r and $r + \delta r$, and let $L(r)$ be the amount of energy crossing radius r per unit time, then energy balance requires that

$$L(r + \delta r) - L(r) = 4\pi r^2 \delta r \rho(r) \epsilon(r) \qquad (10.13)$$

or, in the limit $\delta r \to 0$,

$$\frac{dL}{dr} = 4\pi r^2 \rho \epsilon. \qquad (10.14)$$

Here $\epsilon(r)$ is the rate of nuclear energy generation per unit mass.

This is not yet a complete set of equations, as we have only three equations, (10.10), (10.12) and (10.14) for the five unknowns P, M, ρ, L and ϵ. In order to make further progress, we need to know something about the physical conditions in stellar interiors, which will tell us what forms to take for the pressure and the nuclear energy generation.

10.2.2 The pressure

The gas inside the bulk of a star is highly ionized: essentially all the atoms are split into bare nuclei and electrons. The result is that, although the density is so high that the typical inter-particle spacing is of the order of an atomic radius, the effective particle size is more like a nuclear radius, some 1000 times smaller. This in turn means that, despite the high pressures and densities in the interior, there are very few collisions between gas particles. Stellar material therefore behaves like an ideal gas, with equation of state:

$$P = nkT, \qquad (10.15)$$

where n is the particle density, T is the temperature and k is Boltzmann's constant ($= 1.38 \times 10^{-23}\,\mathrm{J\,K^{-1}}$). We have therefore introduced one new equation, at the expense of two new unknowns, n and T. If the radiation within a star exerts a significant pressure, equation (10.15) can be generalized to read

$$P = nkT + \frac{1}{3}aT^4, \tag{10.16}$$

where a is the radiation density constant, without introducing any further variable. In most stars, the radiation pressure is only a small correction, but it may dominate in very massive stars.

The particle number density is easily related to the mass density, ρ, by introducing the mean particle mass, and astronomers usually write the ideal gas law as

$$P = \frac{\mathscr{R}}{\mu}\rho T, \tag{10.17}$$

where $\mathscr{R}$ is the gas constant ($= 8.3 \times 10^3\,\mathrm{J\,K^{-1}\,kg^{-1}}$) and μ is the mean particle mass in units of the hydrogen atom mass m_H; $\mathscr{R} = k/m_H$. For a fully ionized gas, with a mass fraction X in the form of hydrogen, Y in the form of helium and Z in the form of heavier elements, so that $X + Y + Z = 1$, it can easily be shown that

$$1/\mu = 2X + 3Y/4 + Z/2$$

(e.g. helium contributes three particles, two electrons and one nucleus, for its four units of mass). For solar composition, $X \simeq 0.72, Y \simeq 0.25, Z \simeq 0.03$ and $\mu \simeq 0.6$; allowing for partial ionization, μ is a little higher. Its value is not very sensitive to the exact composition.

10.2.3 Energy transport

10.2.3.1 Radiation and conduction

The appearance of the temperature requires us to introduce some new physics. So far, I have said nothing about the process by which energy flows out from the centre of the star to the surface, but I must now take a brief look at that problem. There are three ways in which energy can be transported: conduction, convection and radiation. Inside a star, conduction and radiation are similar processes, both involving transfer of energy by direct interaction, either between particles or between particles and photons. The energy carried by a typical particle is comparable to that carried by a typical photon (both are about kT, where T is the local temperature), but the number density of

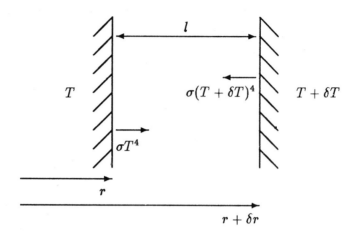

Fig. 10.3. Consider two surfaces at distances r and $r + \delta r$ from the centre of a star, and suppose that they are separated by the photon mean free path l (i.e. $\delta r = l$). The temperatures of the two surfaces are T and $T + \delta T$, where $\delta T = l\, dT/dr$ (< 0, because the temperature decreases outwards). If each radiates like a black body, with emissivity σT^4, where $\sigma = ac/4$ is the Stefan–Boltzmann constant, then the net outward flux across the gap is $F_{\text{rad}} = -4l\sigma T^3 dT/dr$, neglecting squares and higher powers of l. Since the effective cross-sectional area for absorption per unit volume is $\kappa\rho$, and the mean free path between absorptions is l, it is obvious that $l\kappa\rho \simeq 1$. Inserting this into the expression just derived for F_{rad} gives equation (10.20), apart from the factor $4/3$, which comes from a more precise argument.

particles is much greater than that of photons. This means that the energy density in the form of particles is much greater than that in the form of photons (this is equivalent to radiation pressure normally being negligible), so you might expect that conduction would be more important. However, it turns out that the smaller number of photons is far outweighed by the much larger mean free path between collisions: a photon at a typical point inside a star travels about 10^{-2} m before being absorbed or scattered, while a particle travels little more than an atomic diameter, that is, about 10^{-10} m. Conduction is therefore completely negligible in almost all main-sequence stars.

A full treatment of radiative energy transfer is beyond the scope of this book. Fortunately, it is only vital near the surfaces of stars, where the photon mean free paths become extremely long and the photons finally escape, some of them ultimately to be captured by a telescope on the Earth. In the deep interior, the photons' travel can be described by a random walk, similar to the random progress of a drunkard, the direction of whose next step is completely random, but who nevertheless makes a slow progress in a particular direction. A photon is absorbed and re-emitted many times, with the re-emission being in a random direction. If the mean free path between absorptions is l, then after n absorptions the photon will have progressed a distance $\sqrt{n}\,l$ outwards through the star. The net effect can be thought of as diffusion down a temperature gradient. The expression for the flux of energy carried by radiation can therefore be written as

$$F_{\mathrm{rad}} = -\lambda_{\mathrm{rad}}\frac{\mathrm{d}T}{\mathrm{d}r} , \qquad (10.18)$$

where λ_{rad} is the 'radiative conductivity'. Astronomers generally prefer to work with an inverse of the conductivity, known as the *opacity*, κ, which measures the absorptive power of stellar material, averaged in an appropriate way over frequency. If κ is defined as the absorption cross-section per unit mass (dimensions: length^2/mass), then detailed arguments show that

$$\lambda_{\mathrm{rad}} = \frac{4acT^3}{3\kappa\rho}, \qquad (10.19)$$

where c is the speed of light; an approximate derivation is given in the caption to Fig. 10.3. The radiative flux, F_{rad}, can then be written as

$$F_{\mathrm{rad}} = -\frac{4}{3}\frac{acT^3}{\kappa\rho}\frac{\mathrm{d}T}{\mathrm{d}r} = -\frac{c}{\kappa\rho}\frac{\mathrm{d}P_{\mathrm{rad}}}{\mathrm{d}r}. \qquad (10.20)$$

The detailed calculation of the opacity is a very complicated problem in atomic physics, which I will not discuss here. However, since the interior of a star is nearly in thermodynamic equilibrium, with just a slow outward leak of energy, the opacity is to a good approximation a function only of the thermodynamic state variables, being of the general form

$$\kappa = \kappa(\rho, T, chemical\ composition). \qquad (10.21)$$

In restricted ranges of density and temperature, the results of detailed calculations can be represented by power laws of the form

$$\kappa = \kappa_0\rho^\alpha T^\beta, \qquad (10.22)$$

where α, β are slowly varying functions of density and temperature and κ_0 depends on chemical composition and on fundamental physical constants. In the deep interior of many stars, the degree of ionization is so high that the dominant form of opacity is electron scattering, which is independent of ρ and T: $\alpha = \beta = 0$. Elsewhere, the opacity generally increases with density, making α positive. Near the surface, the opacity increases with temperature, so β is also positive, but towards higher temperatures the opacity starts to decrease with increasing temperature and throughout much of a stellar interior β is negative. The main absorption processes in the bulk of a star are bound–free transitions, where an atom is ionized by the absorption of a photon (an electron is freed from a bound state – essentially the same as the photoelectric effect described in Section 3.2), and free–free transitions, where a free electron gains energy from a photon while passing an ion (the presence of the ion is essential for momentum conservation). Both processes can be represented approximately by taking $\alpha = 1$, $\beta = -3.5$.

The reason for discussing the transport of energy was to find another equation for the new variable, temperature, introduced by the equation of state (10.15). We can write equation (10.20) as an equation for T by noting that $F_{\rm rad}$ is the flux of energy per unit area and is therefore related to $L(r)$ by

$$L(r) = 4\pi r^2 F_{\rm rad}. \tag{10.23}$$

Then we find

$$\frac{{\rm d}T}{{\rm d}r} = -\frac{3\kappa\rho L}{16\pi a c r^2 T^3}. \tag{10.24}$$

This equation describes the flow of energy by radiation and we should really write $L_{\rm rad}$ rather than just L. Only if radiation is the sole form of energy transport is the L in equation (10.24) also the total energy flow which, at the surface, becomes equal to the observed luminosity.

You have seen that we may neglect conduction but I have not yet said anything about the third form of energy transport, convection, and I must now turn to that.

10.2.3.2 Convection

Convection is a common form of energy transport on Earth, occurring in porridge, kettles and the atmosphere. However, it is very poorly understood in detail, even under laboratory conditions, because fully developed convective motions are turbulent and are so complicated that there is still no satisfactory mathematical model. This makes it difficult to include convection in a stellar model in a realistic way and the treatment of stellar convection remains one of the least satisfactory features of the theory of stellar structure.

However, it is at least straightforward to derive a condition for convection to occur. Consider a blob of stellar material at a radius r, initially at the same temperature, pressure and density as its surroundings, and imagine moving it upwards through a distance δr. I will make two simple assumptions about this movement: it is fast enough that the blob has no time to exchange

energy with its surroundings, so its properties change adiabatically, but it
is slow enough that the speed of the blob is always small compared with
the speed of sound. The second assumption means that, during the motion,
sound waves have plenty of time to smooth out the pressure differences
between the blob and its surroundings and the blob can be taken to have
the same pressure as its surroundings at all times. However, the adiabatic
assumption implies that the density varies in a particular way as the pressure
changes, which may not be the same as in the surroundings. The adiabatic
relation beween pressure and density is:

$$P/\rho^{\gamma} = \text{constant} \tag{10.25}$$

where γ is the ratio of specific heats in the gas, c_p/c_v. We therefore have that
at $r + \delta r$ the blob and its surroundings have the same pressure $P(r + \delta r)$,
given by

$$P(r + \delta r) = P(r) + \frac{dP}{dr}\delta r, \tag{10.26}$$

but the density inside the blob has changed from $\rho(r)$ to $\rho(r) + \Delta\rho$, where

$$\Delta\rho = \frac{\rho}{\gamma P}\frac{dP}{dr}\delta r, \tag{10.27}$$

while the density in the surroundings has changed to

$$\rho(r + \delta r) = \rho(r) + \frac{d\rho}{dr}\delta r. \tag{10.28}$$

The blob now has a different density from its surroundings and what happens
next depends on the sign of the difference: if the blob is denser than its
surroundings, then it will tend to sink back to its original position and the gas
is said to be stable against convection. However, if the blob is less dense than
its surroundings then it will keep on rising, accelerated by the buoyancy force
$\delta\rho\, g$ arising from the density difference $\delta\rho = \rho_{\text{surroundings}} - \rho_{\text{blob}}$. The gas is
then convectively unstable. The condition for instability is $\rho + \Delta\rho < \rho(r+\delta r)$,
or, from equations (10.27) and (10.28),

$$\frac{d\rho}{dr} > \frac{\rho}{\gamma P}\frac{dP}{dr}. \tag{10.29}$$

The right-hand side of this inequality, as can be seen from equation (10.25),
is the density gradient that would exist in the surroundings if they had an
adiabatic relation between density and pressure, while the left-hand side
is the actual density gradient in the surroundings. Remembering that the
pressure and density decrease outwards, so that both these gradients are

negative, we can use equation (10.17) to rewrite this inequality, for an ideal gas, in the form

$$\frac{P}{T}\frac{dT}{dP} > \frac{\gamma - 1}{\gamma} \; ; \tag{10.30}$$

this can be expressed in words as: a gas is convectively unstable if the actual temperature gradient is steeper than the adiabatic gradient.

If this condition is satisfied, then large-scale rising and falling motions occur in the gas. Rising blobs are less dense than their surroundings and are therefore hotter than their surroundings (this follows, for an ideal gas, from pressure balance – see equation (10.17)), while falling blobs are denser and cooler than their surroundings; both rising and falling motions therefore effectively carry energy upwards. However, in reality the motion is much more complicated than just rising and falling blobs and there is no good estimate for the amount of energy that is carried by convection. The usual approach is to describe the flow in terms of three parameters, the mean speed of a convective element, v_c, the mean temperature excess, ΔT, of an element over its surroundings and the 'mixing length', l, which is the mean distance travelled by an element before it gives up its energy and dissolves back into its surroundings. Then a rough estimate of the energy carried by convection is

$$F_{conv} = 2\rho v_c l c_p \Delta T \tag{10.31}$$

where the factor 2 takes account of both rising and falling elements. The convective velocity can be estimated in terms of the mixing length by equating the mean kinetic energy of a blob with the work done by the buoyancy force, and the temperature excess can be estimated in terms of the mixing length and the difference between the real and adiabatic temperature gradients. However, the mixing length itself is not determinable and is just an adjustable parameter whose value must be fixed empirically by comparing stellar models with observation. It turns out to be approximately equal to the pressure scale height, H_p: the distance over which the pressure changes by a factor e. Formally

$$H_p = |\, dr/d \ln P \,| . \tag{10.32}$$

Fortunately, the lack of a good theory of convection is not so serious as one might expect. We know from the rate of energy generation what energy flux needs to be carried at any radius and equations (10.24) and (10.23) tell us how much can be carried by radiation at a particular temperature gradient. We can then use equation (10.31) to estimate what temperature excess is needed to carry the remaining flux. Except very near the surface of a star, this turns out to be very small. This means that a temperature gradient only slightly in excess of the critical, adiabatic gradient is sufficient for convection to carry all the flux – convection is very efficient at carrying energy. Throughout most of a star the fractional excess over the adiabatic gradient is so small (typically 10^{-6} or less) that to a very good approximation the temperature gradient can be taken to be the adiabatic gradient and we do not need to use equation (10.31) explicitly. The energy balance in a convective region is then determined by the three equations

$$\frac{\mathrm{d}T}{\mathrm{d}r} = -\frac{3\kappa\rho L_{\mathrm{rad}}}{16\pi acr^2 T^3}, \tag{10.33}$$

$$\frac{P}{T}\frac{\mathrm{d}T}{\mathrm{d}P} = \frac{\gamma-1}{\gamma}, \tag{10.34}$$

$$L_{\mathrm{conv}} = L - L_{\mathrm{rad}}, \tag{10.35}$$

where equation (10.34) determines the temperature gradient, equation (10.33) now determines the amount of energy carried by radiation down that gradient and the last equation is used in place of equation (10.31) to tell us how much energy is carried by convection.

10.2.4 Nuclear energy generation

I have not yet said anything about the rate of generation of energy by nuclear reactions, although in Section 10.1.2 I mentioned the main reactions involved. The detailed calculations involve more nuclear physics than I have space for in this book but in equilibrium the rate of energy generation can be written in the general form

$$\epsilon(r) = \epsilon(\rho, T, chemical\ composition). \tag{10.36}$$

Just as for opacity, this general expression can be represented over restricted ranges of density and temperature by power laws. For the two-body interactions that occur in hydrogen fusion reactions, the rate is proportional to the square of the density; thus ϵ, which is the rate per unit mass, is proportional to the density and can be written

$$\epsilon = \epsilon_{\circ}\rho T^n, \tag{10.37}$$

where n is a slowly varying function of ρ and T. For hydrogen-burning, $n \simeq 4$ in the Sun and lower-mass stars and $n \simeq 15$ in more massive stars in which the central temperature is higher and a somewhat different series of intermediate reactions occurs.

10.2.5 The full equations and boundary conditions

We now have the same number of equations as unknowns and it is convenient to summarize them. They consist of a set of first-order, ordinary differential equations:

$$\frac{\mathrm{d}P}{\mathrm{d}r} = -\frac{GM\rho}{r^2}, \tag{10.38}$$

$$\frac{\mathrm{d}M}{\mathrm{d}r} = 4\pi r^2\rho, \tag{10.39}$$

$$\frac{\mathrm{d}L}{\mathrm{d}r} = 4\pi r^2\rho\epsilon, \tag{10.40}$$

$$\frac{\mathrm{d}T}{\mathrm{d}r} = -\frac{3\kappa\rho L}{16\pi acr^2 T^3}, \tag{10.41}$$

where, in a convection zone, equation (10.41) is replaced by the three equations:

$$\frac{\mathrm{d}T}{\mathrm{d}r} = -\frac{3\kappa\rho L_{\mathrm{rad}}}{16\pi acr^2 T^3} \tag{10.42}$$

$$\frac{P}{T}\frac{\mathrm{d}T}{\mathrm{d}P} = \frac{\gamma-1}{\gamma} \tag{10.43}$$

$$\text{and } L_{\mathrm{conv}} = L - L_{\mathrm{rad}}. \tag{10.44}$$

These differential equations are supplemented by a set of relations whose exact form depends on the detailed physics of stellar interiors:

$$P = P(\rho, T, chemical\ composition) \tag{10.45}$$

$$\kappa = \kappa(\rho, T, chemical\ composition) \tag{10.46}$$

$$\epsilon = \epsilon(\rho, T, chemical\ composition). \tag{10.47}$$

Because differential equations are involved, these equations need a set of boundary conditions before they can be solved. It is obvious that at the centre the mass and luminosity must go to zero and two boundary conditions are

$$M(r) = L(r) = 0 \text{ at } r = 0. \tag{10.48}$$

The surface boundary conditions are not so obvious, since we do not know the radius or surface temperature *a priori*. However, the surface temperature is certainly very much smaller than the central one, and a reasonable first definition of the surface is where the density goes to zero; two approximate surface boundary conditions are then

$$\rho = T = 0 \text{ at } r = R_{\mathrm{S}} \tag{10.49}$$

where R_{S} is the radius of the star and cannot be found from the equations. This formulation would in principle enable us to construct stars of a given radius. In practice, since the radius of a star changes by orders of magnitude during its evolution, it is more convenient to specify the total mass, M_S, of a star and to rewrite the differential equations in terms of the mass as the independent variable. The affected equations become:

$$\frac{dP}{dM} = -\frac{GM}{4\pi r^4} \tag{10.50}$$

$$\frac{dr}{dM} = \frac{1}{4\pi r^2 \rho} \tag{10.51}$$

$$\frac{dL}{dM} = \epsilon \tag{10.52}$$

$$\frac{dT}{dM} = -\frac{3\kappa L}{64\pi^2 acr^4 T^3} \tag{10.53}$$

and the corresponding boundary equations are

$$r = L = 0 \text{ at } M = 0 \tag{10.54}$$

and

$$\rho = T = 0 \text{ at } M = M_S. \tag{10.55}$$

With these boundary conditions, the structure of a main-sequence star is uniquely specified by the mass M_S and the chemical composition.

Of course, the surface boundary conditions are not exact. We know, for example, that a better estimate of the surface temperature is $T = T_{eff}$ (see equation (8.7)). Also, the density does not go abruptly to zero – for example, in the Sun the outer atmosphere is detectable well beyond the Earth as the outflowing solar wind. However, experience has shown that the structure of most stellar models is not sensitive to the exact boundary conditions, so if we are interested in the overall structure and evolution, rather than in details of the surface layers, the conditions (10.49) are usually sufficient.

To improve these boundary conditions we need to understand exactly how the radiation leaves the star, a subject whose details are beyond the scope of this book. Roughly speaking, we define the 'visible surface' as the layer from which most of the radiation escapes, or from which a photon is last emitted or scattered, and require the temperature at that depth to be equal to the effective temperature:

$$T(\text{visible surface}) = T_{eff}. \tag{10.56}$$

Detailed models show that the density decreases approximately exponentially outside the visible surface, so $\rho = 0$ is a good enough boundary condition in almost all circumstances. However, the simple $T = 0$ boundary condition can give incorrect results, especially in cool stars where the energy in the

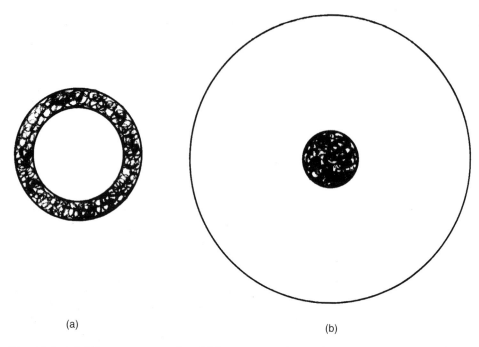

(a)   (b)

Fig. 10.4. (a) The structure of a $1\,M_\odot$ star on the main sequence. Most of the mass of the star is in radiative equilibrium, including the energy-generating regions at the centre, but there is an outer shell in which convection carries most of the energy. For the Sun, the depth of this convection zone is about $0.25\,R_\odot$ (but it contains only 1 or 2% of the mass – see Figs. 10.5 and 10.6). (b) The structure of a $5\,M_\odot$ main-sequence star to the same scale; the radius is about $2.2\,R_\odot$. Convection carries all the energy in the central, energy-generating core, whose radius is about $0.2\,R_{\text{star}} \simeq 0.44\,R_\odot$. Radiation carries the energy from the core to the surface.

surface regions is largely carried by convection. The models with $T = 0$ may then give temperatures at the 'surface of last scattering' (the surface at which a photon is last scattered before escaping from the atmosphere) which are either much greater or much less than the effective temperature and it is necessary to apply condition (10.56) explicitly instead of the $T = 0$ condition.

10.2.6 *The structure of main-sequence stars*

The structure of main-sequence stellar models, based on the above equations and boundary conditions, can be summarized by the sketches in Fig. 10.4 and the detailed graphs for the Sun in Fig. 10.5. The surface temperatures and luminosities of the models increase with mass in a way that satisfactorily reproduces the observed HR diagram and mass–luminosity relationship.

Internally, stars of different mass differ largely in the position and extent

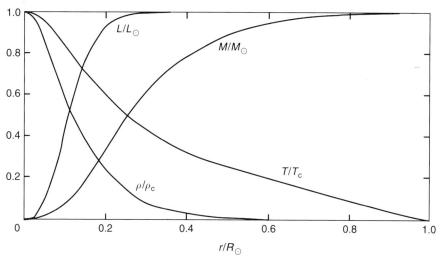

Fig. 10.5. The variation of mass, luminosity, temperature and density with radius in a model of the Sun (data from Böhm-Vitense 1992). The mass, luminosity and radius are given in terms of their surface values, and the temperature and density in terms of their central values. Notice how rapidly the density falls off – it drops to one-thousandth of its central value at about half a solar radius. This is a characteristic of most stars, in which the mass is concentrated towards the centre; for the Sun, 90% of the mass is within half the radius (one-eighth of the volume). The energy generation is even more concentrated in radius, with 99% being within one-quarter of the radius.

of the convectively unstable regions (*convection zones*). Criterion (10.30) for convection to occur can be satisfied in two ways: either the ratio of specific heats, γ, is close to unity or the temperature gradient is very steep. In cool layers, where the gas is partially ionized, much of the heat used to raise the temperature of the gas goes into ionization and hence the effective specific heats of the gas are nearly the same and $\gamma \simeq 1$. In very hot layers, the energy generation rate is sensitive to temperature and the temperature gradient needed to carry the flux away from the site of energy production by radiation is very steep. Where, or whether, such layers occur in a particular star depends on the mass of the star (Fig. 10.6).

In stars with masses less than about $1.3 \, M_\odot$ the surface layers are cool enough to be partially ionized and are unstable to convection for the first reason. The structure of solar-type stars then consists of a hot radiative core, at the centre of which the nuclear energy is produced, surrounded by a cool convective envelope. This envelope is shallow for stars near the upper-mass limit but becomes gradually deeper for lower-mass stars, and stars with masses less than about $0.3 \, M_\odot$ are believed to be fully convective. The

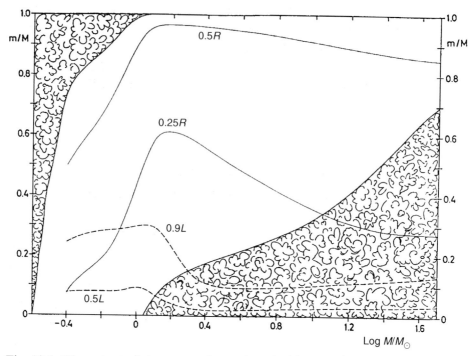

Fig. 10.6. The extent of the convective regions inside stars is shown as a function of stellar mass, M. A vertical line in the diagram at mass M represents the mass distribution inside a star of that mass. 'Cloudy' areas indicate convective regions. The solid lines show the mass values for which the radius is 1/4 and 1/2 of the total radius R; the dashed lines show the masses within which 0.5 and 0.9 of the luminosity L is produced. (From Kippenhahn and Weigert, 1990.)

precise nuclear reactions that transform hydrogen into helium depend on temperature, changing with increasing temperature from a chain involving individual proton–proton reactions to a cycle of reactions in which carbon, nitrogen and oxygen act as catalysts. In stars with masses greater than about $1.1\,M_\odot$ the central temperature is high enough that the 'CNO cycle' becomes more important than the 'pp-chain'. The CNO cycle is much more sensitive to temperature and requires a much steeper radiative temperature gradient, which is unstable to convection. Stars considerably more massive than the Sun therefore have a convective core, which contains the energy-generating regions, and a radiative envelope. Stars just a little more massive than the Sun may have both a small convective core and a shallow convective envelope. The mass in the convective core increases with the total mass of the star and very massive stars, like very low-mass stars, may be fully convective.

The structure of main-sequence stars is fairly well understood, apart from the uncertainties about the theory of convection, because all stars spend some 90% of their lives on the main sequence and there are plenty of observations of main-sequence stars against which to test the theory. However, the slow nuclear reactions at the centre eventually turn all the hydrogen to helium and the star begins to evolve off the main sequence. The last 10% of a star's life is much more eventful than what led up to it, and I now turn to these later stages.

10.3 From main sequence to stellar remnant

The constant theme in the life story of a star is a battle against the tendency to collapse under gravity. In the early stages of star formation, gravity dominates and the initial gas cloud collapses very rapidly until the internal pressure becomes sufficient to oppose the gravitational force. Once the internal pressure is high enough to balance gravity completely (force balance), the centre of the star becomes hot enough for nuclear reactions to start and provide a source of energy to balance the steady leakage of energy from the star by radiation from the surface (energy balance). Thus in a main-sequence star the force balance which keeps the star at a constant radius also produces an energy balance fuelled by the nuclear reactions.

When the nuclear reactions finally cease, because all the hydrogen at the centre has been burned, this energy balance is lost and the star must find an alternative energy source to balance the surface luminosity. Initially, the only available source is gravitational potential energy and the star draws on this by beginning to contract again. Sooner or later, the central temperature increases sufficiently that the helium begins to burn, providing a new nuclear energy source. Equilibrium is again possible, until the helium in its turn is exhausted. Whether further nuclear fuels are available depends on the star's mass, and it is convenient to divide the discussion at this point, concentrating on low- and intermediate-mass stars ($M \leq 8\,M_\odot$). Higher-mass stars, which can go on to burn carbon and other elements, are comparatively rare and I will discuss them briefly at the end of the section.

10.3.1 Central hydrogen exhaustion and the onset of shell burning

10.3.1.1 Low-mass stars: radiative cores

In stars of less than about $1.1\,M_\odot$ there is believed to be no convective core, which means that the central regions of the star are not mixed. Nuclear reactions can therefore draw only on the fuel that is available locally and they cease when the hydrogen there is exhausted. The temperature is

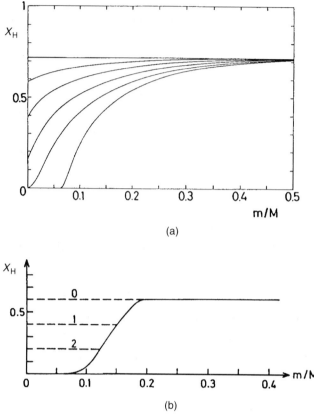

(a)

(b)

Fig. 10.7. The fraction of the mass of the star in the form of hydrogen, X_H, is plotted in these two diagrams as a function of the mass fraction (M is the total mass of the star, and m is the mass out to a particular radius). Initially, the star is supposed to have a uniform composition. (a) In a $1\,M_\odot$ star, X_H decreases smoothly with time, at a rate that is fastest near the centre. The solid lines show the composition profile at six successive times. When $X_H = 0$ at the centre, there is still unburned hydrogen just outside the centre and energy generation continues. (b) In a $5\,M_\odot$ star, convective mixing at the centre keeps the hydrogen content uniform throughout the core at all times, as shown for three successive times by the dashed lines. This ensures that hydrogen exhaustion occurs simultaneously throughout the core; since the region outside the core, with $X_H \neq 0$, is initially too cool to burn hydrogen, nuclear energy generation ceases. Note that the convective region contracts in mass during hydrogen-burning, so that the burnt-out region at the centre is considerably smaller than the initial size of the convective core. (From Kippenhahn and Weigert, 1990.)

highest at the centre, the rate of burning is highest there and it is at the centre that the hydrogen is first used up. However, immediately outside the centre the temperature is only a little lower and there is still hydrogen left to burn. Nuclear reactions therefore continue, but now in a thick shell rather than throughout a sphere (Fig. 10.7(a)). The local rate of nuclear energy production thus decreases gradually as the shell burns outwards into cooler regions, rather than stopping abruptly when the central hydrogen is exhausted. The continuation of nuclear burning in a shell means that there is no need for the whole star to contract, although the star must adjust its structure somewhat in order to keep its total nuclear energy output equal to the star's surface luminosity. Only in the central regions, where there is no longer any energy production but energy is still leaking outwards into the rest of the star, is there a need for a further energy source. A gradual contraction of the hydrogen-free core ensues, releasing gravitational potential energy, some of which goes towards a gradual heating of the central regions and some of which is used to enable a slow *expansion* of the outer envelope of the star, so that the star moves from the main sequence in the HR diagram towards the giant region in the upper right of the diagram (Fig. 9.1).

10.3.1.2 Intermediate-mass stars: convective cores

In stars with masses greater than about $1.1\,M_\odot$, the energy transport in the core is by convection, rather than the radiation that dominates in lower-mass stars. The resultant convective mixing has an important effect on the energy production, since fresh hydrogen is constantly mixed in to the centre and newly produced helium is quickly spread throughout the core, keeping the chemical composition uniform (Fig. 10.7(b)). Thus when the nuclear reactions have used up all the hydrogen at the centre there is no hydrogen left anywhere in the convectively mixed region and energy production ceases throughout the core. Just outside the core hydrogen is still available for burning, but the temperature is too low for fusion to occur and hydrogen-burning ceases altogether when the core burns out.

The star now has no nuclear energy source, and it is forced into a slow overall contraction that draws on its reserves of gravitational energy. Part of that energy is used to balance the continuing radiation from its surface, but about half of it goes into heating up the core as the central density and pressure are increased by the contraction. Eventually, the core is hot enough that hydrogen can start burning in a thin shell just outside the core boundary. The star is now similar in structure to a hydrogen-shell-burning star of lower mass, but it has a much thinner shell. Overall contraction stops, and the energy from the contraction of the core is now fed into an expansion of the envelope, just as for a lower-mass star. The main difference in the evolutionary track in the HR diagram is that the phase of overall contraction causes a hook to the left for intermediate-mass stars before

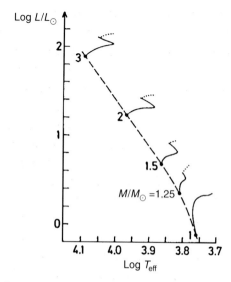

Fig. 10.8. Evolution from the main sequence. Low-mass stars with radiative cores evolve smoothly from core burning into shell burning, and the track in the HR diagram is a smooth curve (e.g. 1 $M_\odot$). Higher-mass stars with convective cores have a period of overall contraction before shell burning starts, and there is a leftward hook in the evolutionary track. (From Kippenhahn and Weigert, 1990.)

progress to the giant region in the upper right is resumed once the shell has ignited (Fig. 10.8).

10.3.2 *Transition to the giant branch*

10.3.2.1 *Isothermal cores*

In both low- and intermediate-mass stars the exhausted core now consists of almost pure helium. Although there is no nuclear energy source, the centre of the star is initially hotter than the edge of the core and heat flows down this temperature gradient, cooling the centre and heating up the outside. For masses less than about 6 $M_\odot$, the smoothing out of the temperature gradient continues until the core is *isothermal* : T_{core} = constant. This has a dramatic effect on what happens next, as you will now see.

10.3.2.2 *Crossing the Hertzsprung gap*

Once an isothermal core has formed, it grows slowly in mass as a result of the nuclear shell source outside it gradually burning its way outwards into fresh fuel, leaving helium 'ash' behind. Because the radius of the core is still contracting, as you saw in Section 10.3.1, its density steadily increases. What happens as we increase the density of an isothermal gas? An isothermal gas in the laboratory has a simple equation of state, $P \propto \rho$. If the density increases, the internal pressure increases and can balance any increase in external pressure. The isothermal gas in a stellar core has an additional property: it is self-gravitating. An increase in density now also increases the gravitational field. Because gravity is attractive, this counteracts the increase in the thermal pressure and is equivalent to introducing a negative 'gravitational pressure'. Detailed calculations confirm the intuitive result that there is a maximum density for such an isothermal gas sphere: at sufficiently high densities the gravitational effects dominate and the effective internal pressure actually decreases as density increases. The stellar core has to withstand an external pressure arising from the weight of the overlying layers of the star. Once it reaches the critical density, it can no longer do so, because further compression decreases the internal pressure of the core. A strictly isothermal core then starts to collapse on an essentially free-fall timescale, as a result of loss of pressure support. The energy released by this rapid collapse quickly heats the centre, restoring a temperature gradient, and hence a thermal pressure gradient, that can again balance gravity, so the initial dynamical collapse rapidly slows to a contraction on a thermal timescale. In a real star, the core develops a small temperature gradient before the critical density is reached, and the contraction is always on a thermal timescale.

The effect of this sudden core contraction is to pour enough energy into the envelope for a rapid expansion of the whole star, and a transition to the giant region occurs on a thermal timescale of some 10^6–10^7 yr. Because this is much shorter than the timescale for nuclear burning, typically 10^9 yr or more, very few stars are observed between the main-sequence band and the giant region in the HR diagram. This accounts for the 'Hertzsprung gap' seen in HR diagrams of open clusters (Fig. 9.13) – stars do evolve through this region, but so rapidly that the chance of catching one in it is very small.

This core instability can be shown to occur when the core mass reaches a critical fraction of the total mass of the star, known as the Schönberg–Chandrasekhar limiting mass fraction. For typical chemical compositions, the critical fraction is about 10%. All stars with isothermal cores are vulnerable to the instability. However, it only occurs for stars with masses between about $2 \, M_\odot$ and $6 \, M_\odot$. For lower masses, the core develops a new source of pressure before the limiting mass fraction is reached, and is able to withstand gravity even at high densities. The new pressure source, electron degeneracy, arises from quantum effects and has profound effects on later stages of evolution.

Before considering degeneracy, an interesting feature of the evolutionary

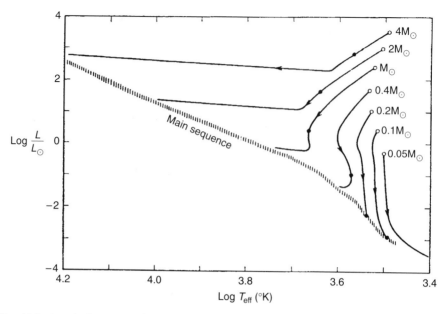

Fig. 10.9. Evolutionary tracks for pre-main-sequence stars. Low-mass stars evolve down the Hayashi line before developing a radiative core and moving nearly horizontally onto the main sequence. High-mass stars become radiative earlier, and spend little time on the Hayashi line. (Reproduced, with permission, from the *Annual Review of Astronomy and Astrophysics*, Volume 4, ©1966, by Annual Reviews Inc.)

tracks to the giant region must be explained: stars first move to lower temperatures in the HR diagram, and then quite abruptly turn a corner and start moving to higher luminosity with little change in temperature. This happens when the surface of the star becomes cool enough that the envelope is fully convective. In 1961, C. Hayashi discovered that there is a limiting temperature for fully convective stars: stars with surface temperatures below a certain value ($T_{\text{eff}} < 3000$–$4000 \, \text{K}$) cannot be in equilibrium. This is essentially because the internal temperature gradient is determined for a convective star by the surface conditions, and for low enough surface temperature the resulting central temperature (and so pressure) is too low to be compatible with hydrostatic equilibrium. This is one situation where it is essential to use the boundary condition (10.56) for the surface temperature; the simple $T = 0$ condition gives unrealistic stellar models. In fact, it was Hayashi's work that revealed the importance of using a proper surface boundary condition for convective stars.

The limiting temperature line on the HR diagram is now known as the Hayashi line, and no stars exist on the cool side of it. Hayashi argued that pre-main-sequence stars must evolve to the main sequence by descending

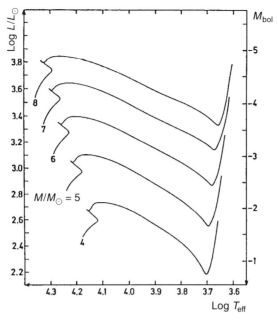

Fig. 10.10. Evolutionary tracks from the main sequence to the giant branch in the HR diagram. When stars reach the Hayashi line, they cannot cross it and evolve up it to higher luminosities. (From Kippenhahn and Weigert, 1990.)

this line, and then swinging onto a more horizontal track as they heat up, develop a radiative core, and approach the main sequence (Fig. 10.9). For post-main-sequence stars, the Hayashi line provides a barrier which cannot be crossed, and the 'giant branch' in the HR diagram (Fig. 10.10) is the result: stars evolve across the Hertzsprung gap and then swing up the Hayashi line, to become giants as they evolve up it.

10.3.3 Degeneracy

So far I have assumed that a star is made of an ideal gas that can be described by classical physics. At sufficiently high densities, quantum effects begin to become important and in particular we have to take account of Pauli's exclusion principle, which says that no two electrons can occupy the same quantum state. Deep inside a star, the matter is virtually fully ionized and the electrons are not in discrete atomic energy levels. None the less, because of Heisenberg's uncertainty principle, the electrons cannot occupy a literally continuous range of energies. For a given electron we can only confine its position and momentum within a certain range given by the uncertainty principle: $\Delta p \Delta x \geq h$. Thus you can think of a quantum state as consisting of a volume h^3 in six-dimensional position–momentum space (phase space), and Pauli's exclusion principle tells us that only two electrons (of opposite spin) can occupy one such elementary volume.

When the density becomes large, electrons are crowded together in position space and, in order to avoid having more than two electrons in a volume h^3 in phase space, some electrons are forced to have higher momenta than they would have at the same temperature in an ideal gas. Since the momentum distribution determines the pressure, the equation of state no longer depends so much on temperature and at sufficiently high densities the pressure is essentially independent of the temperature, being determined only by the density. The electron gas is then said to be *fully degenerate*, and has a finite pressure even at zero temperature. The ions are not degenerate and still satisfy the ideal gas law, but the pressure of the degenerate electrons is much larger and the equation of state can be written as

$$P = K_1 \rho^{5/3} \tag{10.57}$$

when the electrons are moving non-relativistically and as

$$P = K_2 \rho^{4/3} \tag{10.58}$$

in the limit that essentially all the electrons are moving relativistically; for intermediate cases the expression is more complicated, and I will not give it here. The factors K_1 and K_2 are combinations of known constants and of the mass fraction of hydrogen:

$$K_1 = \frac{h^2}{20m_e} \left(\frac{3}{\pi}\right)^{2/3} \left(\frac{1+X}{2m_H}\right)^{5/3}, \tag{10.59}$$

$$K_2 = \frac{hc}{8} \left(\frac{3}{\pi}\right)^{1/3} \left(\frac{1+X}{2m_H}\right)^{4/3}. \tag{10.60}$$

For stars with $M \leq 2\,M_\odot$, the core becomes degenerate before it reaches the critical mass fraction, and the extra pressure of the degenerate electrons prevents core collapse. However, the new equation of state has an even more dramatic effect when the core becomes hot enough for helium ignition.

10.3.4 Central helium-burning

10.3.4.1 The triple-alpha reaction

After the ignition of a hydrogen shell, a star's core continues to contract, whether or not it suffers a sudden collapse. Eventually, the centre of the star reaches a temperature of about 10^8 K, which is hot enough for the helium to start burning. Helium fusion reactions are unusual; the fusion of two helium nuclei forms the highly unstable ^{8}Be nucleus and the only way of forming a stable product is through the much rarer reaction in which three helium nuclei (alpha particles) fuse almost simultaneously to produce a ^{12}C nucleus: the triple-alpha reaction. This reaction is extremely sensitive to temperature, with $\epsilon \propto T^{40}$, and therefore sets in fairly abruptly as the temperature rises through 10^8 K. Because it is essentially a three-body reaction, the rate is proportional to the cube of the density and we can write ϵ, the rate per unit mass, as

$$\epsilon \propto \rho^2 T^{40} .$$

To begin with, only helium-burning is significant. However, the triple-alpha reaction is so unlikely that helium nuclei begin to combine with ^{12}C nuclei once sufficient carbon has been produced, and eventually the reaction

$$^{12}C + \alpha \rightarrow {}^{16}O$$

takes over from the triple-alpha reaction as the main destroyer of helium. The abundance of carbon reaches a maximum, and then decreases again; by the time all the helium has been used up, the amounts of oxygen and carbon in the exhausted core are comparable.

10.3.4.2 The helium flash

In low-mass stars like the Sun ($M < 2\,\mathrm{M_\odot}$), the onset of helium-burning is made even more violent by the fact that the helium core becomes degenerate before it is hot enough for helium to burn. When the helium does ignite, the temperature increases rapidly. In an ideal gas, this would cause the pressure to increase and the core to expand and cool, keeping the temperature just high enough for the nuclear reactions to continue; burning would therefore start in a stable fashion. This is what happens in higher-mass stars, where helium-burning starts quietly. With a degenerate equation of state, the increasing temperature has no effect on the pressure and the core cannot expand to counteract the rising temperature. Instead, the rise in temperature just causes the very temperature-sensitive nuclear reactions to occur faster, increasing the temperature still further. This 'thermal runaway' causes the ignition of helium-burning to occur explosively, in what is called the 'helium flash'. The explosion stops only when the temperature has risen so high that the ion pressure becomes significant and is able to expand the core and cool it. The expansion also lowers the density and lifts the degeneracy, so the star then settles down into quiet helium-burning in a non-degenerate core. The consequences of the helium flash for the rest of the star are unclear;

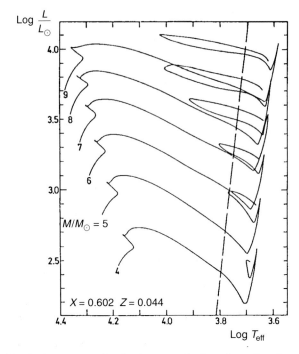

Fig. 10.11. Evolutionary tracks during core helium-burning for stars of intermediate mass. The leftward loop for stars more massive than about $5\,M_\odot$ takes them through the instability strip (marked by a broken line), where they are observed as Cepheid variables. (From Kippenhahn and Weigert, 1990.)

most calculations suggest that at most a little mass is lost, the main force of the core explosion being muffled by the extensive outer layers. After a rapid evolution during the helium flash, the star comes into equilibrium in a position on the HR diagram that corresponds to the horizontal branch in globular clusters (Figs. 9.14, 10.12).

10.3.4.3 Loops in the HR diagram

The effect of helium ignition, whether violent or quiet, is to move a star off the giant branch towards higher surface temperatures. There are now two nuclear energy sources, helium-burning in the core and hydrogen-burning in a shell, and the evolution is more complicated than in the core-hydrogen-burning phase. As the two energy sources vary in relative strength, the star makes a number of excursions to and fro across the HR diagram as it expands and contracts in response to changes in its internal structure. During these loops, the star passes through the 'instability strip', where it appears as a Cepheid variable star (Section 9.4.1) if it is of intermediate

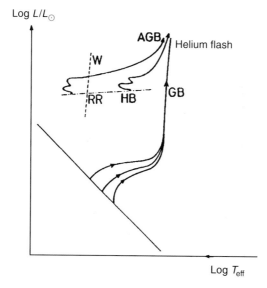

Fig. 10.12. The evolutionary tracks of three low-mass stars of slightly different mass converge on the giant branch (GB). After they have undergone the helium flash (at the top of the giant branch), they become horizontal branch stars (HB) and evolve through the instability strip as RR Lyrae variables. The broken line shows the positions of the RR Lyraes (RR) and of the W Virginis stars (W), which are Population II Cepheids. The tracks, which depend strongly on mass on the horizontal branch, converge again as they tend asymptotically back to the giant branch – the track on this second ascent of the giant branch is therefore known as the asymptotic giant branch (AGB). (From Kippenhahn and Weigert, 1990.)

mass. Theoretical tracks (Fig. 10.11) suggest that most observed Cepheids are in the mass range 5–7 $M_\odot$; for masses less than 5 $M_\odot$ the loops are too short to cross the instability strip, while for masses greater than 7 $M_\odot$ the nuclear evolution timescale is becoming so short that the chance of catching a star in the strip is very low. At lower luminosity, the instability strip crosses the horizontal branch (Fig. 10.12) and stars of 0.5–1.0 $M_\odot$ are seen as RR Lyrae stars or as Population II Cepheids during their core-helium-burning evolution. Observations of Cepheids and RR Lyraes can yield estimates of radii and masses (Section 9.4.1) which provide valuable independent tests of this stage of evolution.

10.3.5 *Up the asymptotic giant branch*

10.3.5.1 *Central helium exhaustion*

The final loop to the left in the HR diagram follows exhaustion of helium in the core. Because the triple-alpha reaction is very temperature-sensitive, the core is convective and burns out everywhere simultaneously, initiating a period of overall contraction similar to that described for the hydrogen-exhaustion phase. The centre of the star now consists of a carbon–oxygen core (Section 10.3.4) surrounded by a shell of almost pure helium; outside this is a vast envelope of unevolved material, mostly hydrogen. The core contains 10% or more of the mass of the star but has shrunk to a very small size: its radius is typically only one part in a hundred thousand of the radius of the star. A scale model of such a red giant star, with total radius 1 km, would have a core only 1 cm in size. The core continues to contract, and becomes degenerate, but for stars with masses less than about $8\,M_\odot$ it never becomes hot enough for any further nuclear fuels to ignite.

10.3.5.2 Double-shell source stars and thermal pulses

Nuclear burning starts again when the base of the helium shell becomes hot enough for the helium to re-ignite and the star moves back to the giant branch. Here it remains for much of the rest of its life, moving slowly to larger radii and higher luminosities as it consumes its helium and becomes a red supergiant. Because the calculated evolutionary tracks approach the giant branch asymptotically, stars at this stage of their lives are often known as *asymptotic giant branch* stars, or AGB stars for short. The hydrogen shell source is still contributing energy and much of the evolution is an interplay between these two shell sources. Initially, an AGB star undergoes a rather turbulent period of evolution, because the helium shell, when it ignites, is unstable and burns in short, violent pulses rather than steadily. During each pulse the hydrogen shell is driven outwards to lower temperatures and goes out, to re-ignite as the helium pulse dies away; because the instability is a thermal one, the interval between pulses is the thermal timescale of the shell, typically 10^5 yr for a $1\,M_\odot$ star. Complex mixing occurs in the region between the helium and hydrogen shells, and in the envelope, and products of the nuclear burning are mixed to the surface of the star. The most prominent results are seen in the carbon stars, red giants with carbon-to-oxygen ratios up to ten times normal, but many other red giants are known with peculiar element abundances, such as excessive amounts of barium, all resulting from mixing during this 'double-shell source' stage of evolution. These stars present direct evidence that our ideas about nuclear processing inside stars are correct.

10.3.5.3 Mass loss: planetary nebulae and white dwarfs

It has gradually come to be realized that all stars lose mass during their evolution. In one sense this is trivially true, since the energy lost by radiation is equivalent to mass loss at a rate

$$\dot{M} = L/c^2 \simeq 7 \times 10^{-14} \, (L/L_\odot) \, M_\odot \, yr^{-1}.$$

However, this rate is small compared to loss of gas by stellar winds (even for the Sun, which loses only a tiny amount of mass per year via the solar wind) and I will ignore it here.

Very massive stars lose mass at a significant rate even on the main sequence, lower-mass stars only when they reach the giant branch. Observations tell us that red giant and supergiant stars lose mass at rates of 10^{-8}–$10^{-9} \, M_\odot \, yr^{-1}$ by stellar winds. The winds are driven off the surface by mechanisms that are not totally clear but probably include radiation pressure on the dust grains that condense in the cool extended atmospheres of these stars; magnetic fields probably also play a role. The mass loss rates are such that the later stages of evolution must certainly be significantly affected, but the details are unclear. Many of the effects are still deduced from observation rather than from detailed theory. Near the top of the red giant branch, the radius of the star becomes so large that the surface gravity is very low and the surface layers are only loosely bound to the rest of the star. It is in this region of the HR diagram that we see the Mira variables, and other long-period irregular variable stars, which seem to be undergoing large-amplitude radial pulsations. Observations suggest that as a star evolves towards the tip of the red giant branch there is a sudden increase in the mass loss rate, referred to as a 'superwind'; this may be connected with a switch from pulsation in the fundamental mode to pulsation in the first overtone. Eventually, some instability seems to occur, possibly when the luminosity exceeds a critical value, when the amplitude of the pulsation increases so much that the star ejects a whole shell of material at a speed of some tens of kilometres per second, just in excess of the escape velocity. The slowly expanding shell reveals hot underlying layers, which ionize the shell, causing it to glow conspicuously in optical and ultraviolet emission lines; we see these shells as *planetary nebulae* (Fig. 10.13). The central star is still a supergiant, but is no longer red – it has moved horizontally across the HR diagram and become a hot, blue supergiant to the left of the main sequence.

Because of the continuous mass loss before the ejection of the planetary nebulae, and especially during the superwind phase, these stars have probably had all their remaining layers of hydrogen stripped off and their structure consists of a carbon core surrounded by a shell of unburnt helium. The temperature is not high enough for these fuels to burn and the star is once again forced to contract. The rise in density causes almost all of the star to become degenerate and the star reaches a new equilibrium in which gravity

Fig. 10.13. The Helix Nebula, NGC7293, is the nearest planetary nebula, about 500 lt yr away. It is a giant shell of gas being blown slowly away from the central star. (Photograph from the Hale Observatories.)

is balanced by the pressure of degenerate electrons. However, there are now no nuclear reactions to balance the loss of energy from the surface, so once the contraction has stopped the star is not in thermal equilibrium and it spends the rest of its life gradually cooling. It is now a *white dwarf* and is very small (typically 1/100 of the Sun's radius) and faint, though quite hot: the surface temperature of observed 'white' dwarfs covers a range from about 5000 K to 20 000 K.

As you will see in Section 10.4, white dwarfs have a maximum mass of about $1.4 \, M_\odot$. We can use this to estimate the mass range of main-sequence stars that can end their lives in this way. One of the characteristics of AGB evolution is that most properties of the star depend only on the core mass. Since the envelope of the star is completely stripped off, the mass of the final white dwarf must be the same as the core mass at planetary nebula ejection, so the final core mass must be less than $1.4 \, M_\odot$. Calculations for various initial masses then suggest that a white dwarf will be produced only if the initial mass is less than about $8 M_{\odot}$, which is why I chose to limit the discussion to that mass range. In practice, observations suggest that once a core mass has grown to about $0.8 M_{\odot}$ the envelope is ejected so quickly that the core has little chance to grow further and the mass of a white dwarf remnant is unlikely to exceed $1.1 M_{\odot}$.

More massive stars may also give rise to compact remnants (neutron stars or black holes), but the route is much more violent: a supernova explosion. I now discuss briefly the evolution of stars that may end their lives as supernovae.

10.3.6 Nuclear evolution beyond helium-burning

Nuclear evolution beyond helium-burning depends on whether or not the carbon–oxygen core ever becomes hot enough for further fusion reactions to occur and whether the core becomes degenerate. For stars with initial masses above about $8 M_{\odot}$, the core is non-degenerate and carbon can ignite quietly, burning first to oxygen and neon. Further reactions are possible and a series of burning episodes builds up successive shells of more and more processed material. Elements produced in these shells include magnesium, silicon and sulphur. For stellar masses greater than about $11 M_{\odot}$, burning can proceed as far as iron and other elements of comparable nuclear mass, principally chromium, manganese, cobalt and nickel. At this point, as I explain below, a star has exhausted all its possible nuclear fuels. It has an 'onion-skin' structure, with successive shells containing the ashes of the various burning stages. Stars less massive than 20–$40 M_{\odot}$ are thought to retain a hydrogen-rich envelope of unevolved material, but in more massive stars all the unevolved material has probably been stripped off by winds before the final iron core is formed. No peaceful end is possible for such stars, and I will discuss their fate in the next section. First I indicate briefly why no further nuclear energy source is available.

If the abundance of a chemical element, as measured in gas clouds, stars and galaxies, is plotted as a function of atomic mass (Fig. 10.14), the major peak is at hydrogen, which is by a factor of ten (by number) the most abundant element in the Universe. However, the general fall-off towards higher masses is interrupted near iron, where there is a strong local maximum – iron and other elements in the 'iron peak' are very abundant relative to both heavier and lighter elements. This is because iron is the most stable element, with the highest binding energy per nucleon: that is,

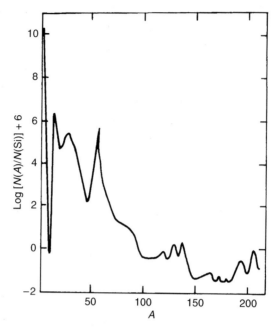

Fig. 10.14. A schematic plot of the relative abundances of the chemical elements against atomic mass, *A*, on a scale where $\log N(\text{Si}) = 6$. The peak between atomic masses 50 and 60 arises from iron and related elements.

more energy is needed to knock a nucleon out of iron than out of any other nucleus.

The *binding energy* of a nucleus, measured by the mass difference between the nucleus and its constituent nucleons, is both the energy needed to break it apart and the energy released in forming it. Elements lighter than iron can be built up out of lighter elements by fusion reactions, and the energy produced by these reactions is what powers stars. Elements heavier than iron can only be produced with the release of energy by the fission of still heavier elements – this is the basis for nuclear power stations, but leaves unanswered the question of where the heavy elements came from in the first place.

What is clear is that to produce elements heavier than iron it is necessary to provide energy – once a star has produced an iron core, no further energy-producing reactions are possible and the star has finally run out of nuclear fuel. As you will see in Section 10.4, it may end its life in a spectacular explosion, leaving behind a compact remnant, but its days as a self-sustaining normal star are over.

None the less, the existence on Earth of elements heavier than iron tells us that nuclear reactions of some kind must build up elements beyond the iron peak. Charged particle reactions require so much energy to overcome the large Coulomb barrier of the iron nucleus that we must look for some other solution. The obvious one is to use neutral particles: neutrons. The difficulty with this is that the neutron is unstable, with a half-life of about ten minutes, so we need a source of neutrons.

Of course, neutrons are produced in some fusion reactions and there seem to be two situations in which enough neutrons are produced to be effective in building up heavier elements. Some of the reactions that occur in AGB stars produce neutrons in fairly small numbers as a side-effect. These cause neutron capture reactions at a rate that is slow compared to the β-decay rate of any unstable nuclei that are formed. These *slow* reactions, which allow time for β-decays, produce stable nuclei, such as barium, known as *s-process nuclei*; the detection of barium and other s-process elements in the surface layers of some red giants encourages us to believe this picture.

The other possible source of neutrons is in supernova explosions. It is now believed that many elements that have been found on the Earth were actually made during the few hours that a blast wave was working its way out through an exploding star. In these conditions, the reactions are far from equilibrium and a large flux of neutrons is available. The neutron capture reactions then occur at a rate much faster than β-decays, and neutron-rich nuclei can be built up through a series of nuclei that would be unstable if given enough time. Because of the *rapid* reaction rate, the final products are called *r-process nuclei* and include such elements as uranium. Detailed calculations of this 'explosive nucleosynthesis' yield relative abundances of the heavy elements that are in remarkably good agreement with those measured on the Earth and in the solar system generally, suggesting that supernovae play a major role in the chemical evolution of the Galaxy. I will return to this topic briefly in Chapter 11.

10.4 Endpoints of evolution

10.4.1 White dwarfs

As you have seen, the final fate of a star, after all its nuclear fuels have been exhausted and it has ejected an envelope, is to become a compact remnant. In particular, stars of less than $8\,M_\odot$ throw off a planetary nebula and reveal a degenerate carbon–oxygen core that becomes a cooling white dwarf.

Because of the degenerate equation of state, white dwarfs have some unusual properties. The most important one is that they have a maximum mass; this was discovered in the 1930s by S. Chandrasekhar, and won him a Nobel prize in 1983. We can understand roughly why there is a maximum mass by using the equation of state of a degenerate gas in the hydrostatic equation, equation (10.10). Let us approximate the pressure gradient by writing

$$\frac{dP}{dr} \simeq -\overline{P}/R,$$

where $\overline{P}$ is some mean pressure within the white dwarf, whose mass and radius are M and R. If we replace r in the hydrostatic equation by R, and ρ by the mean density

$$\overline{\rho} = 3M/4\pi R^3,$$

we find

$$\overline{P} \simeq GM\overline{\rho}/R.$$

Using the non-relativistic equation of state (10.57) in the form

$$\overline{P} = K_1 \overline{\rho}^{5/3},$$

we can eliminate $\overline{P}$ and $\overline{\rho}$ to produce a mass–radius relation:

$$R \propto M^{-1/3}. \tag{10.61}$$

This relationship is valid for low-mass white dwarfs. Now imagine adding mass to such a star – what happens? Initially, it shrinks and the mean density $\overline{\rho}$ increases according to

$$\overline{\rho} \propto M/R^3 \propto M^2. \tag{10.62}$$

As the density increases, so the electrons are pushed ever closer together and more are pushed into high-momentum states. Eventually, the majority of the electrons become relativistic and the equation of state becomes $P \propto \rho^{4/3}$ (equation (10.58)). If we repeat our approximate argument, now using the relativistic equation of state, we find a surprising result: the radius cancels out and we are left with

$$M = \text{constant}. \tag{10.63}$$

In other words, there is a unique mass for which there is a solution with a completely relativistic equation of state. More realistically, as the density increases and the equation of state approaches the relativistic limit, which it can never actually reach, the mass of the white dwarf approaches a limit, above which there are no equilibrium solutions (the limiting mass corresponds to the radius going to zero). For a hydrogen-free white dwarf, Chandrasekhar found that the limiting mass is $1.44\,M_\odot$; a more refined treatment of the equation of state lowers this value to between $1.2\,M_\odot$ and $1.4\,M_\odot$, depending on the chemical composition of the white dwarf.

If a stellar core has a mass less than the Chandrasekhar limiting mass, then it can settle down quietly as a white dwarf. As you saw in the last section, we believe that this is what happens to stars with initial masses less than about $8\,M_\odot$. For higher core masses, the collapsing core cannot reach equilibrium as a white dwarf and must form an even more compact remnant: a neutron star or a black hole. I will discuss the properties of these objects now, and later you will see how they may arise as a result of a supernova explosion.

10.4.2 Neutron stars

Let us consider again what happens as mass is added to a white dwarf near the Chandrasekhar limit. Is it inevitable that as the density increases all the electrons are forced into high momentum states? The answer is no – instead, some of the electrons find it energetically more favourable to be forced into nuclei, turning some of the protons into neutrons. When this process of 'neutronization' is included in the equation of state, it turns out that the mass–radius relationship for white dwarfs is altered; instead of there being a limiting mass, attained only in the limit of zero radius, there is a maximum mass at a finite radius (Fig. 10.15). Models allow a continuous set of solutions at higher densities, but the ones for which the mass decreases as the radius decreases are all unstable.

At sufficiently high density, when the radius has shrunk to a mere 10 km or so, a new, stable equilibrium becomes possible, with mass again increasing for smaller radius. In these models, the density is so high that the nuclei no longer have an independent existence; they have been squashed together to form a sea of free nucleons – because of neutronization, these are largely neutrons. Normally, free neutrons would decay rapidly into protons and electrons. However, because of the extremely high density and the uncertainty principle, all the low-energy electron states are already full. In order to decay, a neutron would have to produce an electron with an energy that is far in excess of the energy released by the decay, and so the decay is prevented; neutrons are stable at nuclear densities, just as they are inside nuclei.

Like electrons, neutrons are Fermi particles and are subject both to Pauli's exclusion principle and to the uncertainty principle. At nuclear densities, the neutrons are degenerate and the internal pressure is dominated, to a first approximation, by the pressure of degenerate neutrons. These 'neutron stars' are therefore an analogue of white dwarfs, but at a much higher density.

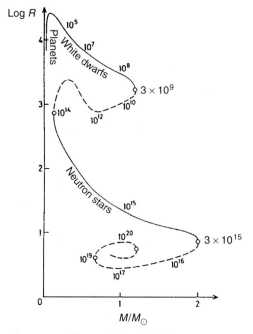

Fig. 10.15. The general mass–radius relationship for degenerate stars has two maximum masses, corresponding to white dwarfs and neutron stars. The models in the range between the maximum mass for a white dwarf and the minimum mass for a neutron star, and beyond the maximum mass for a neutron star (dashed lines), are unstable. The central density (in $g\,cm^{-3}$) is indicated at various points. Planets (which are not fully degenerate) lie at the low density end of the same mass–radius relationship. (From Kippenhahn and Weigert, 1990.)

In particular, they again have a maximum mass. However, it is much more difficult to calculate the maximum mass of a neutron star, for several reasons. The simplest reason is that for solar mass objects with radii of the order of 10 km the gravitational field is far too strong to be described by Newtonian theory and general relativity must be used. This complicates the equations somewhat, and introduces some qualitatively new effects that result from pressure being a source of gravity (this is because of the equivalence of energy and mass in relativity). However, the main uncertainty arises because physicists simply do not know accurately what the equation of state is for matter at nuclear and higher densities. Near the centre of a neutron star, matter probably breaks down into more elementary particles and there may be a 'quark soup'. It also seems likely that the matter has superfluid and superconducting properties. Near the surface, on the other hand, there is probably a strong crystal lattice that forms a rigid crust around the fluid interior. The various uncertainties mean that it is only possible to give estimates for the maximum mass, though most people would agree that it lies somewhere between 2 and $3\,M_\odot$.

10.4.3 Black holes

What is the fate of an object with mass greater than the maximum mass of either a white dwarf or a neutron star? We know of no further source of pressure that can support it against gravity and it must simply collapse indefinitely. According to Newtonian gravity, it would collapse to infinite density at zero radius. As you have already seen for neutron stars, Newtonian gravity cannot give a proper description of behaviour at such densities, where the gravitational field is extremely strong. However, we can use Newtonian physics to estimate the point at which general relativistic effects become dominant. The classical argument makes use of the concept of the escape velocity of a body: the speed that must be given to a particle to enable it just to escape from the body to infinity. If we consider a sphere of mass M and radius R, it is easy to show by considering the total energy of a particle moving radially away from the body that the particle will just reach infinity (i.e. its speed will tend to zero as its distance tends to infinity) if it is given a speed

$$v_{\text{esc}} = \sqrt{\left(\frac{2GM}{R}\right)} \qquad (10.64)$$

at the surface. Special relativity tells us that no object can exceed the speed of light, c, so this classical argument will break down when the escape speed becomes comparable to the speed of light, that is, when

$$v_{\text{esc}} = c \text{ or } R = \frac{2GM}{c^2} \simeq 3\,(M/M_\odot)\,\text{km}. \qquad (10.65)$$

This radius, which I have derived classically, also plays a vital role in the general relativistic treatment of a spherical body, where it is known as the Schwarzschild radius or, more generally, as the *event horizon*. Light cannot escape from a body once it shrinks inside its event horizon, so there is no way of observing its fate at higher densities. General relativity does predict that there is a singularity in the gravitational field at the centre, but the singularity is hidden from view. Because no light can reach us from its surface, a body that is within its event horizon is called a *black hole*: because matter and radiation can fall into the hole but can never emerge again, it is a perfect absorber, just like a black body.

When quantum effects are included, as was shown by Stephen Hawking in 1974, a black hole turns out to be like a black body in another way:

it radiates, with a temperature that is inversely proportional to the black hole's mass. This causes a loss of energy that gradually reduces the mass of a black hole; holes of initial mass less than 10^{12} kg would evaporate completely in about 10^{10} yr, the present age of the Universe. However, the effect is negligible for a stellar-mass black hole.

A proper study of black holes requires general relativity, which is well beyond the scope of this book. However, there is one very important property, that is often known as the 'no hair' theorem. This states essentially that a black hole has no structure, just as a smooth ball has no hair; the only properties of the body that survive its plunge into a singularity are its mass, its angular momentum and its net charge. This makes black hole physics very simple. Because most bodies are uncharged, only two solutions of the relativistic field equations are needed to describe the space outside a black hole: the Schwarzschild solution for a spherically symmetric body, discovered by Karl Schwarzschild in 1916 within months of the publication of Einstein's paper on general relativity, and the Kerr solution (1963) for a rotating body. At large distances, the gravitational field of a non-rotating black hole cannot be distinguished from that of a point mass, and normal Newtonian physics can be used. General relativity only needs to be used when the relativistic parameter $2GM/rc^2$ becomes comparable with unity. In the weak field limit (small mass or large radius), Newtonian physics suffices.

Note, incidentally, that for a large enough mass the density inside a black hole need not be extreme. For example, for a black hole of galactic mass, $10^9\,M_\odot$, say, the density when $2GM/rc^2 = 1$ is only about $20\,\mathrm{kg\,m^{-3}}$ or one-fiftieth of that of water.

10.4.4 X-ray binaries and pulsars

Although the theoretical existence of black holes has been known since 1916, there is still no completely convincing detection of a real example. Because of the no hair theorem, the only hope of detection is by the gravitational effect, which means there is no chance of observing an isolated black hole. Theoretically, we expect black holes to be formed as a result either of the collapse of a massive star or of the accumulation of a large mass at the centre of a galaxy. Although massive black holes, in excess of $10^8\,M_\odot$, are the current favourites as energy sources for quasars and active galaxies (Section 11.5), it is difficult to think of any crucial test that would distinguish them from a dense star cluster and there is no unambiguous detection.

Stellar-mass black holes are detectable if they are in binary systems, since they will still behave like normal gravitating masses. We therefore need to look for binary systems in which one component is invisible but massive (there are plenty of binaries in which there is an invisible companion, but most of these are simply too faint to be seen because they are markedly less massive than their companions). The first serious candidates emerged with the discovery in the early 1970s of X-ray binary stars. Some of these consisted of a massive main-sequence star, of about $20\,M_\odot$, together with an invisible companion and a very strong X-ray source. From the short orbital period, it seemed that the visible star was filling its Roche lobe (Section 9.2.4) and spilling mass onto its invisible companion. The X-ray flux was far stronger than would be expected from a normal main-sequence star and it was suggested that the X-rays came from the accretion of mass onto the companion. To produce photons with such high energies as X-rays required the mass to have fallen down a very deep potential well, which in turn required the companion to be a compact object, much smaller than a normal star. Other evidence for the small size of the companion comes from very short timescale variations in the brightness of the X-ray source. If a source varies on a timescale of a second, the source cannot be more than a light-second in size or the variations would be smoothed out by light-travel-time effects. The X-ray sources exhibit variations on timescales down to a millisecond, ruling out even a white dwarf as the compact object. This leaves neutron stars and black holes as candidates for the invisible companion.

Neutron stars are not very different in radius from the event horizons of black holes of a few solar masses, so size or gravitational potential alone is not enough to decide between these possibilities. However, there is another clue. We can measure the speed with which the visible star is orbiting around its dark companion (what is measured is actually a lower limit to the true speed, unless we know the inclination of the binary orbit to the line of sight). From the orbital period and an estimate of the mass of the main-sequence star (from its spectral type), we can then estimate, or at least obtain a lower limit for, the mass of the compact object. For several massive X-ray binaries, this lower limit is bigger than the maximum mass of a neutron star; the earliest convincing example was the source Cygnus X-1, for which the companion mass is larger than about $8\,M_\odot$. Although this is not a direct detection of a black hole, and other explanations (such as the presence of a third body in the system) cannot be completely ruled out, most astronomers now believe that many massive X-ray binaries contain black holes, which have presumably resulted from the evolution and collapse of a massive star. What is slightly surprising is that the supernova explosion which probably accompanied the collapse did not disrupt the binary completely. Detailed calculations show that survival requires that the explosion ejects less than half the total initial mass of the system.

Neutron stars were not predicted until the 1930s, some 20 years after black holes, but the most convincing examples were found at about the same time as the first X-ray sources were being discovered. It was in 1967 that the Cambridge radio astronomers Tony Hewish and Jocelyn Bell first noticed the objects that were quickly christened *pulsars*. These are radio sources that show extremely rapid pulses of emission, with very regular periods in the range from a few seconds down to a few tens of milliseconds. Later observations showed that the pulses are visible over a very large range of frequency, in one or two cases from the radio through the optical to the γ-ray. Although early explanations included pulsations of white dwarfs, this was quickly ruled out by the discovery of short-period pulsars, the extreme case being the one in the Crab nebula with a period of 33 ms. No white dwarf could pulsate as fast as that, and it became clear that the only possible source was a neutron star. Pulsation periods for a neutron star are much shorter than most of the observed pulse periods, so the strict periodicity is attributed to rotation. Because a neutron star is so small, even a period as short as 33 ms corresponds to 'slow rotation', in the sense that the centrifugal force is small compared to gravity. Only the recently discovered millisecond pulsars, discussed later, are genuinely rotating rapidly.

As you will see in Section 10.4.5, neutron stars are thought to be formed by the collapse of the core of a massive star in a supernova explosion. The discovery of the Crab pulsar in the middle of the expanding remnant of the supernova that was seen to explode in 1054 was hailed as a vindication of this idea, although later searches of supernova remnants have not been very successful in finding other pulsars. If the core contains even a weak magnetic field, then conservation of flux during the core collapse would produce a very strong field in the final neutron star. It is believed that pulsars have fields of 10^7–10^8 T, almost as large a field as is possible without quantum breakdown. The very strong magnetic field causes the formation of a *magnetosphere* around the neutron star, with relativistic particles flowing along field lines and losing energy by radiation, which is beamed along the direction of motion because of the ultrarelativistic speeds. The magnetic axis is inclined to the rotation axis, and the pulses are believed to be caused by a beam of radiation sweeping across our line of sight as the magnetosphere rotates. The details of the structure of the magnetosphere, the emission mechanism and even the exact location of the emitting particles are all still uncertain.

One thing is clear: if the magnetic field is inclined to the rotation axis, the rotation will also give rise to low-frequency electromagnetic radiation from a rotating dipole. This radiation carries away energy, which comes ultimately from the rotation of the neutron star – the rotation rate slows down. The fact that the slowing down of the neutron star causes energy to be emitted is beautifully demonstrated by the Crab nebula. For many years, the energy source of the radiation from the nebula was a mystery. It was known that the nebula shone by synchrotron radiation from relativistic electrons, but the energy loss rate from the electrons was so large that their lifetimes should have been a tiny fraction of the age of the nebula. There must be a continuous source of energy to reaccelerate the electrons. When the Crab pulsar was discovered, it was found that the pulse period was lengthening. This has a natural explanation within the framework of the rotating neutron star model – the rotation rate is slowing down as the kinetic energy is radiated away. The observed rate of slow-down enables the rate of energy loss to be calculated, and it turned out to be just what was needed to supply the radiation from the nebula. The energy in the pulsed radiation can also be measured, and is far too small to be the energy source for the nebula. Presumably, it is the case for all pulsars that rotational slow-down, possibly by emission of low-frequency dipole radiation, is the major energy loss mechanism, with the pulses merely being a very visible froth on top. This cannot be verified independently for pulsars without surrounding nebulae, but in all cases a lengthening of the period is detected, at a rate that cannot be explained by the pulsed flux alone.

For a long time the Crab pulsar, which is known to be young, had the shortest pulse period and it was assumed that all pulsars were born rapidly rotating and slowed down, so that the longest period ones were the oldest. The ratio of the period to the rate of change of the period, $P/\dot{P}$, can then be used as an estimate of the age of a pulsar, and gives ages of up to 10^7 yr, much longer than the ejected shell of a supernova would be expected to remain visible. This explains why most pulsars have no surrounding supernova remnant. (The absence of pulsars in most remnants is more puzzling – perhaps some supernovae leave no remnant core, or the core is not magnetic, or the rotating beam never crosses our line of sight.) As pulsars age, their magnetic fields slowly decay and eventually the radiation ceases when the field becomes too weak.

In the mid-1980s this picture was altered by the discovery of pulsars with periods of 1 or 2 ms – much shorter than the Crab pulsar. However, the periods of these *millisecond pulsars*, instead of changing rapidly, turned out to be completely stable, to the accuracy of the best atomic clocks. If the pulses were still from a rotating neutron star, the rotation rate must be absolutely steady, implying no energy loss by magnetic dipole radiation. This meant that the magnetic field was very weak and that the neutron star was very old. How then could it be rotating so rapidly? After a flurry of speculation, a model emerged which required the millisecond pulsars to be in binary systems. The explanation, then, is that after the pulsar has 'died' (i.e. the magnetic field has decayed to a low value) its companion overflows its Roche lobe and the neutron star begins to accrete mass. This mass has angular momentum, and the neutron star is 'spun-up' by the accretion until it is rotating rapidly again.

This picture emerged quickly because it was already known that some binary systems contained neutron stars. When X-ray binaries were first discovered, some showed rapid X-ray pulsations with periods of about one second. The short period immediately suggested that a compact object was involved, just as for the radio pulsars. The nature of the pulses is different, however, and they are not seen at other frequencies, suggesting that the pulse emission mechanism is different. The current model of the X-ray pulsars is that they are low-mass binary systems with a main-sequence star transferring mass by Roche lobe overflow onto a rapidly rotating, magnetic neutron star, possibly via an accretion disc. The magnetic field channels the accreting matter onto the magnetic poles of the neutron star, and it is the release of accretion energy at the surface of the neutron star that causes the X-ray emission. The pulse originates from a lighthouse effect, just as for the radio pulsars, with the pulse period being the rotation period of the neutron star. Some radio pulsars are also known to be in binary systems. The best-known example is PSR 1913+16, with an orbital period of 7.75 h, in which both components may be neutron stars.

Thus evidence from radio pulsars, millisecond pulsars and X-ray pulsars all testify to the existence of neutron stars. Although all the evidence is as yet indirect – there has been no detection of emission from the surface of the neutron star itself – the reality of neutron stars is established beyond reasonable doubt. Similar confidence in the reality of black holes seems likely to be much more difficult to establish.

10.4.5 Supernova explosions

Observed supernovae fall fairly clearly into two classes, as you saw earlier (Section 9.4.2.2). Apart from a few faint or otherwise peculiar exceptions, all Type I supernovae have the same absolute magnitude ($M_B \simeq -19$ to -20) at maximum, show an exponential decay in their light at late times and show no trace of hydrogen in their spectra. Type II explosions can be recognized by the presence of hydrogen lines in the spectrum, but have a much more variable maximum magnitude ($M_B \simeq -15$ to -18).

The present 'standard models' of these explosions are simple in outline and are quite different for the two types of supernovae. Type I explosions are believed to come from a low-mass progenitor, with no hydrogen. The favourite candidate is a carbon–oxygen white dwarf star with mass close to the Chandrasekhar limit of $1.4\,M_\odot$, accreting mass from a companion in a close binary system. When the mass reaches the critical mass, the central temperature is high enough for carbon to start burning. Because the interior of a white dwarf is degenerate, a thermal runaway occurs, just as in the helium flash, and the ignition is very violent.

Before the carbon flash occurs, various processes are going on that produce neutrinos in large numbers (for example, electron capture by nuclei: $e^- + (Z, N) \rightarrow (Z - 1, N + 1) + \nu_e$). These neutrinos have a very low interaction cross-section and escape directly from the star, removing energy and cooling the centre preferentially because the rate of neutrino production increases with density. The carbon flash therefore starts 'off-centre', in a shell, and propagates outwards as a burning wave. This wave remains subsonic (a 'deflagration' wave, like a flame: in a detonation, the wave is supersonic) and a pressure wave moves out ahead of the burning front and ejects the outer layers of the star, causing the visible explosion.

This model has two particularly attractive features: the progenitor being a C–O white dwarf of a definite mass naturally explains both the lack of hydrogen lines in the spectrum and the great similarity of all Type I explosions. Detailed calculations show that the C–O core burns mainly to ^{56}Ni and ^{56}Fe. Radioactive decay of the ^{56}Ni and its unstable daughter ^{56}Co convincingly explains the uniform exponential decline of the light curve, while iron lines have been detected in the spectra of some supernovae and supernova remnants. However, observations of the unusual Type II supernova 1987A in the Large Magellanic Cloud showed that exponential decline is also a feature of the late-time light curves of Type II supernovae, so the exponential tail does not tie the model down very tightly.

Type II supernovae are believed to come from stars in a broad range of masses:

$$M_{\text{low}} < M_{\text{preSN}} < M_{\text{up}}. \tag{10.66}$$

The lower limit, M_{low}, is about $8\,M_\odot$; stars of lower mass than this can form white dwarfs by non-violent mass loss, as you saw earlier. The upper limit, M_{up}, is less certain, but is probably between 20 and $40\,M_\odot$; higher-mass stars probably lose all their hydrogen before any explosion occurs, and may produce one of the more unusual Type I events.

Throughout this range of mass, carbon can ignite in a non-degenerate core, triggering the complicated series of burning episodes that we discussed briefly in Section 10.3.6 and building up successive shells of processed material. For masses greater than $11 \, M_\odot$, the core can burn as far as the iron peak elements, Cr, Mn, Fe, Co and Ni. After the formation of an iron core, no further energy can be produced by fusion reactions and nuclear burning ceases.

Radiation is still flowing out of the burned-out core, and neutrino processes such as electron capture lead to direct energy loss from the centre. To balance these energy losses, the core starts to contract and, as at earlier stages of evolution, the core temperature rises. However, no new fuel is now available for burning at sufficiently high temperatures. Instead, once $T_{core} > 5 \times 10^9 \, K$, a significant fraction of the photons in the radiation field (which is nearly in equilibrium with the hot core material) have enough energy to cause *photodisintegration* of the nuclei, and a phase change occurs, with the iron peak nuclei being broken down again into small fragments. Schematically:

$$^{56}\text{Fe} + \gamma \rightarrow 13\,^4\text{He} + 4\,\text{n}. \qquad (10.67)$$

This reaction requires the input of a large amount of energy (about 3 MeV per nucleon), comparable to the energy released in the whole of the slow process of evolution that has built up the iron peak elements from helium. The sudden energy demand triggers a catastrophic collapse, on the free-fall timescale of the core. For typical core properties ($M_c \simeq M_\odot, R_c \simeq 10^{-5} \, R_\star \simeq 4 \times 10^6 \, m$), this timescale is $\tau_{ff} \simeq (G\bar{\rho})^{-1/2} \simeq 1 \, s$!

Once started, the collapse continues until nuclear densities are reached. The energy released in the collapse is of the order of the gravitational binding energy of the collapsed core, about $10^{46} \, J$. What happens to this energy? The precise details are still uncertain, but the current model is that the core 'bounces' when it reaches nuclear ($\simeq$ neutron star) densities, because the equation of state of the material suddenly 'stiffens': the core material becomes incompressible ($\rho \simeq$ constant). This causes an abrupt halting of the collapse in the inner core ($M < 0.6 - 0.8 \, M_\odot$), while the outer core continues to fall inwards supersonically. A shock forms outside the inner core and it is this shock, moving back out again, that drives off the outer layers in an explosive expansion: the supernova. The inner core is left behind as a neutron star.

For a long time, detailed difficulties with this model could not be resolved because of the lack of adequate observational tests. This changed dramatically with the discovery of SN1987A in the Large Magellanic Cloud (LMC), the first nearby supernova since Kepler's in 1604 and by far the brightest to be observed in the telescopic era. It was ideally situated, for a number of reasons. The LMC is well out of the obscuring gas and dust of the Milky Way and its distance is known. It is far enough away that there was no danger of the light from the supernova swamping the sensitive detectors of modern large telescopes but close enough that high resolution spectroscopy was possible. Finally, the LMC is circumpolar (in the southern hemisphere) and the supernova could be monitored continuously (although the LMC gets fairly low in the sky when it is below the pole). This led to an unprecedented amount of data being gathered and enabled stringent tests to be applied to the model.

The most important feature of the LMC supernova was that, for the first time, a burst of neutrinos was observed a few hours before the optical outburst. This was a dramatic confirmation of one of the predictions of the core collapse model and was the first time that a direct test of conditions at the centre of an evolved star had been possible. It was also the first time that the main energy loss in a supernova had been observed. As noted earlier, models predict that some 10^{46} J are released by the collapse of the core; the energy emitted in light and kinetic energy is only about 10^{44} J, so about 99% of the energy of the explosion is expected to come out as a burst of neutrinos! The observed neutrino flux was consistent with the model and with known absorption cross-sections and showed that about 7×10^{46} J were emitted from SN1987A in the form of neutrinos. The time delay between the neutrino burst, which lasted only about 10 s, and the optical rise was a direct measurement of the time for the shock wave to propagate from the core to the surface and cause the visible explosion. This enabled various uncertainties in the model to be removed and confirmed that the basic idea of a core bounce was consistent with observations. The decline after the rather unusual rise to maximum followed the exponential decay familiar from Type I supernovae. The explanation in terms of radioactive decay received direct confirmation with the detection both of optical lines arising from cobalt, the unstable product of nickel decay, and of γ-ray lines from the decays of nickel and cobalt. At the time of writing, the final test has still to be made – will a neutron star (and pulsar?) be detected as the expanding shell thins to reveal the compact remnant?

10.5 Evolution in binary systems

You have already seen some of the exciting endpoints of evolution in binary systems, such as X-ray binaries, black holes and Type I supernovae. I will now give a brief outline of earlier stages of evolution.

10.5.1 Formation of binaries

As you saw earlier, binaries are very common and presumably represent a natural outcome of star formation. There are three different ways in which binary or multiple star systems may form: fission, fragmentation and capture. In the first two cases, the binary components form out of a single gas cloud. As the cloud undergoes its initial collapse, two things may happen. The cloud may break into two or more fragments, which then contract independently to form stars with a fairly wide separation. If the initial cloud is large enough, a whole star cluster may form in this way, with some fragments remaining close enough to form bound systems with two or more components. In double-star systems formed by fragmentation, the masses of the two stars tend to be rather similar.

If a sub-fragment in a nascent cluster possesses significant angular momentum, then the protostar will spin faster and faster as it contracts. It will also become steadily flatter, contracting preferentially down the rotation axis but remaining symmetric about that axis. Eventually, it may become unstable and lose its axially symmetric shape. Analytical studies suggest that the most likely outcome is that it first becomes pear-shaped and then hour-glass-shaped, finally splitting into two pieces of rather different mass which settle down into axially symmetric protostars in a close orbit. It is possible that both this fission process and the more general process of fragmentation occur, producing different types of binary; however, some studies have suggested that fission does not occur. Observational studies show that close binaries tend to have components of similar mass, while wide binaries have more disparate masses; this is not what fission would predict. Furthermore, numerical simulations of collapsing, rotating gas clouds suggest that the angular momentum problem may be solved by fragmentation rather than by fission, with spin angular momentum being converted into orbital angular momentum of several fragments.

Capture of a companion is extremely unlikely for a typical star in most parts of a galaxy, because the star density is so low that encounters are very rare; the separations of stars are millions of times their typical sizes. However, there are places where the star density is much higher, such as near the centres of globular star clusters. In these high-density regions, capture is quite probable and may account for the formation of some of the X-ray binaries found in globular clusters. In lower-density regions, capture rates may be enhanced if they occur at the proto-stellar stage, when the protostars are still surrounded by vast envelopes of infalling material, or by discs of high-angular-momentum gas and dust; the capture cross-section would then be much larger because of the larger size of the star.

10.5.2 Effects of mass transfer

Once a binary has been formed, the individual stars will start to evolve. The influence of the companion will depend on the initial separation. If this is sufficiently large, the two stars may go through their whole life cycles independently. This requires the separation to be larger than the largest radius that either star ever reaches, as may be the case for wide binaries with periods of tens of years or more. For shorter-period systems, a time will come when one star becomes a giant or supergiant and fills its Roche lobe (Section 9.2.4). There is then the possibility of mass transfer and new routes are opened up for the later course of evolution of both stars.

Because more massive stars evolve faster, it will always be the more massive component that first becomes a giant. The most likely stages of evolution at which the star first fills its Roche lobe are immediately after hydrogen exhaustion or after helium exhaustion, since the resulting core contraction leads to a corresponding rapid expansion of the outer envelope. Roche lobe filling at these two stages leads to evolution described as Case B and Case C respectively (Case A corresponds to transfer during the rather slower expansion of a star in the hydrogen core-burning phase of evolution).

What happens when a star transfers some mass to a companion? Not only are the masses of the two stars altered, but the orbit is changed, too, because the centre of mass is no longer at the same position along the line of centres. The more massive star is always closer to the centre of mass than is its less massive companion and both stars possess angular momentum about the centre of mass because of their orbital motion. If mass is transferred from the more massive star to the less massive one, the effect is to move mass away from the mass centre and to increase its angular momentum, assuming that it continues to move with the orbital angular velocity. The total angular momentum must be conserved, since there are no external torques, so the whole orbit shrinks slightly to compensate for the mass transfer. Similarly, mass transfer from a less massive to a more massive star causes a slight expansion of the orbit.

These changes in the separation of the two stars react back on the rate of mass transfer. If the mass transfer is from the more massive to the less massive component, the separation of the two stars decreases, causing the more massive star to overflow its Roche lobe further. This leads to an increased rate of mass transfer, which in turn causes a further decrease in separation and an even larger expansion of the star beyond its Roche lobe (strictly, this relative expansion is caused mainly by the shrinkage of the Roche lobe). We clearly have a runaway, with mass transfer stimulating further mass transfer, and a great burst of mass exchange ensues, on a dynamical timescale because the orbital size changes on that timescale. This dynamical mass exchange will continue until the more massive star has lost so much mass to its companion that it has actually become the less massive component. Any further mass transfer will now cause the separation of the two stars to increase, and the mass transfer to be cut off. Further mass transfer will occur on the much longer timescale over which the star expands as a result of evolution.

There is one other complicating factor: how does the star itself react to mass loss? When the Roche lobe shrinks, mass loss will only continue if the star either expands or at least shrinks more slowly than the Roche lobe shrinks. This will generally be the case if the star has a deep convective envelope, and stars that overflow their Roche lobes after they have become giants (Case C or late Case B evolution) will undergo the very rapid mass transfer that I have just described. However, if mass transfer occurs earlier, while the star is crossing the Hertzsprung gap (early Case B) and is still a subgiant with a radiative envelope or at most a shallow convective zone, the initial response to mass loss is such a rapid shrinkage that the star retreats inside its Roche lobe and mass loss is cut off. On a longer (thermal) timescale, the star's envelope readjusts to its lower mass, and expands again. This allows mass exchange to continue, but on a thermal rather than a dynamical timescale.

This evolution, which was first studied in detail in the 1960s, turned out to explain a long-standing mystery. The well-known star Algol shows large dips in its brightness at regular intervals, long attributed to eclipses in a binary system. Closer examination of the system showed it to be rather peculiar: the two stars are a main-sequence star and a subgiant, so it was naturally assumed that the subgiant, being the more evolved, was also the more massive. However, the subgiant is actually the less massive star. This cannot be explained by the evolution of single stars, but finds a natural explanation in terms of mass transfer: the subgiant was indeed originally the more massive star, which expanded to fill its Roche lobe while still a subgiant, and then underwent thermal timescale mass exchange (early Case B) until the mass ratio had been reversed and it became the less massive star. The system is now semi-detached, with the subgiant just filling its Roche lobe and transferring mass at a modest rate. Many such Algol systems are known, most of which have probably formed in this way.

There are clearly many possibilities for interactive evolution of this kind, depending on the initial separation and masses of the two stars. I will give just one more example, of a more speculative nature, which may explain the origin of cataclysmic binaries such as the dwarf novae (Section 9.4.2).

10.5.3 Common-envelope evolution

The evolved star in a cataclysmic binary is a white dwarf. As you saw earlier, single white dwarfs are probably the dead cores of red giant stars, which have lost their hydrogen envelopes in a final puff, producing a planetary nebula. If a white dwarf in a close binary has arisen in a similar way, then the star that is now a compact dwarf of radius 10^7 m must originally have been a giant with a radius of 10^{12} m. For the degenerate core to have formed, evolution to the red giant stage must have occurred before any significant mass transfer took place, so the initial separation of the binary must have been greater than 10^{12} m, corresponding to a period of tens of years. How can a wide binary like that have been transformed into a dwarf nova with an orbital period of less than half a day?

Suppose that the red giant fills its Roche lobe as it approaches the top of the giant branch, leading to late Case B mass transfer. Because the giant is the more massive star, and also expands as a result of mass loss, Roche lobe overflow occurs on a dynamical timescale and an expanding envelope forms, rapidly engulfing the secondary star. The secondary and the degenerate core of the primary now behave like two point masses orbiting inside an expanding gas cloud. Because the cloud conserves angular momentum as it expands, its rotation rate quickly drops below the orbital motion of the stellar pair. Frictional drag between the cloud and the stars then causes the orbit to decay, with the two stars spiralling inward and a huge amount of angular momentum being transferred to the steadily-expanding cloud. Eventually, the complete envelope of the former red giant has been ejected and the white dwarf core is left in an orbit of less than a day. What happens to the red companion during all this is unclear. It may lose little except its outermost layers, stripped off by ablation as it ploughs through the red giant envelope, or it may actually gain some material from the envelope of the red giant, or, perhaps most likely, its surface may contain a mixture of material from both original stars. Certainly, the red dwarfs in cataclysmic binaries seem to be more or less normal main-sequence stars, with hints of anomalous composition (from the red giant?) in at least one case.

Detailed models of this 'common-envelope' evolution have not been constructed, but there is some observational evidence in favour of the scheme:

some planetary nebulae are observed to have very short-period binaries at their centres. These binaries are detached, so an additional process is needed to produce the semi-detached systems that display dwarf nova outbursts. A clue to what may have happened is that the red dwarf, which is transferring mass, is the less massive star. This should lead to an increase in the separation and a cut-off of the mass transfer. Maintenance of a steady mass transfer requires either a slow evolutionary expansion – very unlikely for a star of $0.5\,M_\odot$, which is typical for these systems – or a loss of angular momentum from the system. Since a loss of angular momentum would also explain how the stars evolved into contact from the post-common-envelope detached binary, this is the favoured explanation. One mechanism for angular momentum loss that certainly operates is gravitational radiation: the two stars act as a rotating gravitational dipole and, according to general relativity, act as a source of gravitational waves, which carry away angular momentum.† However, this mechanism is only effective for cataclysmic binaries with periods less than 2–3 h. For longer-period systems, the timescale for significant angular momentum loss is longer than the age of the Universe and we need a more effective mechanism if we are to explain the observed systems.

A candidate can be found by looking at the Sun, which we believe to be gradually slowing down its rotation by *magnetic braking*. This mechanism relies on the star having both a magnetic field and a wind. The wind carries mass away from the star, and the magnetic field forces the wind material to co-rotate with the star until the kinetic energy of the wind becomes comparable to the magnetic energy. This means that when the wind finally breaks free of the field it has a much higher angular momentum than it had at the star's surface, typically by about a factor of 10, and carries away a large amount of angular momentum for rather little mass loss. In a binary system, tidal torques keep the magnetic star rotating at the orbital rate, so the angular momentum is ultimately drawn from the orbital motion, causing a slow decrease in period. This explains both how the post-common-envelope detached systems can become semi-detached and how the red star can continue to transfer mass. More recent calculations suggest that a magnetic wind from the accretion disc may also make a significant contribution.

---

† This effect has been clearly observed in the 'binary pulsar', PSR 1913+16 (Section 10.4.4), in which the orbital period is decreasing at precisely the rate predicted by general relativity.

10.6 Summary

A star's life history can be summarized briefly as follows. After an initially rapid collapse (on a dynamical and later on a thermal timescale) from an extremely tenuous cloud of interstellar gas (Section 10.1), a star settles into equilibrium when the centre becomes hot enough for nuclear fusion reactions to occur. Balance against gravity is maintained by an internal pressure gradient and the energy loss by radiation from the surface is balanced by the internal nuclear energy source. For 90% of its life, it burns hydrogen quietly in its central regions and remains on or near the main sequence in the HR diagram. The timescale for nuclear burning depends in a simple way on the mass of the star, because the fuel supply is proportional to the mass, and central burning ceases when about 10% of the mass has been converted from hydrogen into helium. The rate at which the fuel is burned is, in equilibrium, simply the luminosity of the star. The nuclear lifetime is therefore (cf. equations (9.14) and (10.6))

$$t_{\text{nuc}} \simeq \frac{fuel\ supply}{rate\ of\ burning} \simeq \frac{0.1\,M\,E_{\text{H}}}{L} \simeq 10^{10}\ \text{yr} \frac{M/M_{\odot}}{L/L_{\odot}}\ ,$$

where $E_{\text{H}} \simeq 6.3 \times 10^{14}\,\text{J}\,\text{kg}^{-1}$, the release of binding energy in forming helium from hydrogen.

Once the central hydrogen has been exhausted, the star moves over to the giant branch in the HR diagram, drawing on its gravitational energy reserves until the central temperature is high enough for helium fusion to occur. The star then settles down into a new period of relatively quiet evolution, again on a nuclear timescale but now on the giant branch. The nuclear timescale is still long, but shorter than on the main sequence because the luminosity is higher and the binding energy available from fusing helium to carbon is less than from hydrogen to helium. The helium-burning phase lasts for 5–10% of a star's life.

The timescale for the transition between these two main stages of nuclear evolution depends on the rate at which gravitational energy is used up. This timescale was first estimated in the nineteenth century by Lord Kelvin and H. von Helmholtz and is essentially the gravitational energy of the star divided by the luminosity. This 'Kelvin–Helmholtz' timescale is (equation (10.4))

$$t_{\text{KH}} \simeq \frac{G\,M^2}{2\,R\,L} \simeq 1.7 \times 10^7\ \text{yr} \frac{(M/M_{\odot})^2}{(R/R_{\odot})(L/L_{\odot})}.$$

Because in equilibrium, with the thermal pressure gradient balancing gravity, the internal thermal energy is comparable to the gravitational energy, this timescale is often also called the thermal timescale of the star and represents roughly the time the star would take to lose energy if the nuclear source were switched off and the star continued to radiate at its present luminosity.

Evolution beyond the helium-burning phase in the giant region of the HR diagram is basically on the thermal timescale, interspersed (for high-mass stars) with ever briefer periods of nuclear evolution until all the nuclear fuels have been exhausted. The final evolution to a compact remnant may be on an extremely short timescale if pressure balance is lost and the star collapses in free fall to produce a neutron star or black hole – this is thought to happen in a supernova explosion (Section 10.4). After a white dwarf or neutron star has been produced, the subsequent evolution is just a slow cooling. Since the thermal energy of the star is tiny compared with the gravitational energy, the Kelvin–Helmholtz timescale is not appropriate, and a more complicated calculation is needed. For a helium white dwarf of $1\,M_\odot$, with $L/L_\odot = 10^{-3}$, the cooling time is about 10^9 yr.

A star, then, spends most of its lifetime quietly burning nuclear fuels, drawing on its gravitational reserves from time to time. Finally, there are no nuclear fuels left, and the core of the star contracts to a compact remnant, with the envelope being lost either by steady mass loss or by some violent explosion. In the HR diagram, a star spends most of its life on the main sequence, moves over to the red giant or supergiant region for about the last 10% of its life and finishes as a hot, low-luminosity remnant at the foot of the diagram after a brief excursion through the blue giant or supergiant region. The phases other than hydrogen- and helium-burning, and the final remnant, occupy too little of a star's lifetime to be represented on the HR diagram.

This behaviour explains the shapes of the HR diagrams of globular and galactic clusters. Galactic clusters with a long main sequence are young, with even the hot, massive stars still burning hydrogen at the centre. In extremely young clusters, some low-mass stars are still contracting and have not yet reached the main sequence; the coolest stars in such clusters appear above the main sequence. As a cluster ages, the massive stars move over onto the giant branch and the main sequence becomes shorter – you have seen in Section 9.3 how we can use the luminosity at the top of the main sequence to estimate the age of a cluster. Globular clusters are very old, and only stars of less than a solar mass are still on the main sequence. Because the gas in the Galaxy at the time that globular clusters were born was still dominated by its primordial composition, globular cluster stars generally have a lower heavy element abundance than galactic cluster stars.

Many stars occur in binary systems, and the presence of a close companion affects the evolution of a star, principally by allowing the possibility of mass transfer between the stars. The originally more massive star (the primary) evolves faster and reaches the giant branch first, swelling up until it fills its Roche lobe. What happens next depends on the original masses and separation of the stars. If the primary star fills its Roche lobe not long after hydrogen exhaustion, mass is transferred rapidly to its companion (on a thermal timescale) until the originally smaller star becomes the more massive, after which mass trickles over on the much slower nuclear expansion timescale; this is believed to be the origin of Algol binaries, with a low-mass, Roche-lobe-filling subgiant and a higher-mass main-sequence star. Note that the main-sequence star will generally have been originally of too low a mass to have left the main sequence by the time its companion had filled its Roche lobe. However, it will now be in the anomalous position of being on the main sequence despite having the mass of a star that would have left the main sequence long ago if it had had that mass since birth. If such a star is in a cluster, it will appear on the main sequence of the cluster *above* the turn-off point for single stars; for many years, a few such stars have been known to exist in globular clusters, where they are known as 'blue stragglers' because they are on a blueward extension of the main sequence, and it is now generally believed that these are old low-mass stars which have been 'rejuvenated' by the addition of mass from a companion.

If the mass transfer does not occur until the primary has reached the red giant stage, the mass is transferred on a dynamical timescale and a common envelope may form, which is ejected, leaving behind a much shorter period binary; this may be the origin of cataclysmic variables. These are only two examples of a rich variety of possibilities, which are only just beginning to be explored in detail.

Exercises

10.1 Show that an exact solution of equation (10.1) is given by

$$t = \sqrt{\frac{R^3}{2GM}} \left(\zeta + \frac{1}{2} \sin 2\zeta \right),$$

where R is the initial radius of the cloud (of constant mass M), ζ is defined by $r/R = \cos^2 \zeta$ and the cloud starts to contract from rest at $t = 0$. [Hint: first multiply equation (10.1) by dr/dt and integrate once with respect to time to obtain

$$\frac{dr}{dt} = -\sqrt{2GM}\left(\frac{1}{r} - \frac{1}{R}\right)^{\frac{1}{2}}$$

(the minus sign is needed because the cloud is shrinking); then use the substitution $r = R\cos^2\zeta$ and the identities $\sec^2\zeta = 1 + \tan^2\zeta$ and $\cos^2\zeta = \frac{1}{2}(1 + \cos 2\zeta)$ to write this equation in the form $(1 + \cos 2\zeta)d\zeta/dt = \sqrt{2GM/R^3}$.]

10.2   Find the value of t at which the radius becomes zero in the solution given in exercise 10.1, and compare your answer with the free-fall time given in equation (10.2). [Hint: first find the value of ζ when $r = 0$.]

10.3   Show that for electron-scattering opacity ($\kappa = \kappa_0 = $ constant) the hydrostatic and energy transport equations combine to give:

$$\frac{dP}{dT} = \frac{16\pi acGMT^3}{3\kappa_0 L}.$$

Consider a massive star for which electron-scattering opacity is a sufficient approximation even in the surface layers, and assume that near the surface the mass and luminosity can be taken as constants, equal to their surface values M_s and L_s. Show that, if $P = T = 0$ at the surface, the pressure and temperature near the surface are related by

$$P = AT^4,$$

where $A = 4\pi acGM_s/3\kappa_0 L_s$.

10.4   Assuming that the star in exercise 10.3 is composed of an ideal gas (Section 10.2.2), show that the temperature near the surface is given by

$$T = \frac{GM_s\mu}{4\mathcal{R}}\left(\frac{1}{r} - \frac{1}{R}\right),$$

where R is the radius of the star, μ is the mean particle mass and $\mathcal{R}$ is the gas constant. [Hint: substitute $P = AT^4$ into the hydrostatic equation and eliminate the density using the ideal gas equation.]

11

Properties of galaxies

11.1 Classification of galaxies

In Chapter 9 I discussed the properties of individual stars and you saw that they may have one or more companions in binary and multiple star systems or may be grouped into the larger systems known as open and globular clusters. On an even larger scale, single stars, binaries and clusters are grouped together into galaxies. A first definition of a galaxy is just that it is a large, isolated assembly of stars, bound together by gravitational interactions. However, this is not very satisfactory: how large is 'large'? how far away must other assemblies be if the galaxy is to be 'isolated'? Is this definition not equally applicable to star clusters? The answer to the last question is 'yes', and the difficulty is to refine the definition in such a way as to exclude globular clusters.

The word 'large' is deliberately vague: galaxies may contain as few as a million stars or as many as 10 million million, although in fact we cannot count the individual stars and must estimate these numbers from the measured galactic masses. At the lower end, small galaxies have no more stars than large globular clusters, so there is no clear distinction in mass between small galaxies and large clusters. One difference is that globular clusters are generally much more compact: some dwarf galaxies are so tenuous that their reality can only be established by careful star counts, comparing the region of the sky containing the galaxy with the background sky nearby. However, some dwarf galaxies have a well-defined compact core, so this difference is not definitive.

The word 'isolated' is a reminder that star clusters are usually associated with other clusters and with a large galaxy. However, there is again no sharp distinction – there are intergalactic 'tramp' globular clusters, dwarf galaxies

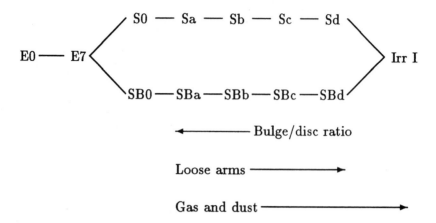

Fig. 11.1. Hubble's 'tuning-fork' diagram for the classification of galaxies. Elliptical galaxies of increasing flatness merge into the barred (B) and unbarred disc galaxies. The lenticular (S0) galaxies show no spiral arms and a large central bulge. Along the Sa to Sd spiral sequence, the bulge decreases in prominence, the spiral arms become looser and the proportion of gas and dust increases. Irr I (irregular) galaxies have the most prominent gas and dust of any galaxies.

may cluster around a larger galaxy and many galaxies are in clusters of galaxies with a dominant central galaxy.

This all suggests that globular clusters would be better to be redefined as compact dwarf galaxies; the clear break comes between open clusters, with masses of less than about $1000 \, M_\odot$, and globular clusters, with masses greater than about $10^5 \, M_\odot$. Galaxies could then be defined as gravitationally bound systems of stars with masses greater than $10^5 \, M_\odot$.

The most obvious property of a galaxy is its structure. There are three basic shapes, elliptical, spiral and irregular, but many variations within these broad classes. The first serious classification of galaxies by their shape (*morphological classification*) was undertaken by Edwin Hubble in the 1920s, and his scheme is still in use, with some extensions and changes of notation. The current scheme is summarized in Fig. 11.1, which is an extension of Hubble's famous 'tuning-fork' diagram, in which the two spiral branches formed the two arms of the fork and the irregulars were omitted. Originally, this was thought to be an evolutionary sequence, but it is now thought that the differences between the various types of galaxy arise primarily from accidents of birth, as you will see later.

Elliptical galaxies (Fig. 11.2) have the simplest appearance, with no pronounced structure, although many have a bright central nucleus. They

Fig. 11.2. The E5 dwarf elliptical galaxy NGC 147 (left) and the E0 giant elliptical galaxy M87 (right). (Photographs from the Hale Observatories.)

appear to contain little gas or dust. The shape is classified in terms of the ratio, b/a, of the semi-minor to the semi-major axis of the apparent elliptical shape: an E0 galaxy appears perfectly round, while the most extreme case is an E7 with $b/a = 0.3$. In general, an En galaxy has $n = 10(1 - b/a)$. Of course, we see only the projection of the true shape onto the sky and the observed b/a is always greater than or equal to the true value for an individual galaxy. However, so many have been studied that if there were elliptical galaxies with $b/a < 0.3$ we should surely have seen at least one at the correct orientation to have measured the true value.

It used to be thought that ellipticals were flattened because of rotation and it was therefore natural to assume that they were axially symmetrical and that the true shapes were oblate spheroids: flattened spheres, or ellipsoids with two equal axes. Measurements of the rotation speeds now makes it clear that most ellipticals rotate far too slowly for rotation to be responsible for their shape and there is strong evidence from the light distribution within the galaxies that many ellipticals are in fact triaxial ellipsoids, with three unequal axes.

Spiral galaxies have the most dramatic shapes and are the best-known type of galaxy. There are three major differences between spirals and ellipticals:

Fig. 11.3. The Sb spiral galaxy NGC 4565 in Coma Berenices, seen edge-on. It is highly flattened and a layer of dust can be seen in the plane of the disc. The Milky Way would look like this if we could see it from outside (cf. Fig. 6.1). (Photograph from the Hale Observatories.)

spirals are highly flattened (Fig. 11.3), contain considerable amounts of gas and dust and display spiral structure, their most obvious feature. Spiral arms were first detected in 1845 by the Earl of Rosse, using his remarkable 6-foot-diameter reflecting telescope at Birr Castle in Ireland. A spiral galaxy as a whole is disc-like, rotates rapidly and is more or less axially symmetrical. However, the brightest objects in a spiral galaxy, such as blue supergiants and luminous gas clouds, are not distributed uniformly across the visible disc but are concentrated into a spiral pattern. The best-known examples – the so-called 'Grand Design' spirals – often have two dominant arms (Fig. 11.4); the majority of spirals have a rather more complicated and less well-defined pattern (Fig. 11.5). There are two distinct types of spiral galaxy, known as normal and barred spirals. The normal spirals possess a more or less spherical nucleus, while the central regions of the barred spirals are

Fig. 11.4. The Sc galaxy M51. This nearly face-on galaxy shows a clear 'Grand Design' spiral structure, slightly distorted by its close companion. (Photograph from the Hale Observatories.)

Fig. 11.5. The Sb spiral galaxy M31 in Andromeda. Spiral arms can be seen, but are not well-defined. The Milky Way probably looks like this. (Photograph from the Hale Observatories.)

Fig. 11.6. The SBc galaxy NGC 7741, showing spiral arms emerging from the ends of the central bar. (Photograph from the Hale Observatories.)

elongated into a bar shape, from the ends of which the spiral arms emerge (Fig. 11.6); the bar often shows a bright nucleus near the centre. Each type of spiral is divided into several sub-classes, a to d, on the basis of two major variations. Sa galaxies have a very large central bulge compared to the size of their disc and the spiral arms are tightly wound around the nucleus and not well broken up into distinct gas clouds and bright stars. Sd galaxies have a tiny nucleus and arms that are very open and quite chaotic. Sb and Sc galaxies show intermediate properties, with the disc-to-bulge ratio increasing as the arms become more prominent and spread out. The classification is somewhat qualitative and our Galaxy is believed to be an Sbc, between Sb and Sc. The fraction of gas and dust also varies along the sequence, with Sd galaxies being bluer than Sa and displaying more gas and dust.

It is this last property that leads naturally into the irregular galaxies of Type I. At first, the irregular galaxies were simply all those that could not be fitted into the Hubble sequence (the term 'peculiar' is still used for galaxies whose classification is doubtful – Section 11.5). They are generally divided into two groups, Irr I and Irr II, whose irregularities are rather different. The prototype of the Irr I systems is the Large Magellanic Cloud (Fig. 11.7), a satellite of our Galaxy, and the Irr I galaxies are sometimes called Magellanic irregulars, or Im galaxies. They contain huge amounts of gas and dust and their appearance is dominated by brilliant gas clouds and luminous stars scattered irregularly across the face of the galaxy. The faint stars, which make up most of the mass, are much more uniformly distributed. The luminous blue stars are very young and it seems that the visual appearance arises from regions of star formation, which is occurring vigorously at the present time but in a very patchy way. Perhaps Irr I galaxies will eventually become more conventional spirals, once more of the gas has been formed into stars. Maybe they are also related to the recently discovered 'starburst' galaxies, which I will mention again in Section 11.5.

The Irr II galaxies, by contrast, seem to owe their strange appearance to gigantic explosions or other violent events. The classic example is M82 in Ursa Major (Fig. 11.8), which has a great cone of hydrogen apparently streaming out from its nucleus and has its disc obscured by enormous streaks of dust. The origin of the explosion is not fully understood. Other galaxies show strange tails and filaments that are caused by tidal interaction with a companion (Section 11.5.1).

Finally, there is the class of spheroidal or lenticular galaxies, known as S0. These appear to be intermediate in structure between ellipticals and spirals, but were not recognized in Hubble's original scheme. Roughly speaking, they

Fig. 11.7. The Large Magellanic Cloud, the prototype Irregular I galaxy and our nearest galactic neighbour. (Royal Astronomical Society.)

are spiral galaxies without the spiral arms. They certainly resemble spirals by being highly flattened, disc-like systems, with a central condensation or bar, and they have similar rotation properties to spirals. As well as having no spiral structure, lenticulars differ from spirals in having an extended envelope. However, there appears to be a clear distinction between lenticulars and ellipticals, with no systems of intermediate flattening. There seems to be

Fig. 11.8. The peculiar galaxy M82, the prototype of the Irregular II galaxies, photographed on blue-sensitive film. The lower picture shows many large sheets and filaments of dust, while the upper picture, exposed for six times as long, shows material streaming away from the central regions in a conical flow; this flow is much more apparent on pictures taken in the red Balmer line of hydrogen at 656 nm. (*Hubble Atlas.*)

a critical flattening of a spheroid beyond which a simple spheroidal structure is unstable and matter arranges itself instead into a disc around a central bulge. Lenticulars may be spirals which have lost their gas. Certainly, they are common in clusters of galaxies, where spirals are rare, and may have had their gas stripped out by interactions with the background gas in the cluster.

11.2 Luminosity, size and mass

Even within individual Hubble types, galaxies display a great variety of properties. The most important global properties are the mass, the size and

Table 11.1. *Visual luminosities for galaxies of different Hubble types*

Giant ellipticals	10^{38} W ($\simeq 2 \times 10^{11}$ L$_\odot$)
Dwarf ellipticals	10^{31} W
Spirals	$10^{36} - 3 \times 10^{37}$ W
Irregulars	$2 \times 10^{35} - 2 \times 10^{36}$ W

the luminosity. None of these is easy to define precisely for a galaxy, as you will see.

11.2.1 Luminosity

Although galaxies were first detected in the visible, it is now clear that many galaxies emit a large fraction of their radiation at other wavelengths, especially radio and infrared. Some galaxies discovered by the infrared satellite IRAS probably emit as much as one hundred times as much in the infrared as they do in the optical band. However, for the great majority of galaxies we still only have information in the optical waveband. The figures given in Table 11.1 are estimated from optical data and are therefore lower limits to the total, bolometric luminosity.

Early studies of galaxies naturally concentrated on the most conspicuous ones, and missed many faint ones – mostly dwarf ellipticals and irregulars. It now appears that most galaxies are faint and that the well-known bright galaxies are rare exceptions. The prominent spiral galaxies whose striking pictures appear in every popular astronomy book are exceptions in two ways – not only are they the commonest type of bright galaxies, they also have no faint counterparts: there are no dwarf spirals. This is illustrated in Fig. 11.9, which shows the luminosity function for galaxies: the number of galaxies in each range of luminosity. The spirals dominate at the bright end, but dwarf ellipticals and irregulars are far more common overall, although they contribute little to the total mass of the Universe.

All luminosities are uncertain, not only because of unknown contributions at other wavelengths but because we need to know the distance of a galaxy before we can deduce the luminosity from the measured flux. We also need to allow for the shape of the galaxy – for example, we cannot assume that a disc galaxy emits the same flux in all directions – and for the fact that the brightness fades gradually towards the edge (Section 11.2.2), so that the total flux depends on how much of the galaxy we include. As you will see

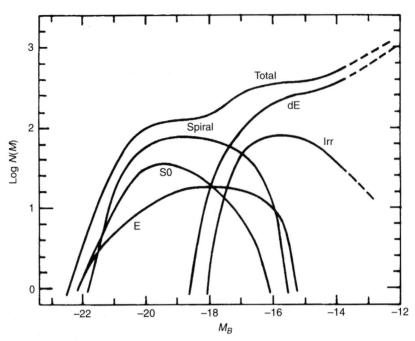

Fig. 11.9. The luminosity function for galaxies in the Virgo cluster; it is probably representative of all galaxies in clusters. $N(M)$ is the number of galaxies of a given absolute blue magnitude M_B; the zero-point of $\log N(M)$ is arbitrary. The dashed parts of the lines are uncertain. There are many more faint galaxies than bright ones. Spirals dominate at the bright end, although the very brightest galaxies are elliptical. Almost all the faintest galaxies are dwarf ellipticals. (Altered, with permission, from the *Annual Review of Astronomy and Astrophysics*, Volume 26, ©1988, by Annual Reviews Inc.)

later, there are uncertainties of up to a factor of 2 in distances to galaxies, and the luminosities of the most distant galaxies are therefore uncertain by at least a factor of 4.

However, because galaxies are extended objects, there is one measure of their luminosity that is independent of distance: the *surface brightness*, or flux per unit solid angle ($\mathrm{W\,m^{-2}\,ster^{-1}}$). This is the analogue of the radio brightness distribution defined in Chapter 4. Its independence of distance is easy to demonstrate. Suppose we have a galaxy at distance d, whose luminosity is L and whose area (projected on the sky) is A. Then the flux measured at the Earth is $F = L/4\pi d^2$, while the solid angle subtended at the Earth is $\Omega = A/d^2$. Thus the flux per unit solid angle is $F/\Omega = L/4\pi A$, independent of distance. The units commonly used in the visual waveband are magnitudes per square arcsecond (to convert to steradians, note that π radians = 180 degrees and that 1 steradian = $(180/\pi)^2$ square degrees).

Table 11.2. *Standard diameters for galaxies of different Hubble types*

Type	D_0 in kpc
dE	0.5–1
Irr I	1–10
S, S0	10–50
E, gE	10–100

11.2.2 Size

The concept of surface brightness is useful in defining what we mean by the size of a galaxy, since there is no well-determined edge: the star density falls off gradually and the measured size of a galaxy depends on how hard one works to distinguish the faint outer reaches from the sky background. We can define a *standard diameter D_0* as the diameter of the surface brightness contour which corresponds to 25 mag arcsec^{-2} after correction for absorption and inclination. The absorption correction allows for absorption of light by dust and gas both in our own Galaxy and within the galaxy being observed; both are uncertain corrections. The inclination correction is mainly relevant to disc galaxies, which are assumed to be axially symmetric and to appear elliptical because the disc is not seen along the axis of symmetry. The effect of these two corrections is to produce circular contours for a face-on galaxy free of all absorption.

There is a great range in standard diameters, especially for elliptical galaxies. Typical values are gathered in Table 11.2.

11.2.3 Mass

The mass of a galaxy is not a directly observed quantity but must be deduced from physical arguments. The simplest rough estimate comes from what is known as the *virial theorem*, which in its simplest form can be written as

$$2 \times kinetic\ energy\ + potential\ energy\ = 0, \qquad (11.1)$$

where the kinetic and potential energies are those of an isolated system in equilibrium. For a self-gravitating system like a galaxy, the virial theorem can be written explicitly as

$$\alpha \, \frac{GM^2}{R} \simeq M \, \overline{v^2}. \tag{11.2}$$

Here M is the total mass of the galaxy, within radius R, G is the gravitational constant and α is a constant that depends on the distribution of mass within the galaxy. The right-hand side is a measure of the total kinetic energy, given by the total mass times the mean square speed of the material moving within the galaxy. It is not easy to decide what radius to use in this equation, since we do not know how far the galaxy extends. One possibility is to estimate the radius from the standard diameter. The constant α can be estimated if it is assumed that the light traces the mass, that is, that the mass distribution is the same as the light distribution; I will discuss this point again later. The mean square speed may be a measure of random motion within the galaxy, in which case the lines in the galaxy's spectrum will show a broadening because of the spread of Doppler shifts present; this is the usual situation for elliptical galaxies. Alternatively, the main contribution may be an organized rotation of the whole galaxy, as in spiral galaxies; this can be measured either from line-broadening or by individual Doppler shifts from different parts of the galaxy.

We therefore have a way of estimating all the quantities that enter equation (11.2), except the mass, and we can rewrite this equation as

$$M \simeq \frac{R\overline{v^2}}{\alpha G}, \tag{11.3}$$

where now everything on the right-hand side is known. This is quite a good first estimate of the mass of a galaxy, and essentially the same method can be used for a cluster of galaxies. For disc galaxies, where the rotation speed can be measured as a function of distance from the rotation axis, more sophisticated arguments can be used to obtain a better estimate, but still only of the mass within some definite radius. (Some simple examples are given in the exercises at the end of this chapter; see also Section 11.2.5.) Because the density falls off outwards fairly slowly, there is always a significant amount of material in the outer reaches of a galaxy, and the estimated mass is only a lower limit. Typical values of these estimates are given in Table 11.3.

As can be seen from Tables 11.1–11.3, the smallest galaxies are generally the least massive and the faintest, and can only be seen if they are nearby. The best-studied galaxies are inevitably those near the upper ends of the ranges of mass and luminosity. Our own Galaxy seems to fall into this

Table 11.3. *Typical masses for galaxies of different Hubble types*

Type	Mass in $M_\odot$
gE	10^{13}
dE	10^6
S	$10^{10} - 3 \times 10^{11}$
Irr	$10^6 - 10^{10}$

category; although it is difficult to estimate the Hubble type from inside, it appears to be an Sbc galaxy, with $L \geq 10^{10} \, L_\odot$ and $M \geq 2 \times 10^{11} \, M_\odot$.

11.2.4 Mass-to-light ratios

Because the most massive galaxies are also the most luminous, the *ratio* of mass to luminosity varies much less from one galaxy to another and is in fact reasonably constant for a given Hubble type. The mass-to-light ratio, as it is usually called, is therefore a more useful characteristic of a galaxy than either mass or luminosity separately. Rough values are:

Hubble type	E	S0	Sa	Sb	Sc	IrrI
$(M/L)/(M/L)_\odot$	$20 - 40$	10	10	10	< 10	< 10

The general decrease down the sequence can be understood qualitatively as due to the higher proportion of bright, blue stars in S and Irr galaxies. Since $L \propto M^4$ for stars, $M/L \propto M^{-3}$ and thus decreases for more massive stars. The more such stars are present in a galaxy, the smaller is its mass-to-light ratio. Ellipticals are dominated by low-mass stars and have a correspondingly high mass-to-light ratio. Note that in solar units $M/L > 1$ for all galaxies. This means that the main contributors to the light in all normal galaxies must be stars fainter and less massive than the Sun; for ellipticals, the dominant stars must be less than $0.5 \, M_\odot$.

11.2.5 Dark matter

The most reliable way to find the mass of a spiral galaxy is to measure its rotation speed as a function of distance from the rotation axis. This is best done in the radio, using the 21 cm emission line of neutral hydrogen, which is not absorbed in the interstellar dust that obscures the disc in the optical. If the mass of the galaxy were concentrated into the central bulge, then the

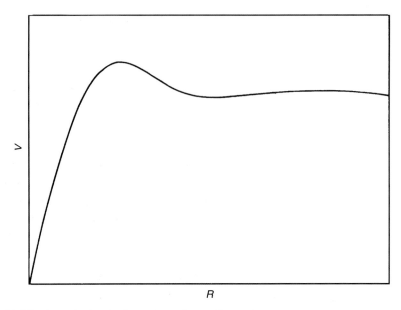

Fig. 11.10. A typical rotation curve for a disc galaxy. Near the centre, the rotation speed increases roughly linearly with distance. Further out, the rotation speed is approximately constant as far out as it can be traced. (R.J. Tayler.)

rotation speed in the outer parts of the galaxy would be expected to fall off outwards, following the same Keplerian law, $V_{rot} \propto R^{-1/2}$, as is found in the solar system. This follows from equating the centrifugal force on a star of mass m in a circular orbit of radius R to the gravitational force from a spherical mass M which is all within this radius:

$$\frac{mV^2}{R} = \frac{GMm}{R^2}. \tag{11.4}$$

(Note that the mass of the star cancels out, and V depends only on M and R.) The observed rotation curves, with rare exceptions, show no sign of a Keplerian fall-off at large radii. To a first approximation, most rotation curves follow the form shown in Fig. 11.10: a steep rise in the inner parts of the galaxy and then more or less flat as far out as the radio signal can be traced – which is usually well beyond the optical limits of the galaxy.

It is clear without calculation that this flat rotation curve is telling us that there is more mass than the eye can see. A simple interpretation of the curve reveals just how much more, as I will now explain. Equation (11.4) is valid for any spherically symmetric distribution of mass, so long as M is interpreted as the mass within radius R, and can be read as giving the mass distribution that corresponds to an observed rotation curve:

$$M = \frac{RV^2}{G}. \tag{11.5}$$

In the inner parts of a disc galaxy, $V \propto R$ (solid-body rotation – uniform angular velocity) and $M \propto R^3$. Thus the central bulge of the galaxy is, to a first approximation, of uniform density. Further out, $V =$ constant and equation (11.5) gives $M \propto R$; although the density now falls off like R^{-2}, the total mass continues to increase well beyond the visible disc. Of course, I have assumed spherical symmetry, which is certainly not justifiable for the optically visible galaxy. However, more realistic flattened models also give $M \propto R$ and in any case it is possible that the 'dark matter' revealed by the flat radio rotation curve has a more spherical distribution than the luminous stars. Note that the radio measurements are only detecting tenuous neutral hydrogen gas, whose density is far too small to account for the hidden mass; the neutral hydrogen is merely acting as a tracer, revealing the underlying gravitational potential in which it is moving. The nature of the 'dark matter' that is producing the potential is as yet unknown, although some of it may be due to dead stars – faint white dwarfs or neutron stars, or black holes – or planetary-size objects. Estimates suggest that as much as 90% of the mass of a spiral galaxy may be outside the visible disc, so our mass estimates in Table 11.3 should probably be increased by a factor of 10. As you will see later, even larger amounts of dark matter may be hidden in clusters of galaxies and in the Universe as a whole.

11.3 Non-stellar contents of galaxies

Although most of the visible mass of a galaxy is made up of stars, they take up very little of the volume. However, this volume is not empty, but contains a variety of other components.

11.3.1 Gas

This fills most of the volume, and is partially clumped into clouds of varying density. The amount varies considerably with Hubble type; a useful measure of the total amount is given by the mass of the most-easily detected component, neutral atomic hydrogen:

Hubble type	E	S0	Sa	Sb	Sc	IrrI
$M_{\mathrm{HI}}/M_{\mathrm{gal}}$	0	0.005	0.03	0.05	0.07	0.2

Until recently, it was thought that E galaxies had no gas at all. Now it is believed that most have gas at their centres. Although the amount is very small, it may have an important effect, as you will see when I discuss active galaxies. Lenticular galaxies also have very little gas, with a few exceptions. Other disc galaxies show a steadily increasing amount of gas towards later Hubble types, culminating in irregular galaxies. Although most of the gas is hydrogen, simply because hydrogen is the most abundant element, there are many other elements present as well, especially in the cool molecular clouds. Molecular hydrogen again dominates in these massive clouds, which may contain up to $10^5 \, M_\odot$, but carbon monoxide is also very prominent and scores of other molecules have been detected in our own Galaxy (Chapter 12).

11.3.2 Dust

In particularly cool regions of galaxies, gas may condense out in the form of interstellar dust grains. Overall, the dust has a mass about 1% of that of the gas, but it is far from uniformly spread. It is concentrated in the plane of disc galaxies and is closely associated with cool, dense, molecular gas clouds and with spiral arms.

11.3.3 Radiation

There is a great deal of radiation traversing a galaxy, coming from the stars, gas and dust, as well as from extragalactic sources. Although it is of course massless (more precisely, its rest-mass is zero), it possesses a significant energy density which can influence the state of the interstellar medium. This can in turn modify the star-formation rate and affect the evolution of the galaxy.

11.3.4 Cosmic rays

These high-energy particles, mostly protons and electrons, can only be directly observed when they reach the Earth and we must assume that the small part of our Galaxy in which we happen to live is typical of galaxies as a whole, in order to estimate the contribution of cosmic rays to the contents of a galaxy. High-energy electrons can be observed indirectly by the synchrotron radiation that they emit when moving in magnetic fields, both in our own Galaxy and in external ones, and these observations confirm local estimates of the cosmic ray number density. The mass density is negligible, but again the energy density is significant, because the particles are relativistic.

11.3.5 Magnetic fields

In our own Galaxy, magnetic fields can be detected directly by Zeeman splitting in some interstellar clouds, and indirectly by a number of methods, including polarization of starlight. An overall field strength of about 10^{-9} T has been deduced, which corresponds to an energy density comparable to that of cosmic rays. The magnetic field of our Galaxy is believed to act to confine the cosmic rays, which in turn expand the field out of the galactic disc. The field is contained within the Galaxy by its interaction with the ionized gas in the galactic disc, which is kept in a flattened distribution by the gravitational field of the stars (and dark matter). The thickness of the gas disc is determined by this balance of magnetic and gravitational forces and the pressure of the gas and the cosmic rays. In external galaxies, the existence of magnetic fields has been deduced from polarization measurements, and it is probable that most gas clouds have some magnetic flux associated with them.

11.4 Chemical composition and stellar populations

I stated earlier (Section 8.2) that all stars have essentially the same chemical composition. This statement now needs some qualification. A better approximation is to say that for (nearly) all stars:

1. Hydrogen and helium are the dominant chemical elements. By number, H $\simeq$ 90%, He $\simeq$ 10%.
2. The other elements are present in the same proportions to one another as in the Sun.

However, the total number of these so-called 'heavy elements' varies considerably with respect to hydrogen. Because one of the more abundant of these elements is iron, astronomers often refer to them collectively as 'metals', although the most abundant elements include such clear non-metals as carbon, nitrogen and oxygen! The overall 'metal abundance' is often measured by the ratio of iron to hydrogen, Fe/H, partly because iron has a very rich line spectrum in the optical and its abundance is easy to measure.

Many stars and much of the interstellar gas seem to have essentially solar composition, with a total abundance of heavy elements of about 2% by mass. Within that total, the most abundant elements are given in Table 9.2 (Section 9.1), which lists all the elements with number densities more than a millionth of that of hydrogen.

A significant number of stars (perhaps the majority) are 'metal-deficient' or 'metal-poor', with a smaller proportion of heavy elements overall, although they are still present in about the same relative proportions to one another. This can be described in terms of different stellar populations, which differ in other properties as well, particularly in their motions. The crudest approximation is to define just two stellar populations, as was first done by Walter Baade in 1944 as a result of his studies of the Andromeda galaxy. These two populations can be characterized as follows:

Population I – these stars have solar composition, the brightest stars are blue, they are found in galactic clusters, spiral arms and the discs of spiral galaxies and they are associated with gas clouds. In our Galaxy, they are found in nearly circular coplanar orbits around the Galactic Centre.

Population II – these stars are metal-poor, by factors of 10–1000, the brightest stars are red and they are found in globular clusters, elliptical galaxies and the haloes of spiral galaxies. In our Galaxy, they are in eccentric orbits, with high inclinations to the galactic plane.

The usual interpretation is that the heavy elements are formed by nuclear reactions inside stars and that the Population II stars were born near the beginning of the life of a galaxy before many heavy elements had been created. In this picture, hydrogen and helium (and a few other light elements) are 'primordial', that is, created in the Big Bang (Chapter 15), while the 'metals' were created inside stars and have been returned to the interstellar medium by mass loss, principally in supernova explosions. The very first stars would then have consisted of almost pure hydrogen and helium, while later stars, formed from an enriched interstellar gas, show an increasing proportion of heavy elements. Thus Population II stars are old, while Population I stars are young.

In fact, as you saw from HR diagrams of galactic clusters, some Population I stars are nearly as old as Population II stars, and nowadays many stellar populations are defined, with a gradation in age. It seems that there must have been a burst of star formation near the beginning of our Galaxy which rapidly created most of the heavy elements that are now observed in the Galaxy amongst the old disc population and in the interstellar gas. Heavy element production did not stop after that initial burst, but has gone on at a reduced rate, gradually building up the proportion of heavy elements in the Galaxy (Fig. 11.11).

The kinematic differences between the two populations can also be explained in terms of the history of the Galaxy, as you will see briefly in the next chapter.

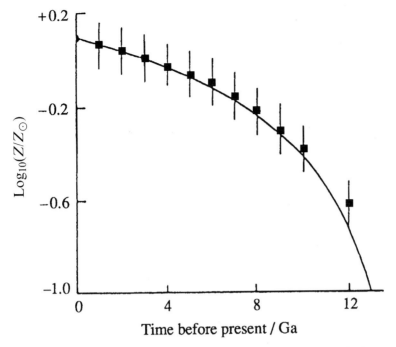

Fig. 11.11. The rate of heavy element production in the Galaxy. Z is the proportion of material in the form of elements heavier than hydrogen and helium. After an initial rapid burst, about 12×10^9 yr ago, heavy elements have been produced at a relatively slow but steady rate. (R.J. Tayler.)

11.5 Peculiar galaxies and active galactic nuclei

11.5.1 Peculiar galaxies

Some galaxies do not fit easily into the Hubble classification scheme, and are given the label 'peculiar' on the basis of their optical appearance. The great majority of these are galaxies that have been distorted by tidal interaction with a passing galaxy or companion galaxy. There is a great variety of strange shapes – tails, warps, filaments and bridges – and it was some time before it was accepted that they could all be produced by purely gravitational interaction. A series of detailed numerical simulations in the 1960s and 1970s made it clear that tidal interaction could produce some remarkable shapes and most peculiar shapes are now attributed solely to gravity. Some examples are given in Fig. 11.12.

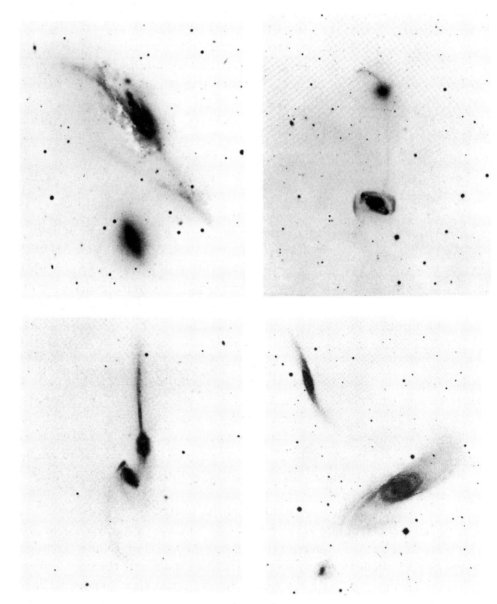

Fig. 11.12. Photographs of four peculiar galaxies, showing the tails and warps caused by gravitational interaction. (Halton Arp.)

11.5.2 Seyfert galaxies, N-galaxies and QSOs

Other unusual galaxies began to be studied in the 1940s by Carl Seyfert and others. These were peculiar in a different way – they looked more or less like normal galaxies, but had very bright nuclei. Some galaxies, which had particularly brilliant star-like nuclei, were christened N-galaxies – in these the nucleus dominates and the rest of the galaxy appears just as a surrounding fuzz. These galaxies were the first of a new class of galaxies with active nuclei, which have become more prominent as observations have been made over more of the electromagnetic spectrum. Since the 1960s, it has gradually been realized that almost all galaxies have activity in their nuclei at some level and the study of active galactic nuclei (AGN) has become a major growth area. I will start by discussing Seyfert galaxies in more detail and relating them to the more famous quasars.

Seyfert galaxies, which are mostly spirals, are defined spectroscopically. Their spectra display broad optical and ultraviolet emission lines whose widths correspond to a spread of velocities of the order of $10^3 \, \mathrm{km \, s^{-1}}$. In some Seyferts, the lines show a double peak that is interpreted as the signature of a disc seen nearly edge-on: the red peak arises from the side of the disc that is moving away and the blue one from the approaching side. If discs are present in the central regions of most Seyferts, then there may be accretion of gas onto some central object, with the accretion occurring through a disc because the gas has angular momentum. Release of gravitational energy by viscous dissipation as the gas flows through the disc could account for the bright nucleus.

The nature of the central object is unclear, except that it must be massive if the gravitational potential well is to provide enough energy for the observed emission, and it must be spinning rapidly because of the accretion of gas with angular momentum. Many astronomers believe that it is a black hole, with a mass of $10^8 \, \mathrm{M_\odot}$ or more. Others have argued that the emission can be caused by massive bursts of star formation, pointing to evidence of prolific star formation in some nearby peculiar galaxies ('starburst galaxies').

Quasars were originally believed to be an unrelated phenomenon. The name is an abbreviation of 'quasi-stellar radio sources', and reflects their original discovery as star-like objects found optically at the positions of strong radio sources. At first they were thought to be stars, but the spectra were quite unlike those of any known star. In 1963, Maarten Schmidt realized that some of the emission lines in the spectra could be interpreted as the familiar Balmer series of hydrogen if the lines had a large redshift, that is, if all the lines in the series were shifted to longer wavelengths, with a shift $\Delta\lambda$ that was proportional to their wavelength λ.

The usual explanation of the redshift is now the same as that of the redshift of distant galaxies – the quasars seem to be receding because they are partaking in the general expansion of the Universe. Their large redshifts then translate into large distances, by Hubble's relationship (Chapter 15).

At these distances, the observed brightness corresponds to an enormous luminosity, in excess of 10^{40} W (about $2\times10^{13}L_\odot$ or 100 times the optical luminosity of a large elliptical galaxy).

Quasars also show variations in their brightness, on timescales of a month or less, showing that the 'central engine' must be less than a light-month across or the variations would be smeared out by time-travel differences from different parts of the source. Like Seyfert galaxies, quasars show broad emission lines, whose breadth probably also arises from a large spread in velocities, this time from clouds close to the central energy source. A few quasars now also show evidence for faint fuzzy patches of light around them; in each case, the spectrum of the fuzzy patch looks like that of a normal galaxy, and it has become clear that quasars are just very extreme examples of galaxies with active nuclei.

For a quasar, the nucleus is completely dominant, enabling it to be seen at enormous distances (several redshifts greater than 4 have been discovered – see Section 15.1). Quasars are rare objects and only a few examples are sufficiently close for the underlying galaxy to be seen, although it is probably as bright as a normal galaxy. As in Seyferts, the central engine in quasars is generally (though not universally) believed to be accretion of gas onto a massive black hole. Since quasars are usually associated with elliptical galaxies, rather than with the spiral galaxies in which Seyfert nuclei are found, the recent discovery of some gas in ellipticals is crucial for this explanation.

Quasars were originally discovered as radio sources and many are very strong radio emitters. Later, other star-like sources without radio emission were found to have the same broad emission lines and large redshift, and the name became less appropriate. Although it has survived in popular usage, a better name now is QSO, or quasi-stellar object, which includes the radio-quiet sources.

Low-redshift QSOs have strong ultraviolet emission, and appear very blue. At larger redshifts, this emission shifts into the optical, or even into the infrared in the most distant objects ($z = 4.4$ corresponds to the Lyman α line of hydrogen, with a rest wavelength of 121 nm in the far ultraviolet, being observed at a wavelength of 656 nm – more or less the rest wavelength of the Balmer α line!). This simplifies the study of intermediate redshift QSOs, which can be observed with ground-based telescopes, but makes it difficult to search for very large redshift QSOs; the usual search technique has been to look for objects with prominent emission lines in low-dispersion slitless spectra on Schmidt plates.

Because of its large range of radio properties and optical colours, a QSO can only be unambiguously identified by its spectrum, which shows the characteristic signature of the hydrogen lines and, in the ultraviolet, various ionized species such as CIV. Because QSOs are so far away, there is a strong probability that the light from a distant QSO will pass through one or more intervening galaxies on its way to us. In several QSOs, absorption lines from the gas in such galaxies are seen.

In 1979, another effect of intervening galaxies was discovered, when an apparently double QSO was discovered: two closely neighbouring QSOs with identical spectra. It was soon realized that these were two images of the same QSO, formed by light taking two paths around a galaxy along the line of sight to the QSO. This dramatic discovery was the first large-scale example of the phenomenon of 'gravitational lensing', which had been predicted long before as a consequence of general relativity (GR). Just as the Sun bends light rays that pass too close to it (one of the classical tests of GR, confirmed by observations at the total solar eclipse in 1919), so a whole galaxy will deflect light from a source behind it. The galaxy acts like a lens, focusing the light from a distant source into an image in its focal plane. Because a galaxy is by no means a perfectly formed lens, the source is usually not exactly on the axis, and the Earth is usually not in the focal plane, the image is not simple and can in fact take many strange shapes. For a point source like a QSO, there are usually several point images, arranged in an asymmetric pattern. For extended sources such as galaxies, the image is more complicated and examples are known of arcs and even complete rings (the so-called 'Einstein rings').

In many QSOs, a 'forest' of absorption lines is seen to the short-wavelength side of Ly α; these are interpreted as Ly α absorption lines arising in intervening hydrogen gas. Because most of these QSOs do not have absorption lines arising from other elements at the same redshift, it seems that most of this gas is in the form of neutral hydrogen clouds between galaxies – if they are really pure hydrogen, then they may be primordial material that has never condensed into stars. Alternatively, the gas density may simply be too low in most cases for less abundant elements to be detected; a few metal lines have been detected which may be associated with the densest hydrogen clouds.

These various pieces of evidence that QSOs are beyond other galaxies and gas clouds tend to confirm that most QSOs are indeed at the large distances suggested by their redshifts, although there is still a minority opinion that the redshifts are not cosmological in origin.

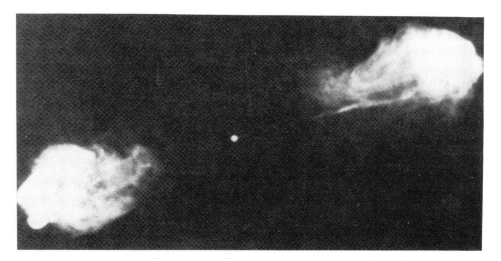

Fig. 11.13. This radio map shows the classical double radio source Cygnus A. There is little radio emission from the central galaxy, but strong emission from the two lobes. A jet can be seen, pointing from the nucleus into the right-hand lobe. (Reprinted by permission from *The Astrophysical Journal*, 1990.)

11.5.3 Radio galaxies

When radio sources were first discovered, astronomers were keen to discover whether they coincided with known objects and careful searches were made for their optical counterparts. Many radio source fields were found to contain galaxies and these sources became known as radio galaxies, because their radio emission far exceeds that of a normal galaxy such as our own – at a typical observing frequency of 408 MHz, the emission is anything from 10^3–10^7 times that of our Galaxy at that frequency.

However, in many cases the central galaxy, which is usually an elliptical, is by no means the strongest contributor to this radio emission. The source is often characterized by a 'double-lobed' structure (Fig. 11.13) with a central source coinciding with the optical galaxy. The two radio lobes stretch out on either side of the central galaxy to many galaxy diameters, often with a jet or bridge of emission reaching out towards them from the central object, which may be quite a weak radio source. The separations of the outer lobes may be anything from 100 kpc to 2 Mpc.

A clue to the nature of these radio sources comes from the radio spectrum, which is a power law:

$$I_v \propto v^{-\alpha}, \ \alpha \simeq 0.8. \tag{11.6}$$

The radiation is also highly polarized. These are both characteristics of *syn-*

chrotron radiation: radiation from ultrarelativistic electrons spiralling around magnetic field lines and so being accelerated. All accelerated particles radiate, and the polarization is imposed by the field direction. The name arises because the radiation was first seen as a bluish light from early particle accelerators, known as synchrotrons. The power law can be shown to imply that the electrons have a power-law energy spectrum: the number of electrons $N(E)$ with energy E is given by:

$$N(E) \propto E^{-(2\alpha+1)}. \tag{11.7}$$

Thus it seems that the double radio lobes have very little mass but are huge tenuous clouds of highly relativistic plasma and magnetic fields. Where these clouds thrust into the intergalactic medium, the surrounding gas piles up and slows the clouds down, causing a concentration of material in confined lobes (and allowing estimates of the density of the intergalactic material). Some lobes are swept backwards into a wake, showing that the central galaxy is moving through the intergalactic gas.

Polarization maps show a relatively ordered magnetic field within the lobes. The discovery of narrow jets, which generally point towards the lobes from the central source, suggested that the plasma clouds are ejected from the central galaxy as part of a continual process of outflow.

In either case, these radio galaxies point again to activity in the galactic nucleus, and there may be no clear distinction between radio galaxies and QSOs – the best-studied QSO, 3C273, has a jet emerging from the centre. Current belief is that the appearance of an active galaxy is largely determined by the direction from which we view it: in radio galaxies, the jet is roughly in the plane of the sky, while in BL Lac objects, a variant of QSOs that show no strong emission or absorption lines, we are looking straight down the jet and the strength of the relativistically beamed continuum radiation from the jet swamps the line radiation from the nucleus that we see in more normal QSOs.

11.5.4 Active galaxies and a unified model

As well as the types of active galaxy that I have mentioned so far, several other classes are now recognized. The terminology is confusing, and not fully consistent, with different authors using different descriptions. The general term 'Active Galactic Nuclei' (AGN) is used for at least the following:

1. *Radio galaxies.* These are strong radio sources, often with radio lobes and possibly jets. The most powerful radio sources are usually

associated with luminous elliptical galaxies. They may also be X-ray sources.

2. *Radio quasars.* The radio spectra are similar to those of radio galaxies, but the optical source is a QSO, which is usually also an X-ray source. VLBI observations (Section 4.4) reveal very compact components which sometimes appear to be moving across the line of sight at speeds in excess of that of light. These 'superluminal' velocities are explained as a projection effect involving relativistically moving jets coming almost directly towards the observer.

3. *BL Lac objects.* These are similar to radio quasars, but show no optical emission lines. They are highly variable in intensity in the radio, optical and X-ray and have strong and variable optical polarization.

4. *Optically Violent Variables (OVVs).* These are similar to BL Lacs in their optical variability, but do show emission lines. BL Lacs and OVVs are sometimes lumped together as 'Blazars'.

5. *Radio quiet quasars.* These are by far the commonest type of QSO, with rather weaker X-ray emission than radio quasars. Many do show some weak radio emission.

6. *Seyfert galaxies.* The main difference between Seyferts and QSOs seems to be that Seyferts are less luminous. Seyfert nuclei are usually found in spiral galaxies; if seen at large redshift, where the galaxy might not be detectable, they might well be classified as QSOs, so the distinction is not clearcut. There are subclasses of Seyferts, based on their emission-line properties. In Seyfert 1 galaxies, the hydrogen emission lines are very broad, with wings extending out to more than $10\,000\,\mathrm{km\,s^{-1}}$ in most cases, while some other emission lines are considerably narrower, suggesting that the lines come from two different regions, the Broad Line Region (BLR) and the Narrow Line Region (NLR). Similar regions are distinguished in QSOs. These names conceal an almost total ignorance of where these regions are, although it is believed that the BLR is probably closer to the nucleus. Seyfert 2 nuclei do not have the extremely broad wings, and all their emission lines have similar widths, suggesting that we may only be seeing the NLR in those cases.

7. *LINERs (Low-Ionization Nuclear Emission-line Regions).* These nuclei are characterized by emission lines from atoms which are neutral or only once ionized. By contrast, most QSO spectra have strong lines from highly ionized species, such as CIV.

Other objects which may be associated with AGN are:

Starburst galaxies. These are galaxies in which blue colours and a strong IR output suggest that star formation is occurring at a high rate. Bursts of star formation may be triggered by interactions between galaxies, or by galaxy mergers (Section 13.1), but there is also some evidence that the galaxies containing AGN may sometimes have high star formation rates.

Strong IRAS *galaxies.* The IRAS satellite (Section 5.1) revealed some galaxies which were extremely luminous in the far IR ($L \approx 10^{12} L_\odot$). This is believed to be re-radiation from dust which has been heated by an intense central source, an AGN or perhaps just a burst of star formation.

Is there any way in which all these apparently different objects can be explained by a single model? As I hinted at the end of the last section, the answer may be that the different appearance is simply because we are looking from different angles.

There have been many attempts to produce unified models of AGN. None is completely successful, but I will outline one plausible picture. In fact, it is really two pictures, since it seems to be impossible to allow all AGN to have the sort of relativistic jet that we certainly see in some radio quasars. The overall picture is of a central, very compact powerhouse, perhaps a black hole, surrounded at a distance of a few parsecs by a ring of absorbing material. Within the ring are the fast-moving clouds that produce broad emission lines (the Broad Line Region), while outside the ring, but still roughly in its plane, are the more slowly moving clouds of the Narrow Line Region.

Suppose, first, that the central nucleus produces a relativistic jet or jets. If the observer looks straight down the jet axis, the luminous and rapidly variable source will look like a BL Lac object or an OVV; the continuum from the jet will completely swamp the emission lines from the BLR clouds. This effect will decrease as the line of sight moves away from the jet axis and a radio quasar will result. When the tiny central source is obscured by the ring, but some of the broad-line clouds remain visible, the observer will see a radio galaxy with broad emission lines. As the observer moves into the plane of the ring, even these clouds are obscured and only the narrow emission-line clouds will be visible in the spectrum. The jet will then be seen more or less edge-on and a normal double radio galaxy will be seen. If there is no jet, the observed systems will be radio quiet. If the nucleus and the BLR are visible, the source will appear as a QSO or a Seyfert 1 galaxy. If the nucleus and the BLR are hidden by the ring, a Seyfert 2 galaxy will result.

Thus this single model of a nucleus, ring and clouds accounts for many of the observed types of AGN.

However, this is certainly not the whole story. There seems to be a genuine range in total luminosity, from QSOs through Seyferts to LINERs, and you have seen that the presence or absence of relativistic jets implies that there are at least two types of system. There is also some evidence that a source may change from one type of AGN to another; for example, from Seyfert 1 to Seyfert 2, or from BL Lac to Seyfert 1. Although I have only outlined very briefly what is known, and the real picture is far more complex, there is clearly still a lot to learn about these enigmatic galaxies.

11.6 Summary

A galaxy is mainly just a huge collection of stars, bound to one another by gravity. Galaxies come in many shapes and sizes, but can be classified into three broad classes: spiral, elliptical and irregular, as was first done by Hubble. The non-circular shapes of ellipticals are unrelated to rotation, and many elliptical galaxies are probably triaxial, but the highly flattened discs in spiral galaxies are dominated by rotation and are axisymmetric. Spirals also contain far more gas and dust than do ellipticals, and display spiral arms, which either emerge directly from the nucleus or, in barred spirals, start from the ends of the central bar. Spirals with very prominent nuclei have more tightly wound spiral arms and less gas and dust than those with small nuclei. The lenticular, or spheroidal galaxies, may be an extreme example of this, resembling spiral galaxies with no spiral arms and essentially no gas.

Galaxies that do not fall within the Hubble scheme are called irregular, and may be young systems with very active star formation, such as the Magellanic Clouds, or genuinely peculiar systems that have undergone some violent event.

Galaxies vary enormously in luminosity, size and mass, with the ellipticals showing the greatest range. As for stars, the small faint galaxies are the most numerous, and the most difficult to study. Mass and luminosity are related for most galaxies, so the *mass-to-light ratio* is often a more useful property, which varies slowly with Hubble type. Even in visible light, the main contributors to this ratio are stars that are fainter and less massive than the Sun.

Mass estimates for spiral galaxies are the most reliable, since they can be deduced by using the 21 cm radio line to look at the orbital motion of neutral hydrogen gas in the gravitational potential of the galaxy. In many cases, the gas can be traced well beyond the visible bounds of the galaxy, and the observations show that there is a great deal of mass in the galaxy that is not detectable at optical wavelengths. The nature of this 'dark matter' is not yet known.

Although stars make up most of the visible mass of a galaxy, they occupy little of the volume, which is filled with other components. The largest contributor is gas, which may account for 20% of the (visible) mass of an irregular galaxy, though only a fraction of a percent for an elliptical. In gas-rich galaxies, dust is also abundant (perhaps 1% of the gas mass) and, like the gas, is distributed in a very clumpy way throughout the plane of disc galaxies. Radiation, cosmic rays and magnetic fields are also present in many galaxies.

Not all stars have the same chemical composition. Although H and He are the dominant elements in almost all cases, and the abundances of the other elements relative to iron are similar for most stars, the overall abundance of the 'heavy' elements relative to hydrogen varies considerably. To a first approximation, stars are either Population I (roughly solar composition) or Population II (metal-poor by a factor of more than 10), and the interpretation is in terms of heavy elements being synthesized by nuclear reactions inside stars and then returned to the interstellar medium, mainly by supernova explosions. New stars can be formed out of the enriched gas, and the heavy element content of stars gradually increased. Detailed studies of the chemical abundances suggest a rapid initial burst of star formation, followed by a more gradual building up of heavy elements over the life of a galaxy.

Some galaxies fit so badly into the Hubble classification, or have such unusual properties, that they deserve a category of their own. The simplest are the peculiar galaxies, whose distorted shapes are witness to tidal interactions with companions, and which may be otherwise normal. More dramatically, the active galaxies show evidence for strange happenings in their nuclei. The Seyfert galaxies, N-galaxies and quasars are all examples of galaxies with unusually brilliant, star-like nuclei, which may outshine the whole of the rest of the galaxy. All show in their spectra evidence for rapid motions near the centre, which testify to the presence of a large, compact mass at the very centre of the galaxy. Most astronomers now believe that these galaxies are powered by accretion of gas and stars into a black hole at the centre, whose mass may be in excess of $10^8 \, M_{\odot}$. The quasars are particularly powerful, and can be detected to immense distances, estimated from the redshift of emission lines in their spectra.

Many active galaxies are also strong emitters in other wavebands, such as radio, infrared and X-ray. Radio galaxies often show gigantic lobes of radio emission, tens or hundreds of times the size of the central galaxy; these are huge tenuous clouds of magnetized plasma, containing relativistic electrons radiating by the synchrotron mechanism. Narrow jets, moving relativistically out from the centre of the optical galaxy, point towards these lobes in many cases and show that the plasma clouds have been ejected from the nucleus.

The term 'active galaxy' covers a great range of different objects, with strong, and variable, emission at many wavelengths. It is possible that they are in fact all essentially the same object, viewed from a range of angles, and with or without a relativistic jet.

Exercises

11.1   Consider a disc galaxy, whose mass distribution can be modelled by a point mass M at the centre. A star of mass m is moving in a perfectly circular orbit in the disc, at a distance R from the centre. If its orbital speed is V, show that

$$V^2 = \frac{GM}{R}.$$

11.2   A spherically symmetric galaxy has a uniform star density and a massless disc of gas. If the gas particles are moving in perfectly circular orbits in the gravitational field of the stars, show that the gas disc has a uniform angular velocity.

11.3   A spherically symmetric galaxy has a mass distribution $M(R)$ and a massless disc of gas. If the gas particles are moving in perfectly circular orbits with orbital speeds that are independent of distance R from the centre, show that $M \propto R$ and that the density follows an inverse square law (use equation (10.12) to relate mass and density).

12

Our Galaxy

12.1 Overall structure

At first sight, our Galaxy looks very different from other galaxies. We view it from within, and we see it spread out around the sky as the Milky Way: one of the most prominent features in a dark sky and certainly known since ancient times. However, the nature of this broad, irregular band of light across the sky only gradually became realized and it was only in the early years of this century, as the true nature of external galaxies was becoming understood, that the true shape and structure of our own Galaxy began to emerge as well.

Students of our Galaxy have the great advantage that it is very close, so it is easy to obtain detailed information. However, there is also a major disadvantage: we are situated in the mid-plane of the disc of a spiral galaxy and have our view blocked by an absorbing layer of dust and gas. Because of this haze, we cannot see far from the Sun except in directions perpendicular to the disc, and early workers believed that the Sun was near to the centre of the system. Sir William Herschel started the quantitative study of the structure of the Milky Way in about 1800, counting stars of different apparent brightness and estimating their relative distances by assuming that all stars were of the same brightness; he called this process 'star gauging'. He found, correctly, that we live in a flattened system of stars, but he did not know about absorption, believing the dark dust clouds in the galactic plane to be actual holes in the star distribution, and his Sun-centred Milky Way was really only a lens-shaped sample of the Galaxy near to the Sun.

A century later, Dutch astronomers such as J. Kapteyn used the huge accumulation of nineteenth-century observations of stars to make a much more careful statistical discussion of the shape of the Galaxy, taking account of the vastly different brightnesses of different types of star. However, the

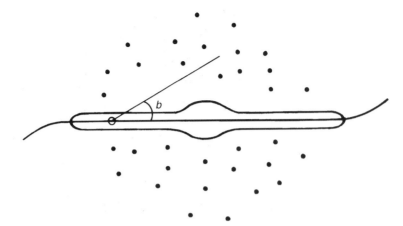

Fig. 12.1. A schematic picture of our Galaxy, seen edge-on. The flattened outline, with a central bulge, represents the distribution of stars, while the central line is the much thinner gas disc, which is warped at galactic longitudes 90° (up) and 270°(down). The spherical distribution of dots represents the halo: globular clusters, field stars, dead stars and dark matter. The Sun is shown as a circle at about the correct distance from the centre. However, in a view which showed the warps the Sun would actually be hidden directly behind the Galactic Centre. Galactic latitude *b* is the angle of view of an object from the Sun, measured north (above, positive) or south (below, negative) of the plane defined by the gas.

presence of absorbing material was still not appreciated and the 'Kapteyn universe' was not very different from Herschel's. It was not until 1930 that R. Trümpler recognized the existence of general interstellar absorption from his studies of distant star clusters and it became possible for the first time to obtain a correct impression of distances in the plane of our Galaxy.

Some ten years earlier, the fact that the Sun was not at the centre had been realized. The American Harlow Shapley noticed that the globular clusters were concentrated in one half of the sky and argued that they were distributed symmetrically about the centre of our Galaxy, which must therefore be a long way from the Sun, in the direction of the constellation Sagittarius. Because many globular clusters are well out of the plane of the Galaxy, there is little absorption between us and them and Shapley's argument was the first step towards our present view of the Galaxy.

We now know that the Galaxy is a fairly normal Sb or Sc spiral: a large, disc-shaped system of stars with the Sun in the disc, some 8–10 kpc from the centre. Figs. 12.1 and 12.2 show, schematically, the appearance of the Galaxy edge-on and face-on respectively. The stellar disc has a radius in excess of 15 kpc and a thickness of about 1 kpc: the density actually falls off

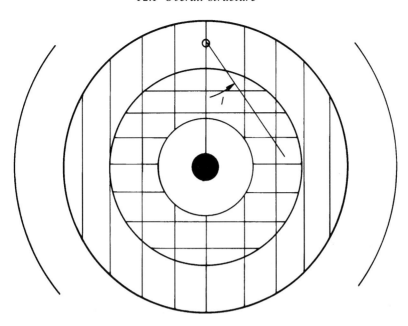

Fig. 12.2. A schematic picture of our Galaxy, seen face-on. The filled circle is the Galactic Centre region (Fig. 12.6). The horizontally hatched ring marks the location of most of the molecular gas in the disc, while the broader, vertically hatched, region contains most of the atomic hydrogen. The Sun is marked by a circle; galactic longitude l is measured anti-clockwise in the plane of the disc, with the Galactic Centre at zero longitude. The broad rings of atomic and molecular gas would not be detectable in an optical picture, which would be dominated by bright stars and ionized gas, forming a roughly spiral pattern (not shown).

exponentially with height above the disc, and the e-folding distance (the *scale height*) varies with stellar population, so an exact value cannot be given.

The dust and gas are confined to a much thinner disc, with a scale height of about 100 pc. The plane defined by the neutral atomic hydrogen is sufficiently precise to be used as the basis of a coordinate system, with galactic latitude being measured from 0 to 90 degrees above or below the plane (positive in the hemisphere containing the north celestial pole). Galactic longitude is measured in the plane, with zero longitude towards the Galactic Centre.

Far out from the centre, at galactic longitudes 90° and 270°, the gas disc displays a warp, which used to be thought to be due to tidal interaction with our nearest galactic neighbours, the Magellanic Clouds. Similar warps are seen in the discs of some external galaxies, several of which also have close companions. However, it now seems that the Magellanic Clouds have never been close enough to the disc to cause such a large warp; current explanations are in terms of a misalignment of the disc axis with the symmetry axis of the halo.

Another interesting feature of the large-scale gas distribution is the Magellanic Stream: a long narrow chain of neutral hydrogen clouds which starts near the Magellanic Clouds and stretches across more than a quarter of the sky, passing nearly through the south galactic pole. It is believed that the Large and Small Magellanic Clouds, which are both in orbit around the Galaxy, came within 3 kpc of one another some 2×10^8 years ago, and that the Large Cloud stripped gas out of the disc of the Small Cloud during the encounter. This material would be expected to stay in orbit around the Galaxy, and the Magellanic Stream is probably the debris of this relatively recent encounter. The orbits of the Clouds have been modelled, and it seems that they are gradually spiralling in towards the centre of the Galaxy, which they will reach in about another 10^{10} years.

As well as the highly flattened disc, the Galaxy has a spherical, or at least spheroidal, component. Around the central nucleus, there is a spheroidal bulge, some 2 kpc in radius, with a very high star density. The contents of this bulge are only gradually becoming revealed as astronomers observe in more detail at infrared and microwave frequencies – the centre is completely hidden at visual wavelengths behind an impenetrable cloak of dust and gas. Beyond the central bulge, there is a much larger and much more tenuous outer halo of material, which is roughly spherical. The most prominent members of the halo are the globular clusters, but there are also many faint, isolated Population II stars, most easily detected as they streak past the Sun on highly eccentric and inclined orbits. There may also be many dead stars, and there is certainly a great deal of dark matter whose nature is unknown.

Most of the visible mass is in the bulge and disc. Within the disc, bright stars and gas are concentrated into spiral arms and rings. Optical tracers, such as bright blue stars and ionized gas (Fig. 12.3), can only be seen within a few kiloparsecs of the Sun and lead to a very partial knowledge of the local spiral structure. The Sun appears to be on the inner edge of the Orion arm, and other sections of arms are seen in the constellations of Sagittarius and Perseus.

Our knowledge of the overall structure of the gas comes from radio and microwave observations. Neutral hydrogen observations at 21 cm show an irregular spiral pattern, rather than the 'grand design' spirals seen in some external galaxies (Fig. 11.4). However, another feature emerges: most of the neutral atomic hydrogen is contained in a broad ring between 4 and 12 kpc from the centre (taking the distance of the Sun from the centre as 10 kpc). Within this ring, microwave observations show a narrower ring

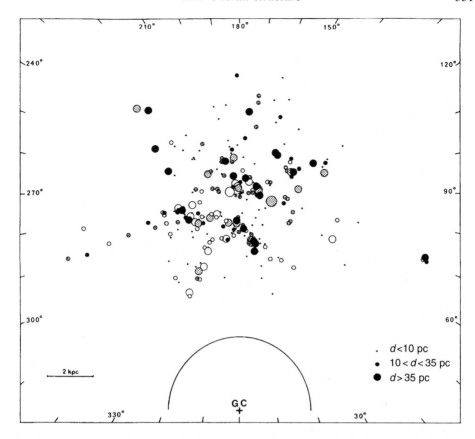

Fig. 12.3. The neutral hydrogen map of our Galaxy shows little sign of the clear spiral pattern seen in the 'grand design' spiral galaxy M51 (Fig. 11.4). Optical tracers, such as the HII regions plotted in this map, show a somewhat more convincing pattern, but can only be seen in the solar neighbourhood. GC marks the Galactic Centre, while the Sun is marked by a small S near the centre of the diagram. The size and density of the symbols represent the size and brightness of the HII regions. (D. Crampton and Y.M. Georgelin; National Research Council of Canada.)

of molecular hydrogen and carbon monoxide stretching from 4 to 8 kpc. There is some dispute about the quantity of molecular hydrogen, whose existence can only be deduced indirectly, but there seems to be at least as much hydrogen in molecular as in atomic form. Carbon monoxide is by far the most stable common diatomic molecule, with a dissociation energy of 11.1 eV (cf. 4.5 eV for H_2), and would be expected to be the next most abundant molecule, especially as C and O are relatively abundant in nature (Table 9.2). It radiates strongly at 2.6 mm and is easily detected, although its total abundance is only about 10^{-5} of that of hydrogen. It is one of the few molecules to have been detected so far in external galaxies, and provides

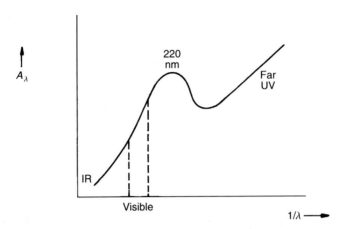

Fig. 12.4. Absorption in the interstellar gas as a function of wavelength, λ. The dependence in the visible is roughly linear with $1/\lambda$; the ultraviolet is dominated by the 220 nm peak. The behaviour in the infrared and far ultraviolet varies from place to place in the sky.

a very valuable supplement to atomic hydrogen as a tracer of galactic structure.

12.2 Dust

The main difficulty in exploring the structure of the Galaxy in the visual is the presence of interstellar dust, which forms a thin haze in the galactic disc and effectively obscures everything beyond a few kiloparsecs. Trümpler discovered this haze when he tried to measure the sizes of open star clusters and discovered that the more distant clusters seemed to be systematically larger. Rather than accept this bizarre result, he argued that he must be systematically overestimating the distances of the more distant clusters and deduced that this was because intervening dust made them appear fainter than they should be at their true distance. By supposing instead that the sizes of open clusters are independent of distance, he was able to deduce an absorption of nearly one magnitude per kiloparsec, although the distribution of dust is by no means uniform and in certain directions (e.g. 'Baade's window', a few degrees away from the Galactic Centre) it is possible to see objects at distances beyond the Galactic Centre.

The dust absorbs preferentially at short wavelengths, and so also causes a *reddening* of distant objects; more red light reaches us than blue. The scattering reaches a strong peak in the ultraviolet at about 220 nm, but otherwise has a fairly smooth dependence on wavelength (Fig. 12.4), which

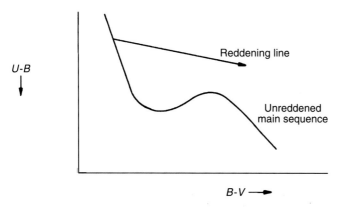

Fig. 12.5. In the two-colour diagram, reddened stars are displaced from the main sequence along a *reddening line*, whose slope is roughly the same for all spectral types. A reddening correction can be made by drawing a line of that slope through the observed colours of the star.

makes it difficult to determine the chemical composition of the dust grains. In the visual, the absorption in magnitudes goes roughly as $1/\lambda$, which is characteristic of particles whose size is comparable to the wavelength of the scattered light. More precise calculations suggest a size distribution in the region of $0.1 \, \mu$m. The observation that distant starlight is polarized as well as reddened suggests that the scattering is anisotropic, and it is believed that many of the grains are elongated. They are then aligned by the interstellar magnetic field, whose direction determines the angle of polarization – indeed, this is one way of estimating the structure of the field.

The composition of the dust grains is unclear and has been a fruitful ground for controversy. On abundance grounds, compounds of C, N and O probably dominate, with significant contributions from Si, Mg and Fe, but the spectrum of scattering provides no clear signature for more precise identification. The 220 nm peak has been ascribed variously to graphite, diamonds and freeze-dried bacteria; some carbon-based compound is almost certainly responsible. In the infrared, an absorption band due to ice has been detected in some directions.

Fortunately, it is possible to correct observed stellar colours for interstellar absorption without knowing anything about the dust composition. To do this, we turn to the two-colour diagram, making use of the reddening effect (Fig. 12.5). As you saw in Chapter 8, main-sequence stars lie on a well-defined line in the $(U - B, B - V)$ plane. Clearly, reddened stars must be displaced from that line, and the direction of displacement can be determined by looking at a number of stars of the same spectral type

but different distances. Using bright O and B stars, the interstellar medium
can be probed to considerable distances and it is found that all stars of a
particular spectral type lie along a line called the *reddening line* that extends
to the right and downward in the diagram. To a first approximation, the
equation of the line is

$$U - B = 0.72(B - V) + Q, \tag{12.1}$$

where Q is a constant for a given spectral type. Once the slope of this
line is known, it is easy to correct the colours of a particular star by
simply drawing a line of slope 0.72 through the position of the star in the
two-colour diagram and reading off the colours where the line crosses the
main-sequence relationship. Ambiguities for cool stars, for which several
crossings may occur, can be resolved if a rough estimate of the spectral type
is known.

12.3 Gas

Because of the strong wavelength dependence of interstellar absorption by
dust, observations in the infrared and radio are hardly affected by absorption,
and it has become possible to explore the structure of the whole galactic disc.
As you saw in Section 12.1, radio and microwave observations have revealed
the overall distribution of gas, in spiral arms and in rings of HI, H_2 and
CO. Microwave and infrared observations are revealing further molecular
rings near the Galactic Centre, and there is a central radio source, Sgr A*,
near the infrared complex of sources IRS 16. A radio and infrared map of
the Galactic Centre region is shown in Fig. 12.6. The radio 'mini-spiral' is
actually a combination of a roughly circular ring, some streamers and a few
individual clouds. This material is within an otherwise almost empty cavity
stretching out to about 1.7 pc, which marks the inner radius of a dust ring
emitting in the far infrared.

Farther out, between about 2 and 8 pc from the centre, there is a thin,
disc-like structure of gas, revealed in emission from neutral oxygen and
hydrogen and from molecular hydrogen and carbon monoxide. Neutral
hydrogen is also present in a much larger rotating 'nuclear disc', on a scale
of a few hundred parsecs to about 1 kpc. Infrared observations have shown
an extensive star cluster of a similar size, centred on the source IRS 16,
which is therefore often taken to be the actual centre of the Galaxy. There
is a strong concentration of mass towards the centre, and it is possible that
there is a relatively small central black hole (possibly as small as $10^2 M_\odot$ or
up to about $10^6 M_\odot$).

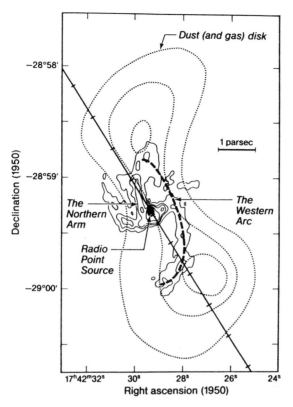

Fig. 12.6. A map of the Galactic Centre. The radio continuum emission at 2 cm shows a 'mini-spiral' centred on a radio point source, Sgr A* (which coincides with an infrared source, IRS 16). Far infrared emission reveals a dust disc, which is actually a ring with a central cavity. The diagonal line marks the galactic plane. (Reproduced, with permission, from the *Annual Review of Astronomy and Astrophysics*, Volume 25, ©1987, by Annual Reviews Inc.)

On a smaller scale, the gas in the disc is clumped into clouds of various sizes. The first to be discovered were the neutral clouds, with temperatures of about 70 K and particle densities of about $3 \times 10^7 \, \text{m}^{-3}$. These are mostly composed of atomic hydrogen, but were first detected from narrow absorption lines of Ca and Na in the visible spectra of stars. The lines were so narrow that it was clear that they could not be produced in the hot atmospheres of the stars (Fig. 12.7(a)), and this was confirmed by their observation in some binary stars, where the stellar lines displayed orbital motion but the narrow interstellar lines showed a fixed velocity, different from that of the binary. This technique of studying the interstellar medium by using the light of background objects as a probe has since been very widely applied, both using ultraviolet spectra of hot stars to probe our own

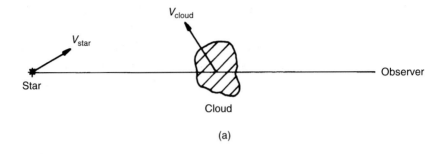

(a)

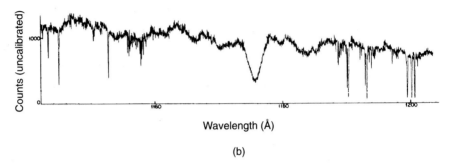

Wavelength (Å)

(b)

Fig. 12.7. (a) When the line of sight to a star passes through a diffuse cloud of gas, the atoms in the gas will absorb some starlight, producing absorption lines in the stellar spectrum that are shifted in wavelength with respect to the star's own lines because of the relative motion of the star and cloud. These interstellar absorption lines are also narrow, because the cloud is much colder than the star. (b) Interstellar absorption lines in the ultraviolet spectrum of a hot star; they are easily distinguished from the broad and shallow features which arise in the star's atmosphere. (Reproduced with permission from *The Astrophysical Journal*, 1975.)

Galaxy (Fig. 12.7(b)) and using the spectra of distant quasars to probe the outer layers of galaxies along the same line of sight.

The neutral clouds are surrounded by a warm intercloud medium, with a temperature of about $5000\,\mathrm{K}$ and a density of about $4\times10^5\,\mathrm{m}^{-3}$, which is partially ionized (10–20%). Remembering that the pressure of a gas is $P = nkT$, where n is the number density, T is the temperature and k is Boltzmann's constant, you will see that there is rough pressure balance between the neutral clouds and their surroundings, so they are expected to be in equilibrium. It can be shown that the equilibrium is stable and therefore fairly long-lived.

The neutral clouds typically have radii of $5\,\mathrm{pc}$ and masses of some $50\,\mathrm{M}_\odot$. More recently, much more massive clouds have been discovered. These are cool, molecular clouds, with $T \simeq 10\,\mathrm{K}$ and $n \simeq 10^9$–$10^{10}\,\mathrm{m}^{-3}$. They are comparable in size to the neutral atomic clouds, though some are larger, but the higher density means that their masses are in excess of $1000\,M_\odot$, and the largest – the 'giant molecular clouds' – have masses of up to $10^5\,M_\odot$. These clouds emit no optical radiation, but show a rich spectrum of molecular emission lines and bands from the infrared to the radio, especially in the microwave part of the spectrum. Some molecules (e.g. H_2) can be detected by ultraviolet absorption lines in the spectra of hot stars.

The main components are H_2 and CO ($\simeq 10^{-5}\,H_2$), but there are nearly 100 other species now known, mainly organic compounds and some of surprising complexity (up to 13 atoms). Not surprisingly in such cool regions, molecular clouds often have dust clouds associated with them.

The third major type of gas cloud found in galaxies is very different. These clouds are fully ionized, very hot and fairly dense, with $T \simeq 10^4\,K$ and $n \simeq 10^8\,m^{-3}$. They form near hot stars, whose ultraviolet radiation acts both to ionize and to heat the surrounding gas. They are generally known as HII regions. They are detected by emission lines that arise as atoms recombine and the electrons cascade down through the various energy levels to the ground state. Hydrogen displays not only the common Lyman and Balmer series in the ultraviolet and visible but also lines in the radio from transitions between much higher levels, with quantum number $n \simeq 10^2$. These radio recombination lines (Section 4.5) cannot occur except in the very low densities of interstellar space; the radius of an atom in such a highly excited state is about $1\,\mu m$, much larger than the typical separation of atoms in a terrestrial gas.

The atomic and molecular clouds and HII regions occupy only a few per cent of the space between the stars. However, the whole volume of the disc is filled with gas. The warm intercloud medium described above occupies some 50% of the volume and the other half of interstellar space is filled with an even hotter and more tenuous gas. This 'coronal gas' has $T \simeq 5 \times 10^5$–$10^6\,K$ and $n \simeq 10^4\,m^{-3}$ and was detected in the ultraviolet spectra of hot stars, which showed lines of OVI, corresponding to a temperature much higher than is consistent with the rest of the star's spectrum. Gas with very similar properties is also seen in soft X-rays from known supernova remnants (SNR). The discovery of this very hot gas led to a new picture of the interstellar medium, in which the structure is dominated by violent events blowing hot gas outwards from stars. On a small scale, HII regions expand because their pressure is higher than that of the surrounding neutral gas, so that an ionized bubble gradually eats its way outwards through a neutral cloud; many HII regions are seen as 'burst bubbles' at the edges of clouds, with hot ionized gas streaming out of the breach in the surface of the cloud. More dramatically, supernova explosions lead to the violent ejection at speeds of up to $10^4 \, \mathrm{km \, s^{-1}}$ of a shell of gas and to the rapid heating and ionization of the growing sphere of gas behind the expanding shell. The expansion of the shell continues for 10^4–10^5 yr, long after the explosion itself has finished. The shell sweeps up the surrounding stationary material and gradually slows down, by conservation of momentum, becoming a supernova remnant: a thick shell surrounding a low-density, ionized bubble. Eventually, bubbles from HII regions and SNRs may merge to form hot, ionized 'tunnels' through the warm intercloud medium, whose shapes are ever evolving as new HII regions form and new SN explosions occur. A whole new vocabulary, with words such as 'worms' and 'chimneys', has grown up to describe the great variety of shapes.

Star formation also plays an important role in the evolution of the interstellar medium. The most obvious place for stars to form is deep in the high-density cores of molecular clouds, where they are initially hidden from sight except as microwave or infrared sources. If a hot star is formed, it will produce a compact HII region around it, which may be detected by radio emission long before it is visible. Eventually, the star or star cluster is likely to blow a hole in the cloud and become visible as it emerges from its cocoon. Thus we expect – and find – that young stars and hot HII regions are closely connected with cool molecular clouds and dust. Spinning protostars seem to form discs of accreting material and to eject jets of molecular gas along the rotation axis. These jets are seen in the microwave and enable us to probe early stages of star formation. In the shells swept up by SNRs, the density may become high enough for new stars to form there as well.

Studies of external galaxies show even more clearly than in our own that young stars, HII regions, molecular clouds and dust tend to be concentrated in the spiral arms, where the general gas and star density is higher than average, and act as tracers for the spiral structure.

In spiral galaxies, the gas generally forms a ring around the centre, just as in our own Galaxy; this is particularly clearly seen in M31 (Fig. 12.8), where the ring shows up in neutral hydrogen, carbon monoxide and the distribution of HII regions; the distribution of the supernova remnants (bottom panel) is more irregular, perhaps because of absorption in the galactic disc. The

---

Fig. 12.8. Various maps of the spiral galaxy M31. The upper two panels show the neutral hydrogen and carbon monoxide distributions, which both reveal a ring-like structure with a central hole. This is also seen in the distribution of HII regions overlaid on the CO map in the bottom panel. The ring is seen to a lesser extent in the IRAS 100 μm and the 11 cm radio continuum maps, whose main features are emission from the centre of the galaxy. (Tom Dame.)

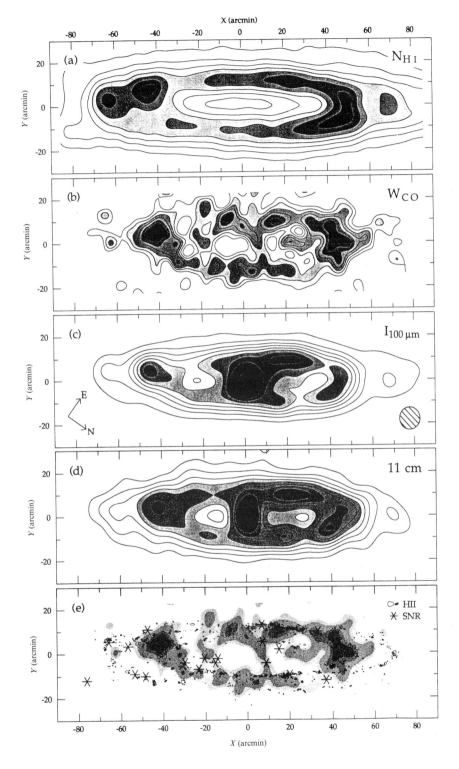

central emission in the infrared and radio continuum is slightly surprising, since there is little evidence for gas and star formation at the centre. Probably the infrared radiation comes from dust heated by cool, luminous giants, while the radio may come from Type I supernovae.

The profusion of gas and dust in irregular galaxies suggests that a great deal of star formation is occurring there. In elliptical galaxies, there is almost no gas, and little current star formation.

12.4 The history of our Galaxy

There is not the space in this book to say very much about the evolution of galaxies, which is in any case much less well understood than the evolution of stars. Here I will just outline some broad principles, largely in the context of our own Galaxy.

The difficulty in studying the evolution of galaxies is that the vast majority of galaxies formed before the Universe was even half its present age, and the only young galaxies easily observable today are some interesting but untypical blue irregulars known as HII galaxies, where star formation is proceeding in a tremendous outburst. Some clues may also be obtained from the active galaxies that I discussed at the end of Chapter 11. But, in general, astronomers can only study young galaxies by looking at extremely distant objects, whose light left them near the time of their formation and which are so faint that detailed study is impossible.

We are therefore thrown back on general theoretical principles and on looking carefully at nearby galaxies, especially our own. There are several important clues. In our Galaxy, which is probably typical of spirals, there is a gradient of chemical composition, with the most metal-rich stars and gas being concentrated towards the centre, where the overall density is higher. The gradient is not large, amounting to perhaps a factor of 2 in Fe/H between the centre and the Sun, but its existence is now well established. A related observation is that even the oldest stars in the Galaxy, the low-mass halo population stars, have a significant amount of heavy elements. Very few stars are known with Fe/H less than one-thousandth of the solar value. These results together tell us that there must have been an initial burst of star formation while the Galaxy was forming and that star formation (and therefore element production) proceeds faster in regions of higher density.

A more general result is that, at least to a first approximation, elliptical galaxies have no gas while spiral ones have a great deal. To interpret this, we must appeal first to general theoretical ideas. Just as stars are believed to start forming by the contraction of an initial gas cloud, so galaxies are thought to start as gigantic clouds of gas. Initially, these are expanding at the same rate as the Universe itself, but if there is a sufficiently large local density peak the expansion rate of a particular region may slow down, causing the relative density to increase further, until eventually the expansion stops and the gas cloud begins to contract. Any departure from a spherical shape is enhanced in the subsequent collapse and virtually all initial gas clouds will flatten as they contract.

The evidence from our own Galaxy and others that there are very few stars with extremely low metal abundance suggests that stars started to form during the initial collapse. What happened to these stars? Because the separations between stars are so much larger than the stars themselves (Table 1.1), the chance of collisions between stars is negligible. As the initial cloud flattened into a pancake shape, the stars that had formed passed straight through the plane of the pancake without collision and the proto-galaxy re-expanded. After several such 'bounces', the stellar orbits would be sufficiently out of phase with one another that only a few stars would be passing through the mid-plane simultaneously, and the galaxy would reach an equilibrium shape not very different from that of the initial gas cloud, with the individual stars moving on random orbits within that shape. The randomization is greatly speeded up by the rapid variation of the gravitational potential during the collapse and subsequent oscillations, a process known as *violent relaxation*. The final stellar orbits have a random distribution of semi-major axes, eccentricities and inclinations, and the total angular momentum of the system is fairly low. This is a fairly good description of an elliptical galaxy, whose ellipticity depends on the precise distribution of stellar orbits.

Why is there essentially no gas in elliptical galaxies? Since the majority of ellipticals are found in clusters, any residual gas left after star formation may simply have been stripped away, either by interaction with background gas in the cluster or by a galactic 'wind' arising from supernova explosions as massive stars reached the ends of their lives. Alternatively, star formation may have been so efficient during the initial collapse that there was essentially no gas left by the time the galaxy attained an equilibrium shape. In either case, it seems that nearly all stars in elliptical galaxies were formed during the initial collapse and must be very old. This is consistent with there being no Population I stars in ellipticals, although the presence of heavy elements at all in the Population II stars that form ellipticals argues for a previous generation of element-forming stars at the very beginning of the collapse.

How can we account for spiral galaxies in this picture? Where do their discs come from and why do they look so different? It used to be thought that an important parameter was the rate of star formation during the contraction phase. If stars form fairly slowly, then a great deal of the proto-galaxy will still be in the form of gas when it flattens into a pancake. Collisions in the gas will dissipate a great deal of the kinetic energy of the collapse, and no bounces will occur. Instead, the gas will continue to contract, spinning ever faster about its rotation axis as a result of conservation of angular momentum. Eventually, contraction in the plane perpendicular to the rotation axis will be stopped by the build-up of centrifugal forces, but contraction along the rotation axis will continue, until the gas has formed into a disc; the final disc plane is determined by the angular momentum vector of the gas, and need not coincide with the initial collapse plane of the proto-galaxy. The gas will stay in a disc, and all subsequent star formation will occur in that disc. Because the disc is younger than the stars formed in the initial collapse, it will contain some material that has already been processed through stars and the disc stars will be more metal-rich than the old stars formed during the collapse. In our own Galaxy, these old Population II stars are seen in the globular clusters and in the halo stars that cross the disc at high inclination and high relative speed – the so-called 'high-velocity stars'. The fact that even these stars show a significant heavy-element abundance shows again that there must have been an even earlier generation of stars that produced the first heavy elements and are now long dead. The disc stars, with a range of age and metal abundance, are the Population I stars seen only in spirals.

This general picture seems to account for the appearance of our Galaxy and, by implication, all spirals and perhaps also the Magellanic irregulars. They differ from ellipticals simply in having retained more gas after the initial collapse. But is it really true that star formation was much less efficient in spiral galaxies, leaving gas behind to form the disc? There is no definite answer to this question, but there is another explanation for the gas in spirals, which now seems more plausible. There is increasing evidence in disc galaxies that gas is still falling onto the disc from outside the galaxy. This suggests that spirals differ from ellipticals in their external environment and happen to have formed in regions where they are able to accrete a lot of gas after the first stars have formed. In this picture, an elliptical galaxy would be the 'natural' state, but subsequent accretion of gas could produce a spiral disc, which did not all form at a particular time but grew gradually and is still growing today. This would be consistent both with the higher metallicity in the central bulge of disc galaxies and with the general similarity between bulges and ellipticals.

I have said nothing in this discussion about the role of dark matter. It is now known from virial theorem estimates of the mass-to-light ratios that giant galaxies are surrounded by a roughly spherical distribution of dark matter which dominates the total mass. If this is also true for most other galaxies, as is strongly suggested for disc galaxies by their flat rotation curves (Section 11.2.5), then the differences between Hubble types may really be quite minor, with the prominent discs and spiral arms just the spinning froth on the surface of the cup. The process of galaxy formation may also have been rather different in detail, because the dominant contribution to the gravitational field would have been from the dark matter rather than from the material that formed into stars. In fact, most present-day studies of galaxy formation start by considering the formation of large-scale structure in the dark matter. The seeds of that structure were grown during the inflationary era, and the pattern of initial fluctuations has now been detected in the microwave background radiation (Section 15.3.2.3). After the dark matter has formed into large concentrations and voids, the small fraction of associated baryonic matter forms luminous galaxies within massive dark halos, which themselves continue to evolve.

My discussion has also assumed that a present-day galaxy formed from a single large cloud of gas. Evidence from clusters of galaxies shows that galaxy mergers may be common, and it is possible that such mergers happen also during the process of galaxy formation, with many small gas clouds forming stars and then merging to form larger galaxies. Observations of large-scale structure are certainly best explained by the hierarchical clustering of smaller masses (predominantly made of dark matter), which were the first objects to form by collapse, and this process may extend to the formation of luminous galaxies. However, the details of galaxy formation are still quite uncertain, and the picture that I have presented here is quite likely to be substantially altered before a consensus is reached.

12.5 Summary

The Milky Way is the visible evidence that our Galaxy is a spiral, with the Sun in the mid-plane. Once the importance of absorption by gas and dust was realized, it became clear that the Sun is also well out from the centre, some 8–10 kpc from the nucleus. Stars dominate the visible mass of the Galaxy, and the stellar disc extends radially to at least 15 kpc from the centre and vertically to about 1 kpc above the plane. The gas and dust, which form some 10% of the visible mass, but occupy most of the volume, lie in a much thinner disc. Outside the disc, there is a spherical halo whose most prominent visible members are the globular clusters. The halo also contains dark matter, whose nature is unknown but whose mass almost certainly dominates the total mass of the Galaxy.

Within the disc, bright stars are concentrated into the spiral arms, while gas is in the arms and in a broad ring, between radii of about 4 and 8 kpc. Much of the gas in the inner parts of the ring is molecular hydrogen and carbon monoxide; further out, and in the spiral arms, most of the gas is neutral atomic hydrogen.

The dust in the galactic disc absorbs much of the visible radiation from distant objects, and causes their light to be reddened. Correction for this reddening effect can be made by using the two-colour diagram. The dust grains also cause polarization of starlight by being aligned with the galactic magnetic field. It is likely that the grains are mainly made of C, N and O compounds.

The gas in the Galaxy occurs in many forms. Clouds of neutral gas, at temperatures of 70 K, are surrounded by warmer (5000 K), partially ionized material in approximate pressure balance with the clouds. Much more massive clouds of cold (10 K) molecular gas are the source of a rich spectrum of molecular emission lines from the infrared to the radio, although they are completely dark optically. The most dramatic clouds in the optical are the hot (10^4 K), fully ionized HII regions found around hot stars, some of which may be newly formed. Between the clouds is the warm intercloud medium and a much hotter 'coronal' component (10^6 K) which provides evidence of violent events in the interstellar medium. The very hot gas arises principally from the expansion of HII regions and of supernova remnants. These form hot, ionized bubbles of gas, which merge and interact to form a complicated network of tunnels throughout the galactic disc.

In general, an observational study of the evolution of galaxies is extremely difficult, since all young galaxies are very distant. However, in our own Galaxy we can obtain clues to its past history by looking at the chemical compositions of stars of different ages and at the motions of stars. There seems to have been an initial burst of star formation while the Galaxy was forming and a slower rate ever since. The remnant of that initial burst can still be seen in the stars of the halo, whose roughly spherical structure reflects the dark matter distribution. The disc, which contains stars with a wide range of age, was probably formed by the gradual accretion of gas from the intergalactic medium. The differences between spiral and elliptical galaxies may be mainly a result of different environments within dominant dark matter halos. There are recent suggestions that galaxies form by the merger of many small gas clouds rather than by the collapse of a single large one.

Exercises

12.1 The Sun is at a distance of about 10 kpc from the Galactic Centre and is moving in a roughly circular orbit round the centre at about $250 \, \text{km s}^{-1}$. Assuming that most of the mass of the Galaxy is concentrated in a spherical bulge centred on the Galactic Centre, estimate this mass. [Take $G = 6.7 \times 10^{-11} \, \text{N m}^2 \, \text{kg}^{-2}$.]

12.2 A supernova explodes in the mid-plane of the Galaxy. Assuming a mean expansion rate of $1000 \, \text{km s}^{-1}$, how long will it take for the expanding shell of the supernova to break out of the galactic gas disc (assumed to extend to a height of 100 pc above the plane)?

12.3 Use the values given in Section 12.3 for the number density and temperature of the various components of the interstellar gas to verify that (a) the neutral clouds are approximately in pressure balance with the warm intercloud medium (WIM), (b) the gas in HII regions has a much higher pressure than the gas in the WIM, and (c) the coronal gas has an intermediate pressure.

12.4 A main-sequence O star in a young star cluster is reddened by dust absorption between the Earth and the cluster. Its observed colour and visual magnitude are $B - V = 0.00$ and $V = 4.84$, and it lies on the line $U - B = 0.72 \, (B - V) - 0.92$ in the two-colour diagram. The main-sequence two-colour relation for O stars is $U - B = 5 \, (B - V) + 0.28$. Find the unreddened colours $(U - B)_0, (B - V)_0$ of the star.

12.5 If the absorption in the visual, A_V, is related to the reddening, $E(B - V) \equiv (B - V) - (B - V)_0$, by $A_V / E(B - V) = 3$, and the absolute visual magnitude of an unreddened O star is -6, find the distance of the cluster in exercise 12.4.

13

The distribution of galaxies

13.1 Clusters of galaxies

When astronomers look at how galaxies are distributed over the sky, the most obvious feature is the 'zone of avoidance': there are essentially no galaxies within about 15° of the galactic plane (i.e. with galactic latitude less than 15°). This is simply another measure of the amount of absorption by dust in the galactic plane, but a band 30° wide around the equator amounts to about a quarter of the sky, so this is a significant gap in our knowledge of the Universe. Unfortunately, it is not easy to fill in this gap even at other wavelengths – there is no dust scattering in the infrared or radio but there is significant emission from the Galaxy at these wavelengths (see Chapter 6). However, the IRAS survey (Section 5.1) has certainly discovered some galaxies in the plane of the Milky Way and there is no reason to suppose that the distribution of galaxies in the zone of avoidance is different from what is observed optically.

A much more fundamental question is how galaxies are distributed in space. To answer it, we need to know the distances of galaxies. You will see in Chapter 14 how to estimate distances; for the moment I will simply assume that it can be done. Note that there are uncertainties of up to a factor of 2 at large distances (these uncertainties are reflected in the range of distance which I quote for a given galaxy or cluster), but that these uncertainties refer to the absolute distance of a particular galaxy. For the purpose of discussing clustering, we are mainly interested in relative distances, which can be determined much more accurately. The scale of the Universe is still uncertain, but its structure can be found independently of its scale.

Most galaxies seem to be members of groups or clusters. Our own Galaxy is a member of a small group of 20–30 galaxies, known as the *Local Group*. Like many clusters, this one is dominated by two galaxies: our Galaxy and the Great Nebula in Andromeda (M31), another large spiral at a distance of about 750 kpc. Each has a number of companions. For our own Galaxy, the nearest and most important satellites are the Large and Small Magellanic Clouds (LMC and SMC), clearly visible to the naked eye (in a dark sky) as fuzzy patches several degrees across near the southern celestial pole and first reported to Europe by Magellan. They are both Irregular Type I galaxies and are at distances of some 50–60 kpc.

The Andromeda galaxy has three main satellites; one is M33, a small spiral (and the only other spiral known in the Local Group), and the other two are small ellipticals, M32 and NGC 205. There are several other dwarf galaxies in the Group which appear to be associated with one or other of the dominant galaxies, and others which are well away from either large spiral but are still members of the Group. The known members are almost all dwarf ellipticals or dwarf irregulars, and there are probably more still to be found.

There are many similar nearby groups, within 15–30 Mpc, such as the Ursa Major group that contains the famous exploding galaxy M82 and its larger companion M81. These nearby groups are not themselves distributed randomly on the sky, but are concentrated in a belt, as was first recognized by Gerard de Vaucouleurs in the 1950s. They are now known to form part of the *Local Supercluster*. This is a large, flattened system, 20–30 Mpc in radius and 15–20 Mpc thick, centred on the large Virgo cluster. The Local Group lies roughly in the mid-plane of the system, 15–20 Mpc from the Virgo cluster, so again we are not at the centre of things.

The Virgo cluster of galaxies, which contains some 2500 known members, is the nearest cluster and dominates the Local Supercluster. However, it is not a particularly massive cluster – many members are known simply because the cluster is close to us. One of the most massive known clusters is the Coma cluster, which is outside the Supercluster at a distance of 70–140 Mpc (Fig. 13.1). Although it is some seven times further away than Virgo, there are still about 1000 known members. Most galaxies seem to be in clusters, from small groups such as our Local Group to massive clusters such as Coma. In fact, there seems to be clustering on all scales, as you will see in more detail in the next section.

I have not yet defined carefully what I mean by a cluster. When can a collection of galaxies be called a cluster? This is actually a very difficult question, with no simple answer, since clusters do not have sharp edges and are really just density maxima in a more-or-less continuous distribution of galaxies. Observationally, they can be defined by picking out regions where more than a given number of galaxies are concentrated within a given radius;

Fig. 13.1. The central region of the Coma cluster of galaxies. It is dominated by a giant elliptical galaxy. (Photograph from the Mount Wilson and Palomar Observatories.)

for example, in the 1950s George Abell defined a cluster as a region where there were more than 50 bright galaxies within a circle of radius 3 Mpc ('bright' meant that they were less than two magnitudes fainter than the third brightest galaxy in the cluster). Theoretically, a cluster can be defined as a group of galaxies in 'virial equilibrium'; that is, they satisfy the virial theorem (equation (11.1)) or, roughly speaking, the gravitational potential energy of the cluster is comparable to the kinetic energy of random motions of the galaxies. The observational definition is complicated by possible contamination by non-members at different distances which just happen to be in the line of sight, and by the necessarily arbitrary choice of numbers such as 50 and 3 Mpc. The theoretical definition is complicated by the fact that many clusters are not isolated but contain infalling matter (intergalactic gas, and other small groups and clusters). Clusters also differ considerably in membership, from a few tens of galaxies in small groups such as the

Local Group to many thousands in large clusters such as Coma, and in composition: some clusters are spiral-rich, while others are dominated by ellipticals. Generally, the more massive, dense ('rich') clusters contain mostly ellipticals and S0 galaxies, while spirals are found in low-density clusters, in the outer parts of rich clusters and in the general field (that is, between clusters). A typical spiral-rich cluster is shown in Fig. 13.2.

A major problem with any attempt to define a cluster is that most definitions concentrate on the optically observed galaxies, which dominate the luminosity of the cluster. X-ray observations reveal large quantities of very hot ($\sim 10^8$ K) gas, which pervades the whole cluster and has a total mass comparable to that of the galaxies (Fig. 6.6). Because it is so hot, the thermal pressure of the gas is sufficient to support it against gravity, and it is in hydrostatic equilibrium (cf. Section 10.2.1). Even more importantly, application of the virial theorem shows that the galaxies are moving around much faster than can be accounted for by the visible mass in galaxies – so if the cluster is really in virial equilibrium the total mass (using equation (11.3)), which amounts to some 10^{14}–10^{15} M$_\odot$ for a rich cluster, is many times that of the visible galaxies. Some of this 'dark matter' is probably hidden in individual galaxies (Section 11.2.5), but there is certainly more that is not: something like 80% of the mass of a cluster is now believed to be in the form of dark matter, whose form is not known, and only about 10% each is in the form of galaxies and hot gas.

Many rich clusters are dominated by one or two supergiant elliptical galaxies at or near the centre (e.g. Fig. 13.1). These galaxies, which often have an extensive outer envelope, are much larger than any other galaxy in the cluster. It is thought that they have probably been produced by mergers of smaller galaxies. Typical galaxy separations in clusters are of the order of 50–500 kpc, not much larger than the typical galaxy size of 30 kpc (cf. Table 11.2), and interactions between galaxies should be quite common. If a large galaxy once begins to swallow smaller galaxies, its gravitational potential well will deepen and it will be more likely to capture further galaxies. This 'cannibalism' is most efficient in small groups, and the large galaxies may form first in small groups, which then merge to form a rich cluster.

Even if galaxies do not actually merge, their structures may be profoundly changed by being in a cluster. For example, they may have their gas stripped out by interaction with the hot gas in the cluster. It is thought that spirals may become lenticulars by this process; certainly, rich clusters are observed to have few spirals, while lenticulars are quite common. In a few clusters, the proportion of spirals is observed to be larger in the outer regions, where the gas density is lower.

Fig. 13.2. A spiral-rich cluster of galaxies in the constellation Hercules. There is a great variety of galaxy types, but no dominant central galaxy. (Photograph from the Hale Observatories.)

If gas is stripped out of galaxies, it will accumulate in the general gravitational potential well of the cluster, joining the existing hot intra-cluster
gas. Detection of X-ray emission lines of highly ionized iron shows that the
chemical composition of this gas is not primordial, suggesting that at least
some of it has indeed been processed through stars in galaxies rather than
being left over from the formation of the cluster.

13.2 Filaments and voids

It is easy to see that nearby galaxies are not distributed randomly, and to
catalogue the many clusters of galaxies and even a small number of superclusters. To what extent does this clustering continue to larger scales? This
is a much more difficult question, to which there is still no complete answer.

On very large scales, represented by the distribution of radio galaxies
(Chapter 11), which are believed to be very distant, there seems to be no
significant clustering and the Universe appears to be smooth. This is borne
out by measurements of the microwave background radiation, which, as you
will see in Chapter 15, comes from distances of the order of the size of the
Universe. Thus cosmologists are justified in assuming, for the purpose of
making models, that the Universe is uniform when averaged over sufficiently
large scales.

How can astronomers study structure on scales intermediate between the
size of the Universe and the sizes of superclusters? A plot of the galaxies in
published catalogues is quite revealing (Fig. 13.3). The distribution does not
look random and the eye seems to be able to pick out lines and chains of
galaxies. Are these real? The eye is notoriously good at picking out linear
structures, even when none exists, as in the famous case of the 'canals' of
Mars, and it is necessary to make careful statistical tests of the reality of any
features which are claimed.

One global statistical test is to ask what the correlation is between the
positions of neighbouring galaxies. More precisely, if the galaxy number
density is N and the distribution is random, then the expected number of
galaxies in an element of volume d^3r at distance r from a chosen galaxy is
just $N \, d^3r$. We can then define a correlation function $\xi(r)$ by the equation

$$N_0(r) = N \, [1 + \xi(r)] \, d^3r, \tag{13.1}$$

where N_0 is the observed density at distance r. If $\xi = 0$, the distribution of
galaxies is random, and a non-zero ξ denotes a departure from randomness.
If $\xi > 0$, there is an excess of galaxies over what is expected for a random
distribution, and there is clustering.

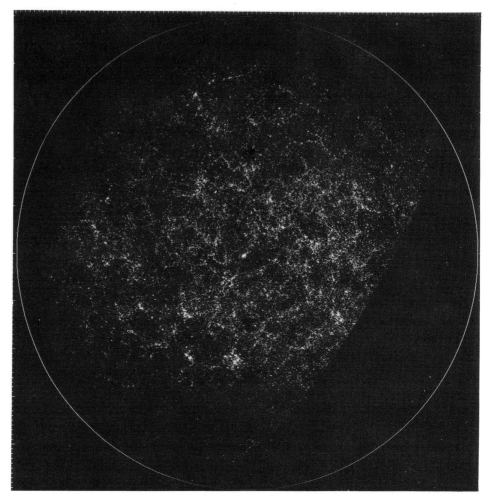

Fig. 13.3. The Lick catalogue of one million galaxies, represented as a map of the sky. The sharp edge to the distribution shows where the survey is cut off by obscuration in the plane of the Galaxy. (Reprinted by permission from *The Astronomical Journal*, 1977.)

To measure ξ directly requires distance measurements, which are difficult to obtain for large numbers of galaxies. First estimates of ξ were obtained by measuring the equivalent angular distribution function, based on angular separation on the sky, and modelling its relationship to the spatial distribution. Now it is becoming possible, with the use of fibre-optic techniques (Section 3.4.2), to measure the radial velocities of many galaxies simultaneously. If it is assumed that the recession velocity deduced from the redshift of the spectral lines is proportional to distance (Hubble's law – Section 15.1),

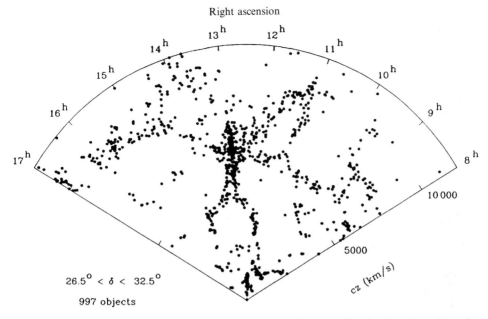

Fig. 13.4. Filaments and voids in the galaxy distribution. The body of the 'dancing man', and his arms and legs, represent long chains of galaxies, and the spaces around the man are largely empty of galaxies. This is a two-dimensional slice through the local neighbourhood, with distance measured radially (in $km\,s^{-1}$, assuming the Hubble relation) and angle on the sky as the other coordinate. (Reprinted by permission from *The Astrophysical Journal*, 1991.)

then it is possible to measure ξ directly. The results are that ξ can be represented by a power law:

$$\xi(r) = \xi_0 \left(\frac{r}{r_0}\right)^{-1.8}, \qquad (13.2)$$

where ξ_0 is a constant that measures the strength of the clustering, and the *correlation length* r_0 is a measure of the scale of the clustering. The value of ξ_0 is larger for the cluster–cluster correlation than for the galaxy–galaxy correlation, showing that clusters of galaxies are more strongly clustered than are individual galaxies.

Redshift surveys recently undertaken by a few groups have enabled them, for the first time, to plot the three-dimensional distribution of galaxies that is measured on average by ξ. So far, the surveys have only been in small regions of the sky and the distributions are only known on narrow cones or slices in particular directions. The results are already dramatic, revealing not only the filaments that are apparent in two-dimensional maps but also large voids, where there are very few galaxies. The best-known example is shown in Fig. 13.4, where recession velocity is used as a measure of distance; this has a slight distorting effect on the distribution, because the galaxies in a cluster have random motions as well as sharing in the general expansion of the Universe. This has the effect of apparently stretching a cluster out along the line of sight when using velocity as a coordinate, because those with larger (smaller) velocities than average are placed further away (closer) in the diagram. This makes the cluster appear to point towards us – the so-called 'fingers of God' effect.

Another dramatic picture was revealed by a redshift survey of galaxies found by IRAS (Fig. 13.5); detailed study shows that the filaments and voids are connected in a sponge-like structure, in which both the filaments and the voids are continuously connected.

The reality of filaments and voids is still a matter of some dispute, but sophisticated statistical tests suggest that at least some are real. If so, there are profound implications for the process of galaxy formation; standard models have difficulty in producing structure on such large scales without violating the constraints imposed by the observed smoothness of the microwave background.

13.3 Summary

Nearly all galaxies appear to be members of groups or clusters. Our own Galaxy and the Andromeda galaxy dominate the Local Group of about 20–30 galaxies, which is about 1 Mpc in diameter. Many similar groups, and some larger clusters, make up the Local Supercluster, dominated by the central Virgo cluster, and this clustering seems to extend to very large scales.

Large clusters often have supergiant galaxies at the centre, which may have grown by mergers with smaller galaxies. Certainly, clusters are crowded places, where interactions are probably common, distorting galaxy shapes and perhaps stripping gas out of disc galaxies. Many rich clusters show strong X-ray emission from very hot gas which has accumulated in the centre of the cluster.

On the very largest scales, the Universe seems to be quite smooth and uniform. Between the scales of superclusters, and the scale of the observable Universe, the strength of clustering is measured by a variety of statistical tools, such as the correlation function. Observations show that clusters of galaxies are more strongly clustered than individual galaxies. Distance measurements for many galaxies are now available, and three-dimensional

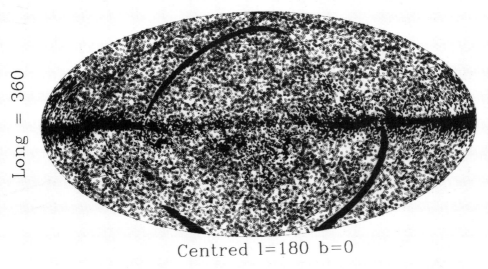

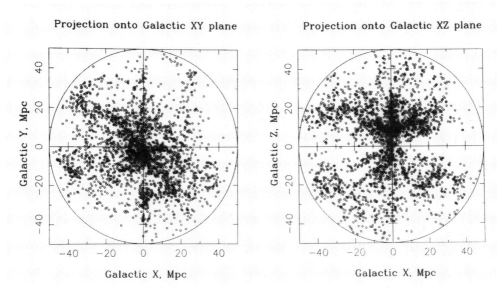

Fig. 13.5. *Top*: The distribution on the sky of galaxies found in the IRAS survey. The map is centred on the direction opposite to the Galactic Centre (which is at the edges of the map). The dark streaks mark gaps in the survey. The spatial distribution found from measuring the redshifts of these galaxies reveals a sponge-like structure on large scales. *Foot*: Projections of the local spatial distribution of the IRAS galaxies onto the galactic plane (left) and an orthogonal plane (right) through the centre of the Galaxy. The right-hand projection shows the incompleteness which results from absorption in the galactic plane. (M. Rowan-Robinson *et al.*, 1991.)

maps of parts of the Universe are becoming available. These have revealed a dramatic structure of interconnected filaments and voids, like a giant sponge.

Exercises

13.1　Show, by integration over the surface of a sphere, that a band around the sky defined by $-15° < b < 15°$ (b = galactic latitude) covers about one-quarter of the sky. Verify that the cruder method of representing the band by a rectangle of angular dimensions $2\pi \times \frac{\pi}{6}$ gives a very similar answer.

13.2　Use the description in the text of the distance and angular size of the Magellanic Clouds to estimate their linear size in kpc.

13.3　The Virgo cluster subtends an angle of about $12°$ in the sky. Assuming its distance to be $15\,\text{Mpc}$, estimate its linear size in Mpc. Assuming that it contains 2500 galaxies, and is spherical in shape, estimate the average distance between galaxies in the cluster in kpc.

14

The distance scale of the Universe

14.1 General principles

Throughout this book I have quoted values for the distances of gas clouds, stars and galaxies. It is now time to explain how these values are found.

The distances of astronomical objects are determined by two basic techniques:

1. *Geometrical methods.* As far as possible, direct measurements are made, using only geometrical arguments which are an extension of methods used on the Earth for measuring the distances of inaccessible objects. Examples are radar and surveying. The great advantage of such methods is that the results are independent of the physical properties of the object. However, because they depend on light-travel effects or on the measurement of angular displacements, the range of these methods is limited.

2. *'Standard candles'.* This method was first used systematically by Sir William Herschel about 200 years ago. If it is possible to identify objects whose intrinsic brightness (or size) is known, then the apparent differences between them can be attributed to their being seen at different distances: if one object appears to be four times fainter than another, it is deduced that it is twice as far away.

   Clearly, this method depends crucially on distant objects being recognizable as the same as nearby ones and does depend on our understanding something about their physical properties. Herschel assumed that *all* stars were of the same intrinsic brightness, an assumption now known to be very far from true. Even if we have correctly selected a true 'standard candle', we can still only obtain *relative* distances unless we can calibrate the method by finding the absolute distance of at least one member of the class – this ultimately

357

depends on using geometrical methods to set up a distance scale. Since this can only be done for nearby objects, the distances of the most remote galaxies depend on a long chain of argument that has many weak links. Remarkably, astronomers now believe that the absolute values of the largest distances in astronomy are known to an uncertainty of only a factor of two. Relative distances are known to a much better accuracy.

We now discuss in more detail some of the important steps in the argument that leads to this conclusion.

14.2 Geometrical arguments for nearby objects

14.2.1 The size of the solar system

The distance scale of the Universe depends ultimately on the accurate deter-mination of terrestrial-scale distances, although it will be a long time before astronomical distances approach the precision of the standard metre. The first method, however, depends on the very precisely known speed of light (now taken to be a defined constant) and on the accurate measurement of time intervals – it is radar, and is used to set up the scale of the solar system.

Since the time of Newton, we have been able to write Kepler's third law in the form

$$\frac{4\pi a^3}{T^2} = G M_\odot, \tag{14.1}$$

where a and T are the semi-major axis and period of a planet's orbit around the Sun (strictly, as you can see by looking at Section 9.2, the right-hand side should contain the sum of the masses of the Sun and the planet, but even for Jupiter this only produces a correction of 0.1% – vital for spacecraft navigation, but unimportant in this context).

Since the product $G M_\odot$ is not known independently, this law provides only a scale model of the solar system. In order to set the scale (and hence to determine the mass of the Sun, if we know G), we must measure at least one distance directly. This used to be done by various surveying techniques, such as measuring the time for Venus to cross the face of the Sun or the distance of an asteroid on a close passage past the Earth from its apparent path against the stars. Since the development of powerful interplanetary radars in the 1960s, by far the most accurate method is to reflect a radar beam off one of the inner planets, Mercury or Venus, measuring the time between transmission and return.

Fig. 14.1 shows the principle behind this method and how it is used to

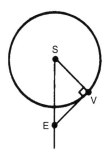

Fig. 14.1. The distance from the Earth to the Sun can be deduced by reflecting a radar beam off the surface of Venus. The distance EV to Venus is then the speed of light times half the time interval between transmission and reception. At a time when the planet is at its maximum angular distance from the Sun, as seen from the Earth, the geometry is particularly easy: the angle EVS is 90° and the angle SEV can be measured. Then the distance ES is just $EV \sec S\hat{E}V$.

find the *astronomical unit* (AU) – this is defined as the semi-major axis of the Earth's orbit around the Sun and has the approximate value 150 million km (Tables 1.1 and 1.2). The AU is the fundamental scale-length in astronomy, and is now known to about 1 part in a million.

14.2.2 Trigonometric parallax

Radar can only be used for objects within the solar system, and even then only if they have solid surfaces that reflect the beam. For more distant objects, we make use of the traditional art of the surveyor – measuring angles. The astronomical unit is used as the baseline and Fig. 14.2 illustrates the principle. If S is the Sun, E the Earth and P a nearby star, and the angles SEP are measured six months apart, then the *parallax angle*, p, can be found. It is formally defined by $p = \sin(a/d)$, where d is the distance of the star and a is the astronomical unit.

In practice, the angle SEP cannot be measured directly, both because the Sun and star cannot be observed simultaneously (that is, the star cannot be observed during the day) and because the Sun is an extended object whose position is difficult to measure to the required accuracy of a fraction of an arcsecond. Instead, the position of the star is measured precisely with respect to fainter stars in the same area of sky, which are assumed to be fainter because they are much farther away. A second measurement six months later will then show an apparent shift in position of the star P with respect to the distant background of faint stars, caused entirely by the change in position of the Earth, and p can again be found.

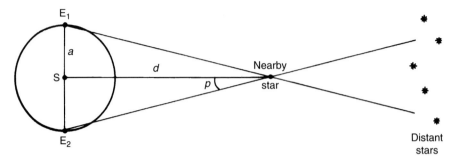

Fig. 14.2. The principle of trigonometric parallax. The Earth, E, moves in a circular orbit around the Sun, S, of radius a (= 1 AU). This motion is reflected in the apparent motion of a nearby star, of distance d, against the background of more distant ones. The maximum apparent displacement occurs for measurements taken six months apart (from E_1 and E_2), and the parallax angle, p, is esssentially half of this displacement.

In practice, additional measurements have to be made to disentangle the effects of the independent movements of the Sun and star through space as a result of their orbital motion around the centre of the Galaxy and to allow for the possibility that one of the faint background stars is in fact an intrinsically faint nearby star with its own apparent motion.

The largest parallax is only 0.76 arcsec (the star Proxima Centauri, a faint companion of the bright southern star α Centauri), so we can approximate the expression for p by:

$$p = \frac{a}{d} \text{ radians}$$

$$= 206\,265\,\frac{a}{d} \text{ arcseconds,} \tag{14.2}$$

since 1 radian = 206 265 arcseconds. This large factor makes the astronomical unit inconvenient for measuring stellar distances, and we introduce the *parsec* by

1 parsec = the distance at which 1 AU subtends an angle of 1 arcsecond.

Then equation (14.2) yields

$$1 \text{ parsec} = 206\,265 \text{ AU} \simeq 3 \times 10^{13} \text{ km} \tag{14.3}$$

and we can write $d = 1/p$, where d is in parsecs and p is in arcseconds.

The measurement of parallax is difficult, and typical errors are about 0.01 arcsec. The method can thus only be used at present for $p > 0.01$ arcsec or $d < 100$ pc. This will change dramatically when the results are available

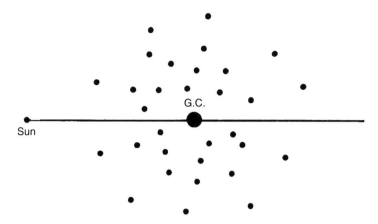

Fig. 14.3. Shapley's model of the distribution of globular clusters about the centre of the Galaxy (G.C.).

from HIPPARCOS (Section 7.5), which is capable of extending parallax measurements to distances of up to 1000 pc.

14.3 The centre of the Galaxy

The distance to the centre of our Galaxy is 8–10 kpc, which is much too far to use trigonometric parallax. The earliest realization that the Sun was not at the centre was due to the American astronomer, Harlow Shapley, who noticed that the globular clusters are heavily concentrated in one half of the sky, centred on the constellation of Sagittarius. He argued that the most likely explanation was that the clusters were symmetrically placed about the Galactic Centre (Fig. 14.3) and that the Sun was well out from the centre.

This immediately provides a way of estimating the distance to the Galactic Centre:

$$distance\ to\ centre\ = mean\ distance\ to\ globulars. \qquad (14.4)$$

So the problem reduces to finding the distances of the globular clusters.

There are several ways of doing this, but the most reliable is to use the fact that they contain RR Lyrae variables. As you saw in Chapter 9, these are giant stars, on the horizontal branch in globular cluster HR diagrams, which show the same regular pulsations as Cepheids. Because they always appear on the horizontal branch, they must all have the same absolute magnitude, M_V: they are 'standard candles'.

The value of M_V can be found by looking at the many RR Lyrae stars in the solar neighbourhood. None is close enough for a direct distance

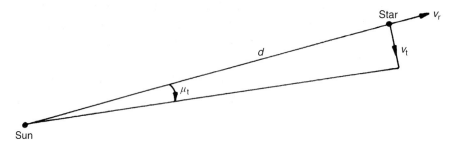

Fig. 14.4. Radial and transverse motions of a star at distance d. The radial velocity, v_r, is the component of the star's motion in the line of sight from the observer to the star. The transverse motion, v_t, is the component perpendicular to the line of sight. The corresponding angular motion, μ_t, is known as the 'proper motion'.

measurement by trigonometric parallax, but we can use the method of *statistical parallax*, based on the motions of the stars. This method depends on one crucial assumption – that the RR Lyraes have random motions with respect to the Sun (more precisely, with respect to a local standard of rest at the Sun, defined by the motion that the Sun would have if it were in a perfectly circular orbit around the centre of the Galaxy). If the motions are indeed random, as various tests suggest, then for a large enough sample of RR Lyraes the mean motion along the line of sight from the Sun should be the same as the mean motion perpendicular to the line of sight. That is (Fig. 14.4):

$$\overline{v_r} \simeq \overline{v_t}. \tag{14.5}$$

The radial component, v_r, can easily be measured using the Doppler effect. For the transverse component, v_t, we actually measure the *angular* velocity, μ_t; this is usually called the 'proper motion' of the star, to distinguish it from the periodic apparent motions arising from the orbital motion of the Earth. In practice, a baseline of many years is often necessary before a proper motion can be detected. If d is the distance of the star, it is clear that

$$v_t = \mu_t \, d \tag{14.6}$$

and equation (14.5) then gives

$$\overline{d} = \overline{v_r}/\overline{\mu_t}. \tag{14.7}$$

We then measure the apparent magnitudes, m_V, of the stars in our sample, take a suitable average (exercise 14.3) and deduce the absolute magnitude. The value usually quoted is $M_V = 0.6$, but there is still an uncertainty of several tenths of a magnitude in this figure.

Measurement of the apparent magnitudes of the RR Lyraes in a particular globular cluster then gives a good estimate of the distance of that cluster. Since most globular clusters are well out of the plane of the Milky Way, they are not much obscured by dust and gas, and Shapley's original estimate of the distance to the Galactic Centre was close to today's figure of between 8 and 9 kpc. (For many years, 10 kpc was used as the standard figure, but more recently there have been several arguments in favour of a smaller distance.)

14.4 Distances to nearby galaxies

14.4.1 The period–luminosity relationship

RR Lyrae variables are extremely useful as distance indicators within our Galaxy, and can even be seen in the nearest external galaxies, the Large and Small Magellanic Clouds. However, they are too faint to be detected in most galaxies and we use the much brighter Cepheid variables instead.

As you saw in Chapter 9, the Cepheids are regularly pulsating stars whose luminosities are directly proportional to their pulsation periods. Because Cepheid periods range up to 100 days, the brightest Cepheids are some 10 000 times brighter than RR Lyraes and can be seen 100 times farther away. The main difficulty in using this method is the calibration of the period–luminosity relationship, that is, finding the constant of proportionality in the relationship $L \propto P$. This is complicated by the fact that there are two types of Cepheid, corresponding to stellar populations I and II, with the Population I Cepheids being about four times brighter at a given period. We therefore need two methods of calibration, since the two types of Cepheid are found in different kinds of star cluster:

1. Population II Cepheids are found in globular clusters, whose distances are known from RR Lyrae stars. It is therefore fairly simple to tie the Population II Cepheid P–L relationship to the absolute magnitude of the RR Lyrae variables; as can be seen in Fig. 9.18, the two relationships are essentially one continuous curve.

2. Population I Cepheids are never found in globular clusters. They are found in open clusters, but such clusters contain no RR Lyrae stars. We therefore need another way of calibrating the relationship for Population I Cepheids. This is carried out in a two-step procedure, the first step again involving geometrical arguments, which I will describe in the next section.

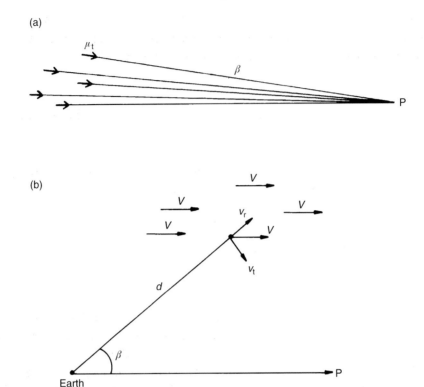

Fig. 14.5. (a) The motion of a star cluster away from (towards) the Earth causes the apparent motions across the sky to converge to (diverge from) a point P at an angle β from the position of the cluster. (b) The direction of the convergent point in the sky is parallel to the motion of the cluster, V, and β is also the angle between the direction to the cluster and the cluster's direction of motion. By measuring β, the radial velocity v_r and the proper motion μ_t, the distance, d, of the cluster can be found.

14.4.2 *Moving cluster parallax*

Stars in a galactic cluster are believed to have been born from the same initial gas cloud and are now moving together through space on parallel tracks. Seen from the Earth, they therefore appear to converge to, or diverge from, a particular point on the sky, just as parallel railway tracks appear to converge in the distance. The point where the tracks appear to meet is known as the *convergent point,* and is in a direction from the Earth that is parallel to the stars' direction of motion (Fig. 14.5). The position of the convergent point is found by measuring the *proper motions* of the stars: their angular motions across the sky, corrected for parallax effects. These measurements give values of the proper motion μ_t and the angular distance from the convergent point,

β, for each star. If we also measure the radial motion of a star, v_r, using the Doppler shift of the lines in its spectrum, we can find the total space motion, V, from the geometrical condition (Fig. 14.5(b))

$$V = v_r / \cos \beta. \tag{14.8}$$

The transverse motion v_t, perpendicular to the line of sight, is then

$$v_t = V \sin \beta = v_r \tan \beta. \tag{14.9}$$

But, if d is the distance of the star, the transverse component of the space motion and the proper motion are related by

$$v_t = \mu_t d \tag{14.10}$$

and we can deduce the distance of the star from

$$d = \frac{v_r \tan \beta}{\mu_t}. \tag{14.11}$$

This can be done for all stars in the cluster with measurable proper motions, enabling both a test of the assumption of common space motion and a measure of the spread in distance of the stars in the cluster.

This method is in principle a very powerful technique for finding distances to star clusters. However, it cannot be applied to globular clusters, which are too distant for proper motions to be measured, and unfortunately only a few open clusters are close enough for the convergent point to be determined at all accurately. The best example – indeed, the only cluster with a really well-determined distance by this method – is the Hyades cluster, near Aldebaran in the constellation of Taurus. Its distance is about 44 pc. Since the Hyades cluster contains no Cepheids, it cannot be used directly to calibrate the period–luminosity relationship.

14.4.3 Main-sequence fitting

Although the moving cluster method can only be applied directly to a few clusters, we can use the distance to the Hyades to find the distance to other clusters by making another assumption, this time with a good theoretical basis. Theoretical models show that all unevolved stars of the same chemical composition have the same locus in the colour–magnitude diagram: the Zero Age Main Sequence (ZAMS). If we assume that all clusters have the same chemical composition (and unfortunately the Hyades may be an exception to this!), then their cluster colour–magnitude diagrams should be the same, at least for the low-mass stars which have not yet begun to evolve.

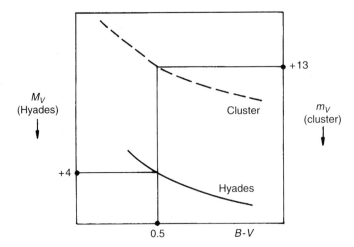

Fig. 14.6. An artificial example of determining the distance to a cluster by fitting its main sequence to that of the Hyades. For the Hyades is plotted the absolute visual magnitude as a function of colour. For the other cluster, only the apparent visual magnitude is known. The *difference* between the apparent and absolute magnitude at a given colour is independent of colour and is the distance modulus of the cluster (in this case it is $13 - 4 = 9$).

For most clusters, we observe only $(m_V, B - V)$, and obtain the *shape* of the main sequence. For the Hyades, we know $(M_V, B - V)$, and we can determine the distance modulus, $m_V - M_V$, for each cluster by fitting its ZAMS to that of the Hyades (Fig. 14.6). The vertical shift needed at a given $B - V$ to fit the two ZAMSs together over their region of overlap is just the distance modulus of the cluster. In practice, because the more massive stars in a cluster are partially evolved, the Hyades ZAMS is extended to a larger range of colour by fitting together the main-sequence relationships for a small number of well-observed clusters which cover a range of age (Fig. 14.7). This standard ZAMS can then be used to determine the distance of almost any cluster.

In particular, there is a small number of open clusters that contain Cepheids, and we can use the main-sequence fitting method to find the distances of these clusters and hence the absolute magnitude M_V for cluster Cepheids. With a correction for the difference between the visual and bolometric luminosity, this allows calibration of the period–luminosity relationship. (At a given luminosity, there is also a smaller dependence of period on the effective temperature, or colour, of the star, so the relationship is strictly a period–luminosity–colour relationship. This introduces a small scatter into the P–L relationship, which has to be allowed for in accurate work.)

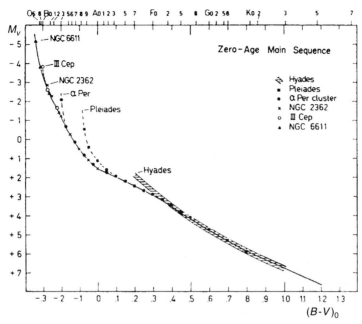

Fig. 14.7. The standard Population I Zero Age Main Sequence is constructed by fitting together the unevolved portions of the main sequences of several well-observed clusters of different ages. (©1963 by The University of Chicago. All rights reserved.)

The main-sequence fitting method can also be used to find the distances of globular clusters. However, because globular clusters certainly have different chemical composition from open clusters (the metal content is lower by a factor of 100 or more), the open cluster ZAMS cannot be used to calibrate the globular cluster distance scale. Instead, a standard Population II ZAMS is constructed from nearby Population II subdwarfs whose distances are known from trigonometric parallax. These stars, which belong to no obvious cluster, but move randomly in the galactic gravitational field, are known as *field subdwarfs*. This method is less widely applicable than the use of RR Lyraes, because the main sequences of many globular clusters are faint and near the limit of detection, but for the nearer, brighter clusters it provides a useful check.

Because Cepheids are very luminous stars, they can be seen at large distances, and this method allows us to determine distances to external galaxies, both in our Local Group and in other nearby groups. The more luminous Population I Cepheids are found only in spiral galaxies, so distance measurements for nearby groups of galaxies are taken from the distances of their spiral members.

14.5 Beyond the help of Cepheids

Although Cepheid variables are very luminous, even they fade into invisibility beyond about 4 or 5 Mpc with present telescopes. This range will certainly be extended as the new generation of telescopes (Section 2.6) becomes active, but even with the 8 m Keck telescope or the newly corrected optics of the Hubble Space Telescope (HST) the Virgo cluster is close to the limit of what is possible. At 15–20 Mpc, that is not yet far enough away for the pure Hubble expansion (Chapter 15) to be determined directly, although the Virgo cluster distance is an important stepping stone whose value is becoming increasingly well determined. There is no single method that is as useful and reliable as Cepheids for larger distances, but there is a large number (about 30) of more or less reliable methods that can be used to estimate distances larger than 4 Mpc, and some do extend well beyond the Virgo cluster. Some of the more reliable ones are:

1. *Brightest stars.* The brightest supergiants in bright Sc galaxies all have $M_V \simeq -8$ to -9. This method is useable up to a range of about 20 Mpc.

2. *Novae and supernovae.* It appears that both novae and supernovae have maximum brightnesses that are the same for all galaxies of the same type. For Sc galaxies: Novae have $M_V(\text{max}) \simeq -8$ [range about 20 Mpc], SN (Type I) have $M_V(\text{max}) \simeq -20$ [range about 200 Mpc].

3. *Supernovae.* A variant of the Baade–Wesselink method discussed in Section 9.4 can be used if the effective temperature, T_{eff}, of the supernova and its expansion velocity, v_r, are known; it is usually applied to Type IIs. Because of the relationship $L = 4\pi R^2 \sigma T_{\text{eff}}^4$ (equation (8.7)), the bolometric luminosity, L, of the supernova can be found if the effective temperature and radius, R, can be estimated. The radius at any time t is given by $R(t) = R_{\text{star}} + \int_0^t v_r \, dt$ where R_{star} is the radius of the star before it exploded as a supernova. In general that is not known. However, the expansion of a supernova shell is so fast that its radius is soon enormously larger than that of the original star and the stellar radius can be ignored. The radius can thus be found simply by integrating the observed radial velocity function or, more simply, by replacing the integral by the product $v_r t$ – a good approximation in the initial explosion, for which the expansion speed is roughly constant. Thus $R(t) \simeq v_r t$. The effective temperature at time t is found from the colour of the supernova at that time, and then the bolometric luminosity and the absolute

bolometric magnitude M_{bol} are known. If the apparent magnitude of the supernova can be measured over a sufficient range of wavelength, the apparent bolometric magnitude m_{bol} can be estimated, and the distance modulus found, yielding the distance d from the equation $m - M = 5\log_{10}(d/10)$. This method has the great advantage of giving distance estimates that are independent of any other method.

It can be used to distances of 100 Mpc or more, provided only that measurable spectra can be obtained. The main uncertainty is that the method tacitly assumes that the spectral lines used to measure the expansion speed are formed in the same part of the shell that produces the continuum radiation used to estimate the effective temperature. Corrections for departures from this assumption are model-dependent and somewhat uncertain. None the less, this is probably the best single method available for estimating large distances. Of course, it requires there to be a supernova explosion in a suitable galaxy at a suitable time!

4. *HI line width.* The width of the 21 cm neutral hydrogen emission line measured in spiral galaxies is well correlated with the luminosity of the galaxy, and is fairly easy to measure (up to about 25 Mpc). The correlation is easy to understand theoretically. The line-width is a measure of the spread of velocities in the galaxy, because gas moving towards the observer has its emission Doppler-shifted to shorter wavelengths while gas moving away has the emission shifted to longer wavelengths – we say that the line is *Doppler-broadened*. As I explained in Section 11.2.3, the range of velocities present is a measure of the mass of the galaxy, so it is not surprising that it is related to the galaxy's luminosity, which arises mainly from stars with a well-defined mass–luminosity relationship. The luminosity–line-width correlation is known as the Tully–Fisher relationship after the astronomers who pioneered its use; it is most effective when used with infrared magnitudes, which are less affected than optical magnitudes by absorption in the galaxy and between us and the galaxy. It is calibrated from nearby galaxies, normally using Cepheids, and is one of the most reliable ways of extending the Cepheid distance scale to more distant galaxies.

A similar correlation between velocity-width and luminosity exists for ellipticals, for the same reason, but the line-width measurement is based on optical data for stars. It represents the range of velocities present in the central regions of the galaxy, known as the *central veloc-*

ity dispersion, σ. The correlation with the absolute blue magnitude of the galaxy is known as the Faber–Jackson relation. An even better correlation exists with the *diameter* of the galaxy, as has been shown by Dressler and others for the Virgo and Coma clusters. The diameter used, D_n, is (like the standard diameter defined in Chapter 11) measured out to a particular surface brightness contour. This D_n–σ relation is interesting, because it is one of the few reliable examples where a standard 'yardstick' is used instead of a standard candle.

5. *Brightest galaxies.* At very large distances, all purely stellar standard candles fail, and we must seek brighter objects, such as entire galaxies. In rich clusters of galaxies, the so-called 'first-ranked', or brightest, galaxy, which is usually a supergiant elliptical galaxy, turns out to have about the same absolute visual magnitude for all clusters of the same degree of richness. This very useful result allows galaxy distances out to about 500 Mpc to be determined; beyond that, although the galaxies can still be seen, we are looking so far back in time that the galaxies are very young, and we can no longer be sure that they have the same properties: estimates of distance become inextricably entangled with estimates of evolutionary effects. First-ranked galaxies have also been used (Section 15.3.1) to establish that the Hubble relation between redshift and distance remains valid out to large distances. There are no nearby clusters of sufficient size, so this method can only be used for distances larger than about 50 Mpc and must be calibrated by one or other of the above methods.

With a combination of these methods and others, astronomers have found moderately reliable distances to a significant number of galaxies and have been able to obtain an idea of the scale of the Universe. But these methods are slow and laborious and yield at most a few thousand reliable distance estimates (known to 20% or better). Our current ideas about the large-scale structure of the Universe depend on a new and much more rapid way of estimating large distances, made possible by the dramatic discovery in the 1920s that the Universe is expanding. In the next chapter I will discuss briefly the evidence for the expansion and how we can use it to find distances. This will lead into a discussion of simple models of the expansion, using Newtonian gravity.

14.6 Summary

Nearby objects can have their distances determined by purely geometrical methods, which are independent of the physical properties of the objects. For more distant objects, the principle of the 'standard candle' is used: the apparent difference in brightness between a nearby object and a distant object is attributed purely to their difference in distance. All 'standard candle' methods must be calibrated using geometrical arguments.

Geometrical methods include radar (within the inner solar system) and trigonometric parallax (for nearby stars). Statistical parallax can be used for slightly more distant stars, and the convergent point method can be used for open clusters, such as the Hyades.

For galactic-scale distances, pulsating variable stars can be used. The constant mean absolute magnitude of RR Lyraes, in globular clusters, is calibrated from field RR Lyraes, using statistical parallax, while the period–luminosity relationship for Cepheids is calibrated from Cepheids in galactic clusters, using main-sequence fitting.

At large distances, Cepheids are undetectable and brighter standard candles are used, such as the brightest stars in galaxies, novae and supernovae at maximum, and the brightest galaxies in clusters. Another method using supernovae is a variant of the Baade–Wesselink method for determining the radii of variable stars, and involves measuring the temperature and expansion speed of the ejected shell. This method is in principle independent of any other distance and is therefore potentially very powerful. The (infrared) Tully–Fisher method can be used for galaxies which are rich in neutral hydrogen, using the correlation between 21 cm linewidth and (infrared) luminosity. For gas-poor galaxies, the Faber–Jackson relation between luminosity and central velocity dispersion, or the alternative D_n–σ relation between linear diameter and velocity dispersion, can be used.

Exercises

14.1   The radii of the orbits of Venus and the Earth about the Sun (assumed circular) are 0.72 and 1.00 AU respectively. Taking 1 AU $= 1.5 \times 10^8$ km and the speed of light as 3×10^5 km s^{-1}, find the time for a radar beam to travel from Earth to Venus and back when the planets are in the configuration shown in Fig. 14.1.

14.2   An RR Lyrae star (absolute visual magnitude +0.6) is observed to have an apparent visual magnitude of +7.2 in a direction in which interstellar reddening causes a visual absorption of +0.3 magnitudes. What is the distance of the star?

14.3 The proper motions μ_i, radial velocities v_i and apparent magnitudes m_i are measured for a large number of RR Lyrae stars, all of which are believed to have the same absolute magnitude M. Write down an expression (in terms of sums of the μ_i and v_i) for the mean distance $\bar{d}$ of the group of stars. Hence show that

$$M = 5 - 5\log_{10}(N\bar{d}) + 5\log_{10}\left(\sum_i 10^{0.2m_i}\right).$$

14.4 Equation (14.11) requires the proper motion to be in 'natural' SI units of radians s^{-1}, and the distance and the radial velocity both to involve the same units of length. More usually, d would be measured in pc, v_r in km s^{-1} and μ_t in arcsec yr^{-1}. Show that in that case

$$d = \frac{v_r \tan \beta}{4.74\mu_t}.$$

14.5 Relative to the local circular motion the Sun is moving at a speed of about 20 km s^{-1}. What would be its proper motion seen by an observer at a distance of 50 pc in a direction orthogonal to the Sun's motion if that observer were stationary in a frame moving with the local circular velocity?

14.6 A Population I Cepheid variable in an external galaxy is observed to have a period of ten days and a mean apparent bolometric magnitude of 22.2. If the period–luminosity relationship for Cepheids is written as

$$L/L_\odot = 400(P/1\ \text{day})$$

and $M_{bol}(\odot) = +4.7$, find the distance of the galaxy in megaparsecs.

14.7 From the measurement of proper motions of stars in a galactic cluster in Coma, the cluster is found to have a convergent point 93° from a star X in the cluster. The proper motion of star X is 0.02 arcsec yr^{-1}, while its radial velocity is 0.4 km s^{-1} towards the observer. Calculate the distance of the cluster in parsecs. [You may either assume 1 radian $= 2.1 \times 10^5$ arcsec, 1 pc $= 3.1 \times 10^{13}$ km, 1 yr $= 3.2 \times 10^7$ s, or use the formula in exercise 14.4.]

14.8 The supernova 1987A in the Large Magellanic Cloud reached a maximum V magnitude on about 1987 May 20. As viewed from the Anglo-Australian Telescope, it was at a large zenith distance and so atmospheric absorption was significant (see Chapter 1). Given the following data for May 20, estimate the absolute visual magnitude of SN 1987A at maximum light:

Apparent V magnitude	Air mass
3.56	3.8
3.38	2.9

You may take the distance modulus of the Large Magellanic Cloud to be 18.5.

15

The Universe

15.1 Expansion of the Universe

Our present understanding of the nature of galaxies started with the work of Hubble in the 1920s. You have already seen in Chapter 11 how he classified galaxies into ellipticals and spirals. In 1924 he finally put an end to the controversy over the size and distance of galaxies by discovering Cepheid variables in the Andromeda galaxy, M31. This showed conclusively that Andromeda was a vast, independent collection of stars, at least as large as our own Milky Way, and disposed of the idea, current only a few years earlier, that it might be simply a star cluster within the Milky Way. The presently accepted value of the distance to Andromeda is about 2 million light years.

This was the first of a large number of galaxy distance measurements made by Hubble, which confirmed the view that galaxies are independent star systems at large distances from us. Hubble also began to study the

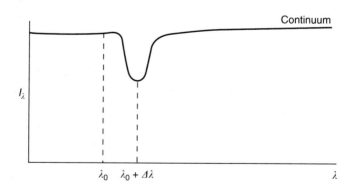

Fig. 15.1. An absorption line in the spectrum of a galaxy. If λ_0 is the rest wavelength of the line, the redshift $\Delta\lambda$ is a measure of the recession speed of the galaxy.

spectra of galaxies. For normal galaxies, the spectra represent the integrated light of thousands of millions of faint stars. They thus resemble stellar spectra, but with some important differences of detail. Fig. 15.1 shows, schematically, an absorption line in the spectrum of a galaxy. The first obvious difference from the spectrum of a single star is that the line is much broader. This is because the stars in the galaxy have large relative motions, with some stars moving towards us, some moving away and some moving across the line of sight. There is therefore a large range of Doppler shifts and the absorption lines from individual stars are spread out over a range of wavelengths. For spiral galaxies, the dominant motion is rotation, and the width of the line is a measure of the rotation speed of the galaxy, while for ellipticals, the linewidth measures the amplitude of the random motions. In both cases they allow an estimate of the galaxy's mass (see Chapter 11).

The motion of the entire galaxy can be found from the shift of the central wavelength of the line from the rest or laboratory wavelength. For stars within our Galaxy, this shift may be either to the blue or to the red, depending on whether the star is moving towards us or away from us. The first measurement for an external galaxy was made by V.M. Slipher in 1913, who found the remarkable result that the Andromeda nebula was approaching the Earth at about $300 \, \mathrm{km \, s^{-1}}$. This was before the nature of these nebulae was agreed, and was another step along the road towards our present view that they are vast systems of stars outside our own Galaxy. Slipher and others continued to measure radial velocities for galaxies, and found that nearly all the shifts were to the red – the only exceptions were nearby galaxies, such as Andromeda. In 1917, shortly after the publication of Einstein's general theory of relativity, W. de Sitter suggested a solution of Einstein's equations that predicted a linear redshift–distance relation for distant objects. Although this was not yet a model for an expanding universe, it did prompt observers to begin looking for evidence for a relation between redshift and distance. In 1924 K. Lundmark produced the first tentative evidence that the redshift increased linearly with the distance of the galaxy, but definitive confirmation did not come until 1929. By that time Hubble had measured accurate distances for 24 galaxies, and his colleague M. Humason had used the 100 inch (2.5 m) telescope at Mt Wilson to measure recession velocities more precisely than was possible with Slipher's 24 inch refractor at the Lowell Observatory. The measurements were planned to test de Sitter's model of the Universe, and the results were quickly accepted.

If the redshift is interpreted as a Doppler shift, this means that all sufficiently distant galaxies are moving away from us at speeds that increase with their distance, and that is the modern interpretation. We can express

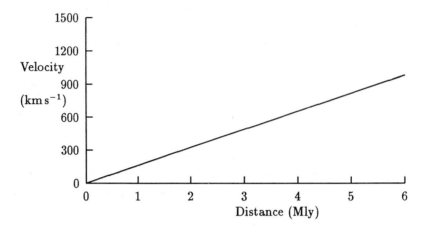

Fig. 15.2. A schematic representation of Hubble's original data. Hubble's values for the distances are used – to fit modern data, the numbers on the horizontal axis need to be increased by a factor of between 5 and 10.

this simply if we write the redshift as

$$z \equiv \frac{\Delta\lambda}{\lambda_0} = \frac{v}{c}, \tag{15.1}$$

where we have used the non-relativistic form of the Doppler shift.† Then Hubble's relationship $z \propto d$ becomes

$$v = H_0\, d, \tag{15.2}$$

where d is the distance to the galaxy and H_0 is known as Hubble's constant (Fig. 15.2). Hubble's relationship is interpreted as arising from an expansion of the whole Universe, since it seems to be a universal phenomenon at large enough distances. It breaks down for nearby galaxies because the random motions of the galaxies within their groups or clusters are comparable to the expansion speed. The expansion becomes dominant only for distances in excess of 10–20 Mpc.

Because of the simple linear relationship between velocity and distance, and the ease with which redshifts can be measured, redshift is now used as a measure of distance for galaxies at sufficiently large distances (≥ 20 Mpc).

† For very large distances, the speeds are so large that we must make relativistic corrections to the relation between redshift and velocity; the *velocity*–distance relation (15.2) remains true (see the next section for a simple derivation) but there is no longer a simple relation between *redshift* and distance – Hubble's relationship $z \propto d$ is valid only for small redshifts.

The application of Hubble's law is essentially the only way of estimating the distances to galaxies beyond about 100 Mpc. The value of the Hubble constant is only known to about a factor of two (Section 15.2), so the distance *scale* is not very well-established for these distant galaxies. However, the redshifts can be measured with an uncertainty of less than 10% so the *relative* distances of different galaxies are well-determined, at least for $z < 0.1$ ($v < 30\,000\,\mathrm{km\,s^{-1}}$, $d < 300$ to 600 Mpc).

At the time of writing, the most distant normal (radio-quiet) galaxy has a redshift of about 3.4, while radio galaxies are known with redshifts of up to 4 and QSOs with redshifts of nearly 5. At these extreme redshifts, relativistic corrections to the redshift–distance relationship are very important, and a factor of 2 in redshift does *not* correspond to a factor of 2 in distance. None the less, we can use the measured redshifts to estimate the distances to about the factor of two uncertainty that corresponds to the uncertainty in the Hubble constant.

Until the 1980s, redshifts had to be measured one at a time: each spectrum of a galaxy required a separate observation, and only a few distances could be found during a night of observing. With the advent of fibre-optic spectrographs (Section 3.4.2), more than a hundred galaxy redshifts can be measured in a single exposure, allowing many thousands of redshifts to be measured in a few years. This hundred-fold increase in observing efficiency has transformed our view of the Universe, allowing the three-dimensional structure of the Universe on a large scale (several hundred Mpc) to be probed for the first time (see Section 13.2).

The ability to measure so many velocities has also revealed large-scale streaming motions superimposed on the local Hubble flow. To probe the structure of these streaming flows, redshift-independent distance measurements for the galaxies involved are needed, and many of the methods discussed in the last chapter play a vital role.

15.2 Simple models of the expansion

How can we understand the expansion of the entire Universe? Before we can begin to tackle that problem we need some way of describing the Universe as a whole. On the very largest scales, all the detailed structure of stars, galaxies and clusters of galaxies is blurred out, as if we were looking at the ocean from a high-flying aircraft. The Universe on that scale appears very smooth, and looks the same (as far as we can tell) for any observer. The simplest models of the Universe take advantage of this apparent lack of structure on very large scales, and assume that the Universe is *homogeneous* and *isotropic*,

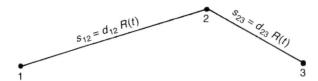

Fig. 15.3. In an isotropic and homogeneous universe, the distance between a pair of galaxy clusters, i and j, say, can be expressed as $s_{ij} = d_{ij}R(t)$, where d_{ij} is a constant for a particular pair and the scale factor $R(t)$ is a universal function of time which represents the expansion of the universe.

that is, we assume that it is of uniform density and looks the same in all directions. Despite the local clustering of galaxies, this assumption seems to be a good description of the Universe as a whole; the best evidence is the microwave background radiation, which we will discuss in Section 15.3.2.3.

If we make this assumption, we can see that Hubble's relationship is a natural result of expansion and we can construct simple expanding models of the Universe. First we need to describe the expansion more carefully. In an isotropic and homogeneous Universe, we can represent all distances as a dimensionless number multiplied by a universal time-dependent scale factor $R(t)$ (Fig. 15.3). Then, as the Universe expands, the *relative* distances between galaxy clusters remain the same, but the *scale* changes.

Hubble's law follows at once from the existence of a universal scale factor. For if we write $s_{ij}(t) = d_{ij}R(t)$ then

$$v_{ij}(t) \equiv \frac{ds_{ij}}{dt} = d_{ij}\frac{dR}{dt} = \left(\frac{\dot{R}}{R}\right)(t)\, s_{ij}(t). \tag{15.3}$$

Thus, at a particular time t, the measured speed of recession of one galaxy from another (which is what we mean by the 'expansion speed' of the Universe) is proportional to their separation at that time. Hubble's law is simply this general relationship applied at the present time, and Hubble's 'constant' is not really a constant at all but just the present value of the ratio $\dot{R}/R$:

$$H_0 = \frac{\dot{R}_0}{R_0}, \tag{15.4}$$

where $R_0 = R(t_0)$ and $t_0 =$ now (the time since the expansion started). Hubble's 'constant' is constant in space, because of the uniform density of our model universe (the assumption of homogeneity), but not in time: it changes as the Universe expands.

A proper treatment of that expansion involves general relativity and is beyond the scope of this book. However, we can see what is involved by

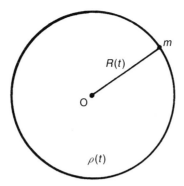

Fig. 15.4. A Newtonian model of an expanding universe. The observer is at the centre, O, of an expanding sphere of uniform density $\rho(t)$. A galaxy of mass m is at a distance $R(t)$ from the observer.

using a Newtonian argument. This is not strictly valid, but leads to the same equations as the rigorous treatment and is much easier to follow.

Consider an isolated galaxy of mass m at distance $R(t)$ from an observer O (Fig. 15.4). Because we are assuming isotropy and homogeneity, the galaxy can be considered to be on the surface of a sphere of uniform density $\rho(t)$ centred on the observer. Newton showed that the gravitational force at radius r within a spherically symmetric distribution of mass is equal to that of a point mass, at the centre of the distribution, whose mass is equal to that contained within the sphere of radius r; the mass outside radius r exerts no net force within it, so long as it is spherically symmetric. Using this result, we can neglect the effect of the Universe outside the radius $R(t)$ and we find the equation of motion of the galaxy to be

$$m \frac{\mathrm{d}^2 R}{\mathrm{d}t^2} = -\frac{GMm}{R^2}, \qquad (15.5)$$

where $M \equiv 4\pi \rho(t) R^3(t)/3$ is the mass within the sphere of radius R. This mass remains constant as the sphere expands: it contains the same galaxies and clusters, whatever the size of the sphere. It is as if the clusters were attached to the surface of an expanding balloon – as the balloon expands the clusters merely get further apart. (It is at this point that I have cheated in using a Newtonian treatment. Can we really assume that the whole Universe is spherically symmetrical about one observer, but yet looks the same to all observers? This assumption cannot in fact be made logically self-consistent within a non-relativistic framework. However, I use it because the answer is the same as that given by relativistic theory.)

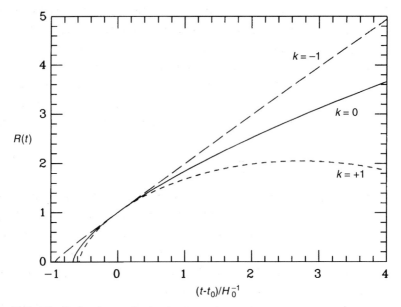

Fig. 15.5. The behaviour of $R(t)$ for $k = -1, 0, +1$ (equation (15.8)). The curves have been normalized in such a way that they have the same value and slope at the present time, t_0. The origin of the time axis has been taken to be the present time, and the time is measured in units of the inverse Hubble constant (see equation (15.10)). The age of a model is the time since $R = 0$; all the models have ages less than H_0^{-1}, with the closed model ($k = +1$) having the smallest age and the open model ($k = -1$) the largest one. (Courtesy D.G. Wands.)

Equation (15.5) can be written as

$$\frac{d^2 R}{dt^2} + \frac{GM}{R^2} = 0. \tag{15.6}$$

Multiplying by $\dot{R}(= dR/dt)$, we can integrate this equation once to obtain

$$\frac{1}{2}\dot{R}^2 - \frac{GM}{R} = \text{constant}, \tag{15.7}$$

which is usually written in the form

$$\frac{\dot{R}^2}{R^2} - \frac{8\pi}{3}G\rho(t) + \frac{kc^2}{R^2} = 0, \tag{15.8}$$

where k is a constant and c is the speed of light. This is the fundamental equation that governs the evolution of the scale factor R as the Universe expands.

The behaviour of $R(t)$ as a function of time depends on the density of the Universe and on the constant k, which can be positive, negative or zero. By convention, the units of R are chosen in such a way that k takes only the values 0, −1 and +1. (This makes the treatment consistent with the relativistic one, in which R is the curvature scale of the universe and k automatically has only these three values.) In all three cases, there is a solution for which R is initially zero: the Universe started from an initial singularity with infinite density. However, the behaviour for large values of t depends qualitatively on k. Fig. 15.5 shows the three solutions for different k. Their nature can be understood by considering equation (15.8) in the form

$$\frac{\dot{R}^2}{R^2} = \frac{8\pi}{3} G \rho(t) - \frac{kc^2}{R^2}. \tag{15.9}$$

Because the density is always positive, this shows that, if $k \leq 0$, $\dot{R}^2 > 0$ always. Thus $\dot{R}$ never changes sign and a universe that starts expanding remains expanding for ever. If $k = 0$, equation (15.9) can be written as $\dot{R}^2 = 2GM/R$, so $\dot{R} \to 0$ as $R \to \infty$. For $k = -1$, $\dot{R}^2 = 2GM/R + c^2$ and the expansion speed never goes to zero. (It is in fact always larger than the speed of light! However, this only means that distant parts of the Universe are separating at speeds faster than light; no signal can propagate *through* the Universe at such a speed, so there is no violation of special relativity.)

If $k = +1$, there is the possibility of $\dot{R}$ changing sign, and in fact $\dot{R} = 0$ when $8\pi G \rho R^2 = 3c^2$, that is, at a finite radius and density. So for $k > 0$ the model universe expands until it reaches some critical density and then starts to contract again, vanishing eventually back into a singularity. Such a universe is finite in extent and we say that it is *closed*. The geometry of space is not the normal Euclidean geometry that we are used to on small scales on the Earth; if we were to measure the area of a large circle of radius r in such a universe, we should find that it was less than πr^2. We say that such a space has a *positive curvature*. The two-dimensional analogue is the surface of a sphere – if we were to measure the area of a large enough circle on the surface of the Earth, we should obtain the same result.

The $k = 0$ universe does have Euclidean geometry. We say that space is *flat*; the two-dimensional analogue is a plane, and the area of a circle is exactly πr^2. In this case, the universe is infinite and expands for ever, but the expansion slows to zero after an infinite time.

The universe with $k = -1$ is also infinite, but never stops expanding. We say that it is an *open* universe, and that the geometry has a negative curvature. The two-dimensional analogue in this case is more difficult to visualize – it is a hyperboloidal surface with a saddle-point. If we were to measure the area of a circle in this geometry, it would be greater than πr^2.

You can see from equation (15.4) that the inverse of Hubble's constant, H_0^{-1}, defines a characteristic expansion time, which would be the age of the

Universe if the Universe had been expanding at a uniform rate since the initial singularity. In all the models we have just described, the expansion has in fact been slowing down (Fig. 15.5) and

$$\text{age of Universe} \equiv t_0 < H_0^{-1}. \qquad (15.10)$$

To put better bounds on the age of the Universe, we must find not only H_0, which gives the rate of change of R at the present time (the *slope* of the graph), but also the deceleration: the rate of change of slope, or the *curvature* of the graph. This is usually written in terms of the so-called *deceleration parameter*:

$$q_0 = -\frac{\ddot{R}_0/R_0}{(\dot{R}_0/R_0)^2} = -\frac{\ddot{R}_0/R_0}{H_0^2}. \qquad (15.11)$$

But $\ddot{R} = -GM/R^2 = -4\pi G\rho R/3$ and we can write

$$q_0 = \frac{4\pi G\rho_0}{3H_0^2}, \qquad (15.12)$$

where ρ_0 is the present mean density of the Universe.

This shows that if the present mean density of the Universe is large enough the expansion will be halted and the Universe will re-collapse under its own gravity. The $k = 0$ model marks the transition between infinite expansion and re-collapse. For that model

$$\frac{\dot{R}^2}{R^2} - \frac{8\pi}{3}G\rho = 0. \qquad (15.13)$$

Evaluating this at t_0 gives

$$H_0^2 = 8\pi G\rho_0/3, \quad \text{or } q_0 = \frac{1}{2}. \qquad (15.14)$$

The density given by equation (15.14) is known as the *critical* or *closure* density, ρ_c, say, and it is common to write the mean density in terms of ρ_c, defining

$$\Omega_0 \equiv 2q_0 = \rho_0/\rho_c, \quad \text{where } \rho_c = 3H_0^2/8\pi G. \qquad (15.15)$$

Can we say anything about the values of all these quantities in the observed Universe? Clearly, Hubble's constant can in principle be measured by observing the redshifts of galaxies of known distance. The methods that I described in the last chapter enable distances to be estimated to within an uncertainty of a factor of two. The uncertainties in the redshifts are small by comparison, so H_0 is also known to within a factor of two:

$$H_0 = (50 - 100)\,\text{km s}^{-1}\,\text{Mpc}^{-1}. \qquad (15.16)$$

The eccentric-looking units arise naturally by dividing the velocity of expansion found from the redshift by the distance. Remembering that $1\,\mathrm{Mpc} \simeq 3 \times 10^{19}\,\mathrm{km}$, this range of values for H_0 yields a characteristic expansion time

$$H_0^{-1} = (1-2) \times 10^{10} \,\mathrm{yr}. \tag{15.17}$$

We can also find, from equation (15.15), that the critical density is

$$\rho_{\mathrm{c}} \simeq (0.5 - 2.0) \times 10^{-26}\,\mathrm{kg\,m^{-3}}. \tag{15.18}$$

Estimates of the actual mean density at the present time are very uncertain. The visible matter in galaxies gives

$$\Omega_{\mathrm{visible}} \simeq 0.01 \text{ to } 0.03. \tag{15.19}$$

If this were all the matter there were then the Universe would be open.

However, this is not the whole story. As you have seen, many galaxies show evidence for dark haloes, so Ω_0 must certainly be larger than $\Omega_{\mathrm{visible}}$. Furthermore, observations of large-scale structure, and of the motions of galaxies relative to the Hubble flow (their so-called 'peculiar motions') are very difficult to understand if the Universe has Ω_0 as small as 0.03. There is also the question of whether all the matter in the Universe is in the form of ordinary matter, made up of protons and neutrons and similar particles, known collectively as *baryons*, or whether it is in the form of some other kind of particle. The current situation is that there seems to be strong evidence for some baryonic dark matter (dead stars?) but that baryons alone do not close the Universe. If some of the present attempts to unify the forces of physics are valid (Grand Unified Theories, or GUTs), then the Universe may have undergone a period of very rapid expansion at a very early stage in its history (inflation) and there are then reasons, which I will discuss briefly in Section 15.3.2.1, for expecting $\Omega_0 = 1$. No-one seriously suggests at the moment that $\Omega_0 > 1$.

We can then put bounds on the age of the Universe by integrating the equation for R in two limits that correspond to an empty universe and to one with critical density.

1. $\Omega_0 = 0, k = -1$. This is the extreme case of an empty universe. Since there is no matter, there should be no decelerating forces and we expect that the universe will simply expand for ever at a constant speed. This is confirmed by equation (15.8), which for $k = -1$ and $\rho = 0$ takes the form

$$\frac{\dot{R}^2}{R^2} = \frac{c^2}{R^2}.$$ (15.20)

This immediately reduces to $\dot{R} = c$, a uniform expansion. The solution which starts from zero radius is then simply $R = ct$ and we find

$$t_0 = \frac{R_0}{c} = \frac{R_0}{\dot{R}_0} = H_0^{-1}.$$ (15.21)

2. $\Omega_0 = 1, k = 0$. This is the opposite extreme of a flat universe with the closure density. The deceleration is the maximum possible without re-collapse occurring. The appropriate equation is (15.13), which can be written in the form

$$\frac{\dot{R}^2}{R^2} = \frac{2GM}{R^3}.$$ (15.22)

Recasting this as $R^{1/2}\dot{R} = \sqrt{(2GM)}$ enables immediate integration, yielding

$$R = (9GM/2)^{1/3}t^{2/3}.$$ (15.23)

Then $\frac{t}{R}\frac{dR}{dt} = \frac{2}{3}$ and if we evaluate this at the present time:

$$t_0 = \frac{2}{3}\left(\frac{R}{\dot{R}}\right)_0 = \frac{2}{3}H_0^{-1}.$$ (15.24)

Thus the range of possible lifetimes for the Universe is not very sensitive to the mean density. Taking into account also the range in the observed values of Hubble's constant (equation (15.17)), the range of lifetimes is given by

$$t_0 \simeq (7 - 20) \times 10^9 \text{ yr.}$$ (15.25)

The lower end of this range is uncomfortably short, since the current best estimate of globular cluster ages is about 16×10^9 yr! Unfortunately, the fashionable value of Ω_0 is 1 rather than 0, which pushes the value towards the lower end of the range. Astronomers would therefore be happier with a small value of Hubble's constant, and indeed recent evidence from supernovae is suggesting a value nearer 50 than 100, and possibly even less than $50 \, \text{km s}^{-1} \, \text{Mpc}^{-1}$. On the other hand, evidence from other methods still implies a value more like 70 or $80 \, \text{km s}^{-1} \, \text{Mpc}^{-1}$.

One possible resolution of this problem is to introduce a non-zero value of the *cosmological constant*. This constant, which was first introduced by Einstein in an attempt to create static models of the Universe , is essentially a constant of integration of the equations of general relativity. It has the effect of altering the law of gravitation at large distances into an attraction or repulsion that is directly proportional to distance: $\ddot{R} = \Lambda R$. Since no such effect is seen on small scales, Λ must be very small, so that the force is significant only on cosmological scales. However, if $\Lambda > 0$ the cosmological term opposes gravity and can have the effect of lengthening the time since $R = 0$. I will not go through the mathematics, except to say that equation (15.8) is altered by the addition of the term $\Lambda/3$ on the right-hand side (some relevant calculations are given in the exercises at the end of this chapter). However, if Λ has just the right positive value, and $k = +1$, it can be shown that there is a set of models, known as the Lemaître models, which have a long period of very slow expansion; the age of the Universe would then be much older than the Hubble time and there would be no conflict with globular cluster ages. Only rather weak upper limits to the value of Λ are currently available from observation, and this remains a possible model for the Universe.

15.3 Observational constraints

When the first cosmological models were constructed in the 1920s and 1930s, the Hubble relationship was essentially the only observational fact in cosmology. This situation has changed dramatically over the last few decades and there are now several types of observational evidence that can be used to put constraints on models. In this section we will discuss just two pieces of evidence, which relate to what are perhaps the two most obvious questions to be asked about the models we have discussed so far. These are:

1. Given that distances are uncertain by a factor of about 2, how well established is the Hubble relationship?
2. Is there any direct evidence that the Universe was ever very dense?

15.3.1 The Hubble relationship

The relationship $v = cz = H_0 d$ can be rewritten, using the distance modulus which we defined earlier (equation (8.4)): $m - M = 5\log_{10}(d/10) = 5\log_{10}(cz/10H_0)$. This gives

$$m = 5\log_{10} z + M - 5\log_{10} H_0 + \text{constant}. \tag{15.26}$$

This expression can be used to *test* the Hubble relationship *without knowing* H_0, so long as the absolute magnitude M is nearly the same for all the objects used in the test. In that case, if the Hubble relationship holds (i.e. if

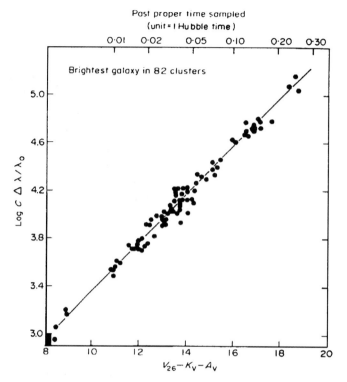

Fig. 15.6. Sandage's test of the Hubble law: the observed values of $z(= \Delta\lambda/\lambda)$ and apparent magnitude cluster tightly about a straight line of slope 5 (see text). (A.R. Sandage.)

H_0 is really a constant) the graph of m against $\log z$ should be a *straight line*, of slope 5, irrespective of the value of H_0 – we only require it to be a constant. The value of H_0 determines only the zero point of the graph, not its slope.

The classic test is that made by Allan Sandage at Mount Palomar, using the first-ranked galaxies in clusters of galaxies, over the redshift range $0.003 \le z \le 0.46$. The results are shown in Fig. 15.6. The correlation between apparent magnitude and $\log z$ is excellent. Unless some malevolent conspiracy is present, whereby variations in absolute magnitude are exactly compensated by variations in H_0, this tight relationship confirms simultaneously that (a) first-ranked galaxies really are good standard candles, with the same absolute magnitude in all clusters; (b) the Hubble relationship holds accurately.

Of course, we cannot expect the Hubble relationship to hold for arbitrarily large distances. When we consider distances that probe a significant fraction

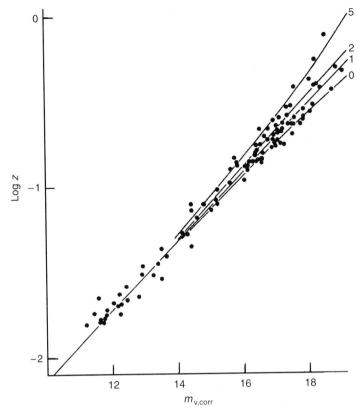

Fig. 15.7. The Hubble diagram for a range of deceleration parameter, q_0. The curves for different q_0 begin to diverge significantly just beyond the redshift of the faintest galaxy available for testing the Hubble relation. (M. Rowan-Robinson.)

of the observable Universe, we are also looking back in time and we can no longer regard Hubble's 'constant' as a constant. This introduces curvature into the relationship at large distances (faint apparent magnitudes), which can in principle be used to determine the deceleration parameter, q_0. Unfortunately, as indicated in Fig. 15.7, the observations are not yet good enough to distinguish between the various curves for different q_0.

There is another problem which makes it difficult even in principle to use this method to determine q_0. At the very large distances involved we are looking far back in time, to a stage near the birth of galaxies. There is no guarantee that galaxies had the same properties when young as they have now. If, as seems likely, galaxies evolve with time, the standard candles may well have had different luminosities in the distant past. This evolutionary effect, whose size and nature are uncertain, would also introduce a curvature into the relationship, and the two effects cannot easily be distinguished. We must turn to other methods, such as direct estimates of the mean density, to determine the deceleration parameter.

Estimating the mean density might seem relatively straightforward, since we are assuming that the Universe is homogeneous and we only need to estimate the mean density locally. Naïvely, all we need to do is to count the local density of galaxies and estimate the masses of typical galaxies. If this is done by estimating the mass of *luminous* material in galaxies, the mean density comes out to be only a few per cent of the critical density, suggesting that the Universe is open and will go on expanding for ever. However, as we saw in the last section, this conclusion is put in doubt by the increasing evidence that the mean density of the Universe is dominated by dark matter, whose amount is very difficult to estimate. The deceleration parameter is thus a very poorly known quantity.

15.3.2 *The hot Big Bang*

In order to answer our second question, 'is there any direct evidence that the Universe was ever very dense?', it is useful to start by giving a brief description of the present best guess at what happened in the very early Universe.

According to the so-called 'Standard Model', the Universe started from a singularity, and at very early times was both very hot and very dense. At times earlier than the 'Planck time', a mere 10^{-43} s after the initial 'Big Bang', we have essentially no understanding of the properties of the matter in the Universe – the gravitational field would have been so strong that quantum effects need to be taken into account in describing it. Quantum gravity is a fashionable subject, but no-one has yet produced a generally accepted way of unifying the theory of gravitation with the fully quantized theories of the other force fields, despite many attempts. What we do believe is that the Universe has been expanding and cooling ever since the initial singularity.

15.3.2.1 *Inflation*

Immediately after the Planck time, conditions in the Universe were so extreme that the strengths of the electromagnetic force, and the strong and weak nuclear forces, were comparable, according to Grand Unified Theories. The physics involved is well beyond the scope of this book, but there is one very important related phenomenon, that of *inflation*, which was suggested in the early 1980s and which has significant observational consequences. It is best to begin with two observational puzzles which the idea of inflation helped to solve (there are others, which I will not discuss here).

One feature of the microwave background (Section 15.3.2.3) is that it is extremely smooth: areas of the sky which are far apart have the same temperature. It is possible to estimate how far apart such regions would have been when the radiation was emitted, and the curious result emerges that regions more than a few degrees apart now would have been out of contact with one another at the time of emission. That is, a light signal would not have had time to get from one to the other since the Big Bang. If we define the *horizon* as the size of region within which communication is possible, then the horizon size is about ct at a time t after the Big Bang, and is much smaller than the size of the whole Universe at that time. So how can regions which have never been in contact agree to have the same temperature? This is known as the *horizon problem*.

Another problem arises from the fact that, whatever the precise value of Ω_0, the Universe is quite close to being flat at the present time. It turns out that this requires it to have been quite amazingly close† to being flat at early times, otherwise the kc^2/R^2 term in equation (15.8) would completely dominate today instead of being just a bit smaller than the $G\rho$ term. This need for a very fine tuning of the initial conditions is known as the *flatness problem* and is one argument for Ω_0 being identically equal to 1.

Both problems can be solved by inflation, which is essentially a short period of extremely rapid expansion in the very early Universe. The horizon problem is solved because in the inflationary picture a single region expanded to fill the entire Universe today – the uniformity we see in the microwave background was imposed within a single tiny region in causal contact. (The implication, of course, is that there are vast regions of the Universe today which are beyond our horizon, and always have been.) The flatness problem is solved because the very rapid expansion in the inflationary era 'stretched out' the curvature, tending to make the kc^2/R^2 term small, and so imposing the fine tuning naturally.

15.3.2.2 Primordial nucleosynthesis

In its initial very hot and very dense phase, the radiation and matter were in thermodynamic equilibrium, that is, the matter and radiation had a common temperature, and particles and their anti-particles were constantly annihilating and being re-created out of the radiation. For a given particle, this continued so long as $kT > mc^2$, where T is the temperature and m

---

† The density at the Planck time would have had to be equal to the critical density at that time to within one part in 10^{57}.

the rest mass of the particle. After one second, the temperature had cooled to 10^{10} K and re-creation of baryons was becoming very slow. Because there was a slight initial asymmetry between the numbers of baryons and anti-baryons, all the anti-baryons disappeared, leaving a small number of protons and neutrons (about 1 for every 10^9 photons). For a few minutes, the temperature and density remained high enough for the protons and neutrons to undergo nuclear fusion reactions, producing deuterium, helium-3, helium-4 and traces of other light elements. There was no time to produce elements with atomic mass greater than 7, because there are no stable elements of masses 5 and 8 and reactions that by-pass these elements take much longer than the few minutes for which conditions were suitable for fusion to occur.

The predictions of the hot Big Bang model for the relative abundances of the light elements produced in this primordial nucleosynthesis are quite precise, and can be tested if any primordial matter is left today. Although no-one has identified any totally unaltered material, it is possible to estimate the primordial abundances by examining, for example, the value of helium abundance by mass, Y, as a function of the abundance of heavier elements, Z. By extrapolating this relationship to $Z = 0$, an estimate of the primordial helium abundance can be found. There is undoubtedly *some* helium that was not produced in stars, and that alone is enough to confirm that the Universe was once at least hot and dense enough for nuclear fusion to have occurred. Detailed observations of metal-poor galaxies yield an estimate of the primordial helium abundance ($Y = 0.229 \pm 0.004$) which is in remarkably good agreement with the prediction from the Big Bang model.

This is not only one of the best pieces of evidence that the Big Bang theory is correct. It also places a strong constraint on the baryon density in the Universe, for agreement with observation requires the baryon density to be in a small range. The current extreme limits on the present-day value of Ω_{baryon} are about 0.011 to 0.11, depending on the value of Hubble's constant, and somewhat on the details of the cosmological model. Probably, in fact, $\Omega_{\text{baryon}} < 0.1$: this is the strongest piece of evidence that baryons alone cannot close the Universe.

15.3.2.3 *The microwave background*

Long after inflation and nucleosynthesis, the material of the Universe remained fully ionized, giving an abundant supply of electrons. The photons that made up the radiation were constantly scattered by these electrons, and this process of *electron scattering* ensured that there was a sharing of energy between the radiation and the electrons; collisions between the electrons and the ions distributed the energy amongst all the material particles, and the radiation remained in equilibrium with the matter. Of course, the ion–electron collisions would occasionally lead to the capture of an electron into a bound orbit, and the formation of a neutral atom of hydrogen if the ion was a proton. But at the very high temperatures in the early Universe these electrons didn't stay bound for very long and the material was essentially fully ionized.

As the Universe expanded, however, its temperature fell, collisions between particles involved less energy and some neutral atoms began to survive for significant lengths of time. Eventually, when the temperature had fallen to about 3000 K, most of the collisions lacked sufficient energy to re-ionize the neutral hydrogen that was the dominant chemical element present and, at a time some half a million years after the Big Bang, the Universe came for the first time to be mostly made up of neutral atoms. This is usually called the *recombination era*, although in fact the electrons and protons had mostly not been combined before (at least for any significant length of time). The effect of this on the radiation was quite dramatic. Recombination drastically reduced the number of free electrons available to scatter the photons, which quite suddenly found themselves free to travel large distances between interactions with matter. To a good approximation, the radiation became de-coupled from the matter and subsequently cooled separately.

Although the matter and radiation were no longer in thermodynamic equilibrium, the radiation retained the same spectral distribution as in equilibrium – the black-body or Planck spectrum, characterized by a single parameter, the temperature, T. As the Universe continued to expand, the energy density of the radiation fell and the corresponding temperature decreased, with $T \propto 1/R$. Since the time of decoupling, the Universe has expanded by a factor of about 1000, so we predict that today a general background of radiation, with a black-body spectrum characterized by a temperature of about 3 K, should be detectable.

A 3 K Planck spectrum has a peak at a wavelength near 1 mm, which was for a long time rather a difficult region of the spectrum to observe, both because the detectors are hard to build and because it is near the atmospheric cut-off for short-wave radio signals. However, it also happens to lie neatly in a gap between two local sources of background radiation, arising in our own Galaxy (Fig. 15.8), and the universal background radiation was eventually discovered in 1965 by two Americans working at Bell Telephone Laboratories, Arno Penzias and Robert Wilson. It had been searched for initially after a prediction (which gave a slightly higher temperature) by George Gamow some 15 years earlier, but it was not found. The accidental discovery by Penzias and Wilson just preceded an independent, and more

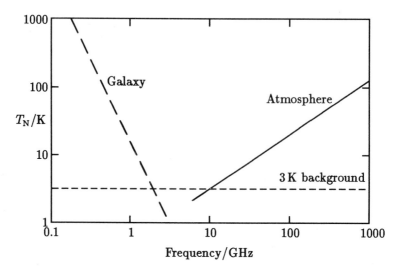

Fig. 15.8. This logarithmic plot of the 3 K microwave background radiation shows that it can be observed in a narrow range of frequency between radiation from our Galaxy and radiation from the Earth's atmosphere. T_N is the noise temperature of the signal (Chapter 4).

precise, prediction by the Princeton physicist Robert Dicke, who immediately recognized the significance of the new discovery.

One of the interesting by-products of this discovery was the fact that the microwave background is so uniform, as I will explain below, that it can be used as an absolute frame of rest. Any motion will cause an apparent rise in the temperature of the background in the direction of motion, and a decrease in the opposite direction, by a physical process analogous to the Doppler effect. In fact, there is a slight effect in all directions, with higher temperatures in the hemisphere towards which the observer is moving, and negative temperatures behind the observer, in a dipole pattern of the form $T = T_0 + T_1 \cos \theta$, with $T_1 \simeq 3 \times 10^{-3}$ K. This dipole anisotropy was one of the first features of the microwave background to be measured, and reveals that the Sun (and the whole Galaxy) is moving relative to the background radiation. Recent measurements by the COBE satellite have resulted in a velocity of the Galaxy of about $550 \, \mathrm{km \, s^{-1}}$ towards the direction with galactic coordinates $l = 260°, b = 30°$.

It is now known that the microwave background radiation has two particularly important properties:

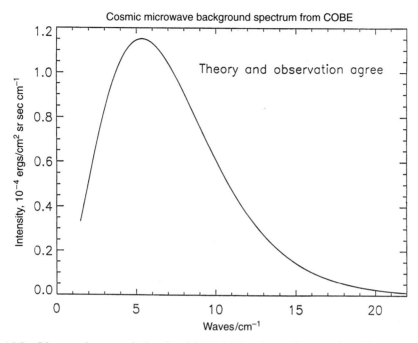

Fig. 15.9. Observations made by the COBE (COsmic Background Explorer) satellite, showing the very exactly black-body nature of the microwave background radiation. The best-fit black-body spectrum has a temperature of $2.726 \pm 0.010\,\mathrm{K}$. The 34 equally spaced data points fit the curve so precisely that it is impossible to distinguish the data from the theoretical curve; the uncertainties are smaller than the width of the plotted curve. (NASA/Goddard Space Flight Center and the COBE Science Working Group.)

1. The spectrum is accurately that of a black body, with a temperature of $T_0 = 2.726 \pm 0.010\,\mathrm{K}$ (Fig. 15.9).

2. After correction for the dipole anisotropy arising from our motion, the intensity is highly isotropic (the same in all directions), to about one part in 10^5.

The first property gives strong support for the idea of a hot Big Bang, and it is very widely accepted that the microwave background photons are indeed messengers from the very early Universe, last scattered by electrons at the era of decoupling. As we saw earlier, this era occurred at $t \simeq 500\,000\,\mathrm{yr}$, that is, about t_0 ago. The photons have therefore travelled a distance of about $ct_0 \simeq 10^{10}$ light years. The second property then tells us that the Universe must be isotropic on these scales, as is required for the simple cosmological models described above.

The COBE data, which made front-page news in many papers when they were published in 1992, suggest that there are slight anisotropies ('ripples') at about the level of one part in 10^5. These ripples in the temperature distribution are also expected, and reflect corresponding slight fluctuations in the density of the Universe at the era of recombination, which are the seeds from which galaxies have grown to produce the large-scale structure observed in the Universe today. The precise size of the fluctuations, combined with observations of the large-scale structure of the Universe (Chapter 13), puts severe constraints on models of galaxy formation, and on the nature of any non-baryonic dark matter.

Before decoupling, the Universe was opaque to electromagnetic radiation, because of the scattering of photons by free electrons. The microwave background therefore represents the limit of our horizon – we can never see further back in time. The only possible way round this fundamental barrier is to look for something other than photons. Neutrinos interact much less strongly with matter than do photons, and they decoupled from it at a much earlier stage. However, they are correspondingly more difficult to recapture, as you have seen in Chapter 5, and there is no immediate prospect of detecting neutrinos from these earlier phases. Our discussion of the microwave background has therefore brought us to the very edge of the observable Universe and is a good place to end this brief introduction to astrophysics.

15.4 Summary

The discovery that the Universe is expanding must rank amongst the most important of this century. It has become part of popular culture and changed profoundly the way that we think about the Universe. For a long time, it was essentially the only observational fact in cosmology, based on a relatively small number of spectra of faint galaxies.

Simple models, assuming that the Universe is isotropic and homogeneous and using purely Newtonian gravity, show that such a universe has two possible behaviours: it may continue to expand for ever, or it may slow to a halt and then re-collapse to a singularity (the Big Crunch). These open and closed models are separated by a unique model, corresponding to normal, 'flat', Euclidean geometry, which expands for ever but slows to zero expansion rate after an infinite time. The distinction between the models can be described either by their curvature (negative, positive or flat) or by their mean density. The flat model has the closure or critical density, while the 'open' model that expands forever has a lower density and the 'closed' model has a higher density. Observations of luminous matter suggest that the real Universe is open, but evidence for dark matter clouds the argument. There are theoretical reasons for believing that the Universe may be close to being flat.

Estimates of the time since the initial Big Bang range from 7 to 20 thousand million years, depending on the model and the current scale of the Universe. Greater ages are possible if the cosmological constant is non-zero.

The Hubble relationship, describing the expansion, is now very well established up to distances well beyond those Hubble was able to measure, although the value of Hubble's constant is still uncertain to about a factor of two.

The major change in cosmology over the last 30 years has been the expansion in observational tests – I mainly discussed the most dramatic, the discovery of a universal microwave background which is a remnant of the original Big Bang, but measurements of the primordial helium abundance also provide a powerful constraint. Other tests include detailed studies of the clustering of galaxies and clusters of galaxies (Chapter 13), which have led to the detection of filaments and voids on the scales of clusters and superclusters and are leading to clues about the formation of galaxies.

Understanding of the very earliest phases of the universe is still largely theoretical, and now builds heavily on new theories of fundamental particles and the unification of the forces of nature, but these ideas take us well beyond the scope of this book.

Exercises

15.1 Consider equation (15.8) with the term $\Lambda/3$ added to the right-hand side (Λ is the cosmological constant). Show that if $\Lambda < 0$ there is a finite value of R, for any value of k, for which $\dot{R} = 0$ (this gives rise to oscillating universe models).

15.2 For equation (15.8) with the cosmological term, with $\Lambda > 0$ and $k \leq 0$, show that the universe always expands and that for large R the growth is exponential, with $R \propto \exp(\sqrt{\Lambda/3}t)$.

15.3 With the cosmological constant, equation (15.6) can be written in the form: $\ddot{R} = -4\pi G\rho_0 R_0^3/3R^2 + \Lambda R/3$, where ρ_0 and R_0 are the present density and scale factor of the Universe. Using equation (15.8) with the cosmological term, and $k = 1$, show that there is a critical (positive) value of Λ, Λ_c, say, for which $\ddot{R} = 0$ and $\dot{R} = 0$ can be satisfied simultaneously (for the Lemaître models, $\Lambda = \Lambda_c(1 + \epsilon)$, where $\epsilon << 1$).

15.4 The temperature, T, of the radiation in the Universe decreases linearly with the scale factor, R, as the Universe expands (i.e.

$T \propto 1/R$), while all lengths increase linearly ($\propto R$). Hence show that the radiation retains its initial black-body shape at all times. (A black body has the Planck radiation spectrum

$$B_\nu = \frac{2h\nu^3}{c^2} \frac{1}{\exp(h\nu/kT) - 1}.)$$

16

Postlude

This book started with the Earth's atmosphere, and has ended with a glimpse of the farthest reaches of the Universe, beyond the wildest dreams of the first humans who looked up at the night sky.

The last half-century has seen more changes in our view of the Universe than the previous half-millenium, and it has not been possible in this book to do justice to more than a few of them. I have tried to concentrate on those things about which astronomers are most certain, and on the main observational techniques needed to establish them. Even so, there are many gaps and I make no claims for a comprehensive coverage. I can only hope to have whetted your appetite – and to have provided enough pointers to other books to enable you to begin to satisfy it.

Appendix 1
Encore: A summary of the Universe

Here is a final round-up of the contents of the Universe, in summary form, starting with what can be seen with the naked eye (Section 1.1), progressing to what is revealed by instruments and then looking at the whole Universe in a series of steps, from the solar system to the farthest galaxies.

A1.1 The naked-eye sky
A1.1.1 A snapshot

On a clear, **dark** night, a snapshot view shows:

- stars;
- some planets (sometimes);
- a comet (occasionally);
- the Milky Way (a faint band across the sky).

Stars and planets can be distinguished because:

- stars twinkle more than planets;
- planets move against the star background (hence the name, from the Greek word *planetes*, wanderers).

The Milky Way is composed of many faint stars, which cannot be resolved with the naked eye.

If the Moon is visible at night, it dominates the night sky. The scattered light makes the sky background so bright that only the brightest stars and planets are visible.

In the daytime sky, the Sun has an even more drastic effect, and only the very brightest objects are visible:

- the Sun;
- the Moon (sometimes);
- Venus (if you know where to look!);
- the sky itself.

The daytime sky is bright because of *scattered sunlight*, just as the moonlit sky is full of scattered moonlight. Because blue light is scattered more strongly than red, the sky is blue (if it is clear: clouds scatter all colours equally, so a cloudy sky is grey).

Thus most naked-eye observations are restricted to the hours of darkness, preferably away from a full Moon. However, solar (and some lunar) observations can be made at more social hours!

A1.1.2 A longer view

The sky is not static, and a longer study reveals changes on many timescales.

Timescale	*Phenomenon*
Seconds	'Shooting stars', or meteors: these are debris of dead comets, and other interplanetary dust particles, burning up in a few seconds as they enter the atmosphere. Typically, one can expect to see about 6 per hour; occasional showers may display 1 per minute or more.
Minutes	Artificial Earth satellites are now an extremely common sight, and cross the sky in 15–20 minutes.
Hours	On this timescale, the rotation of the Earth becomes apparent: the star pattern (by night) or the Sun (by day) appear to move across the sky at about 15° per hour (15° is approximately the width of an outspread hand at arm's length).
Days	The (orbital) motion of the Moon against the star background becomes apparent; it moves about 13° per day from W to E across the sky, in the opposite direction to its more obvious daily motion arising from the Earth's rotation.

Weeks to months The motions of planets, comets and the Sun with respect to the stars become detectable:

- Planets: from about 30° per week (Mercury) to about 1° per month (Saturn);
- Comets: the rate depends on how close they approach the Earth, but is in the same range as that of the bright planets;
- Sun: about 1° per day, or 30° per month.

A1.2 The hidden sky – the growth of modern astronomy

There are only about 6500 'naked-eye stars', even in good conditions. Modern astronomy has progressed beyond the naked-eye astronomy of the ancients in four main ways:

- invention of the telescope (~1600);
- use of the spectroscope (~1800);
- invention of photography (~1850);
- observations outside the optical waveband (radio: ~1930; IR, UV, X-ray, etc: ~1960 onwards).

All the above enabled qualitatively new discoveries to be made and led to great bursts of activity:

Telescope – discovery of sunspots; Jupiter's satellites; Saturn's rings; etc.

Spectroscopy – discovery of spectral lines in Sun and stars: beginning of astrophysics.

Photography – first permanent, objective record; ability to make measurements at leisure, after the observing session; long exposures for faint objects.

Other wavebands – new understanding of familiar objects; interstellar hydrogen; quasars; pulsars; complex molecules; X-ray binaries; etc.

These developments also led to a change in emphasis away from the solar system:

Naked-eye – restricted to night-time and bright objects: largely solar system motions.

Optical telescope – still mainly night-time, but:

- able to resolve planets ⟶ study of surface features;
- able to detect faint objects ⟶ more emphasis on stars and nebulae.

Other wavebands – Sun less dominant source as move away from optical band, so can observe by day as well:

>   **Radio**
>   - Sun a weak source;
>   - observe 24 h per day from Earth's surface;
>   - main emphasis: our Galaxy and external galaxies.
>
>   **γ-ray, X-ray, UV, far-IR**
>   - observable only from *above* atmosphere (balloons, rockets, satellites);
>   - no scattered sunlight in space, so can again observe 24 h per day (but not too close to Sun!);
>   - main emphasis again *outside* solar system.

Interest in solar system studies was restored and maintained when *space exploration* began: the direct exploration of planets by unmanned spacecraft. This has led to the development of *planetary science* – a bridge between astrophysics and geophysics. Because of the very different skills involved, most astronomers work either in planetary science or in stellar and galactic astrophysics (outside the solar system).

There has been another important change of emphasis in astronomy. When planets and stars were just points of light, the main interest was in their *motions*, so their *positions* were of great importance. With the rise of astrophysics, interest in positional astronomy declined. However, this interest, like that in the solar system, has revived recently, for two main reasons:

- to guide spacecraft to other planets, their positions must be known extremely accurately at all times;
- when observing objects in many wavelengths, how do we know when (for example) an X-ray source is the same as a radio source? – only by accurate, absolute measurements of their *positions*, to see whether they coincide.

A1.3 Contents of the Universe

Finally, let us take seven steps out into the Universe:

Nearest 50 AU

The *astronomical unit* (1 AU) is the semi-major axis of the Earth's orbit around the Sun ($\simeq$ mean Earth–Sun distance $\simeq 150 \times 10^6$ km) and is the unit of distance used within the solar system and in binary star systems.

All planets are within $\sim$40 AU of the Sun.

Pioneer 10 was the first spacecraft to go beyond 50 AU (1990).

Nearest 5 parsecs

The *parsec* (1 pc) is the distance at which 1 AU subtends 1 arcsec:
$$1\,\text{pc} \simeq 2 \times 10^5\,\text{AU} \simeq 3 \times 10^{13}\,\text{km} \simeq 3\tfrac{1}{4}\,\text{light years.}$$
Nearest star (at $\sim$ 1.3 pc): Proxima Centauri, currently the nearest member of the triple system α Centauri.

Other stars:
– well spaced out: interstellar space is very empty. Mean distance between stars $\sim$ 1 pc $\sim 4 \times 10^7\,\text{R}_\odot$ ($\Rightarrow$ stars don't collide within lifetime of Galaxy, $\sim 10^{10}$ yr);
– majority are fainter, cooler and of lower mass than the Sun;
– about half are double or multiple systems.

Space between stars:
– neutral and ionized hydrogen (each $\sim 2 \times 10^4$ particles m^{-3}).

Nearest 20 kpc (1 kiloparsec $= 10^3$ pc)

– includes most of Milky Way (our local Galaxy);
– mostly *stars*: single, double, multiple and *clusters*, containing from a few hundred stars (*open* or *galactic* clusters) to 10^5–10^6 stars (*globular* clusters). Total: $\sim 10^{11}$–10^{12} stars;
– gas and dust ($\sim$ 10% by mass);
– high-energy particles (*cosmic rays*) and magnetic field.

The structure of the Milky Way is a flattened, smooth star distribution (disc) plus a low-density, roughly spherical halo. The Sun is in the mid-plane, about 8 kpc from the centre, and *dust* hides the centre in the optical (the centre, in the constellation Sagittarius, is visible in the infrared and radio). Bright stars and gas are concentrated in *spiral arms*, which are poorly defined in our own Galaxy, though prominent in many external galaxies. The gas is clumped into clouds, though there is also a diffuse background of gas. The composition of the gas, which may be ionized, neutral or molecular, is mainly hydrogen (H$^+$, H and H$_2$) but there are also complex (organic and other) molecules, for example, CO, H$_2$O, HC$_{11}$N, C$_{60}$. Many of the gas clouds are concentrated in a *ring* (4–8 kpc from the centre), and at the centre of the Galaxy.

Nearest megaparsec (10^6 pc)

Most galaxies form part of a group or cluster of gravitationally bound galaxies. Our *Local Group* (radius $\sim$ 1 Mpc) contains two dominant galaxies, the Milky Way and Andromeda, both spirals and each of mass $\sim 10^{11}\,M_\odot$. Each has several satellites:

Milky Way – closest satellites are Large and Small Magellanic Clouds (LMC, SMC), at $\sim 55\,\text{kpc}$; both are irregular in shape;

Andromeda – main satellite is a small spiral (M33: M $\equiv$ Messier, Frenchman who constructed a catalogue of 'nebulae' around 1787).

There are about 20 other galaxies, mostly small (10^6–$10^9\,M_\odot$) and either *irregular* or *elliptical* in shape.
Irregulars – lots of gas, dust and young stars;
Ellipticals – almost no gas or dust, old stars, smooth shape.

Note that galaxies are much closer together, compared to their size, than stars:

$$\text{Milky Way to Andromeda} \sim 600\,\text{kpc} \sim 50\,R_{\text{Gal}}.$$

However, like stars, most galaxies are small and faint.

Nearest 20 Mpc
Beyond the Local Group, other small groups, and larger clusters, are clustered together into the *Local Supercluster*, centred on the *Virgo cluster* with about 2500 members, at a distance of some 15 Mpc. The Local Supercluster appears in the sky as a flattened system spread out in a belt roughly at right angles to the plane of the Milky Way.

Nearest 100 Mpc
Many groups and clusters appear, some in other superclusters. We begin to detect:

– large-scale structure of the Universe: galaxies found in *filaments* and *sheets*, at the boundaries of *voids*, which are empty of galaxies, or nearly so;

– a general expansion of the Universe, following Hubble's law

$$v_{\text{expansion}} = H_0 d,$$

where d is the distance and H_0 is Hubble's constant, whose value lies in the range 50–100 $\text{km}\,\text{s}^{-1}\,\text{Mpc}^{-1}$ (uncertain by a factor of two, because distances are uncertain by that factor);

– increasing numbers of *active galaxies* (often in rich clusters).

Nearest 10^4 Mpc ($\simeq$ radius of observable Universe)

Because of expansion of Universe, we use *redshift z* as a measure of distance:

$$1 + z = \frac{\lambda_{observed}}{\lambda_{rest}} \quad \text{or} \quad z = \frac{\Delta\lambda}{\lambda_{rest}}.$$

All redshifts with $z \geq 1$ correspond to distances of the order of the size of the observable Universe, and 'look-back times' (light-travel times) of the order of the age of the Universe ($1–2 \times 10^{10}$ yr). At high redshift we see:

   (i) Galaxies, to $z \sim 4$;
  (ii) Quasars, with $0.1 \leq z \leq 5$; these are probably extreme cases of Active Galactic Nuclei (AGN);
 (iii) Background radiation; this is detectable at almost all wavelengths, from various sources. The strongest is the microwave background with a thermal spectrum, corresponding to a temperature $T \sim 2.7 K$, and believed to be the relic of the Hot Big Bang. If so: it originated at $z \sim 1000$, when radiation decoupled from matter. For $z \geq 1000$, the Universe is opaque, so the microwave background forms a boundary to the observable Universe.

Appendix 2
Solutions to selected exercises

1.1 $215\,R_\odot$ (Table 1)

1.2 $263\,000\,\mathrm{AU}$ (Table 1)

1.3 M increases slowly at first as z increases, then rapidly as $z \to 90°$ (Section 1.3)

1.4 $m_{\mathrm{vis}} = 8.803$ above atmosphere (equation (1.6))

2.1 $m_{\mathrm{lim}} = 14.5$ (equation (2.6))

2.2   Magnification $= 62.5$; angular size of image of 1 arcsec seeing disc $= 62.5$ arcsec – big, fuzzy and faint (Fig. 2.2 and Section 2.4)

2.3 $\lambda = 0.55\,\mu\mathrm{m}\,(5D/1\,\mathrm{m})^{5/6}$; $\Delta\theta_\mathrm{s} < \Delta\theta_\mathrm{d}$ for $\lambda > 11.9\,\mu\mathrm{m}$ for an 8 m telescope and for $\lambda > 2.1\,\mu\mathrm{m}$ for a 1 m telescope (Section 2.3)

2.4   0.0075 arcsec (UV) and 0.025 arcsec (IR) (equation (2.9)); both need space-based telescopes

2.5   Nine 2 m mirrors (Section 2.6)

2.6   A 4.4 m single telescope would be 2.45 times more expensive (Section 2.6)

2.7   370 mm × 370 mm (linear size = focal length × angular size (in radians); Section 2.4)

3.1 5.07×10^9 pixels on the plate; 4×10^4 pixels per square mm (Section 3.1)

3.2 2.23×10^5 pixels on the CCD; 108 square mm for CCD, 1.27×10^5 square mm for Schmidt plate; 4.19×10^6 pixels on large CCD, of total area 944 square mm (Section 3.3.3)

3.3   Gain $= 1.05 \times 10^6$ (equation (3.2))

3.4   Thickness of layer $= 360$ nm; neighbouring maxima at $\lambda_1 = 972$ nm, $\lambda_3 = 324$ nm (equation (3.4)) – cut out using coloured glass filter

3.5   0.027 arcsec (equation (3.8))

3.6   10 050 sec (equation (3.20) with $f = \frac{1}{2}$)

4.1 $V = V_1 \sin(\omega t - \phi_1) + V_2 \sin(\omega t - \phi_2)$ (Section 4.1; for $V_1 = V_2 = A$, use the identity $\sin A + \sin B = 2 \sin \frac{1}{2}(A + B) \cos \frac{1}{2}(A - B)$)

4.2 $\phi = \phi_1 + \phi_0 + 90°$; $\psi = \phi_1 - \phi_0 + 90°$ (use the identity $2 \sin \alpha \sin \beta = -(\cos(\alpha + \beta) - \cos(\alpha - \beta))$) (Section 4.1)

4.3   Take $\lambda = 1$ m and use equation (4.16) in difference form:
        $\Delta\phi = 2\pi \sin \alpha \, \Delta D / \lambda$ (Section 4.2)

4.4   (a) $\frac{(305)^2}{27 \times (25)^2} = 5.5$; (b) VLA maximum separation $= 36$ km and $\frac{36\,000\text{m}}{305\text{m}} = 119$; (i) Arecibo (ii) VLA (Section 4.3)

4.5   About 300 K (equations (4.7), (4.10), (4.15))

5.1 γ-rays: 3×10^9 K; X-rays: 3×10^6 K; UV: 3×10^4 K; optical: 6000 K; IR: 30 K; radio: 0.03 K (equation (5.1))

5.2   Area $= 1.57$ m^2; effective area $= 1.57 \times \sin 1° = 0.027$ m^2 (Section 5.3.3)

7.1   South point is below the pole, north point is on the meridian, etc.

7.2   Plane formulae: $a/\sin A = b/\sin B = c/\sin C$; $a^2 = b^2 + c^2 - 2bc \cos A$

7.3   ZA $=$ ZB $= 90°$ and use equation (7.1) in triangle ZAB

7.4   At midday, altitude of Sun $= 90° -$ latitude of observer $+$ declination of Sun (latitude, declination both negative if south, positive if north) (Figs. 7.1, 7.5, Section 7.4)

7.5 $\phi = 90° - (18° + 23°27') = 48°33'$ (Sun has altitude $-18°$ at midnight) (Section 7.2)

7.6   (a) Pole: whole visible sky circumpolar, but only half the celestial sphere ever visible; (b) Equator: none of the sky circumpolar, but whole celestial sphere visible at some time (only half at any one time, of course!) (Fig. 7.1)

7.7   0.29 km s^{-1} ($\simeq$ 1050 km h^{-1} $\simeq$ 660 m.p.h.) (Section 7.1)

7.8   ZT $= 15^\text{h}12^\text{m}$ (Section 7.3)

7.9   LST 0^h (by definition) (equation (7.3))

7.10 $\delta =$ altitude $- (90° -$ latitude$) = 0°$
        RA $=$ LST $= 5^\text{h}$.
        Altitude for observer at $-30°$: still 60°, but now north of zenith instead of south of it. (Draw the Earth, with the directions to the zenith marked for both observers. Since the star is on the equator, the situation is symmetrical about the equator.) (Equation (7.9), Section 7.2)

7.11 Use cosine rule in $\triangle$ZPX (X = star), with ZX = 90° (equation (7.1))

7.12 0^h (First Quarter implies direction to Moon is at 90° to the direction to the Sun, which is due west at sunset if on the celestial equator)

7.13 RA = LST = 0^h, $\delta = 49° - (90° - 51°) = 10°$ (equation (7.9), exercise 7.4)

7.14 $a = 51.3°$ (cosine formula in $\triangle$ZPX); overhead when LST = RA = 6^h; overhead at local midnight on December 21 (RA$_\odot = 18^h$)

7.15 $\phi = 30°$; LST = RA(star) = 12^h; UT = 0^h on March 21

7.16 $\delta = 14.5°$ (cosine formula in $\triangle$ZPX); RA = 1.77^h (sine formula for HA plus equation (7.9))

7.17 ZT = 11^h25^m (Section 7.3)

7.18 δ(south) = 50° − latitude(north) = 20° (Section 7.4)

7.19 0^h (Full Moon 180° $\equiv 12^h$ away from Sun, therefore on meridian at midnight)

8.1 1000 pc (equation (8.2) for star and Sun, plus equation (8.4))

8.2 $9.3 \times 10^{-12}\,\mathrm{W\,m^{-2}}$ (equations (8.5), (8.6))

8.3 $R = 100\,R_\odot$ (equation (8.7) for star and Sun)

9.1 Use equations (9.9) and (9.10)

9.2 $a = 1\,\mathrm{AU}$; $M_1/M_2 = 51/36$; $M_1 = 5.0\,M_\odot$, $M_2 = 3.5\,M_\odot$ (Section 9.2.2)

9.3 $R_1 = 25\,R_\odot$, $R_2 = 2.5\,R_\odot$ (Section 9.2.3)

9.4 5.75 kpc for nova, 724 000 kpc for supernova (equation (8.4) for visual magnitudes)

9.5 3.7×10^8 yr (equation (9.14))

9.6 $M_1/M_2 = 36/18 = 2$; $M_2 \simeq M_\odot$ (same spectral type as Sun), so $M_1 \simeq 2M_\odot$ (equation (9.9) and Section 8.2)

9.7 $R(t_1) = 1.73 \times 10^7$ km, $R(t_2) = 1.88 \times 10^7$ km (equations (9.16), (9.18))

10.1 Divide equation (10.10) by equation (10.24); integrate with respect to T

10.2 $t(r = 0) = \frac{\pi}{2}\sqrt{\dfrac{R^3}{GM}}$

10.3 —

10.4 —

11.1 Equation (11.5)

11.2 Section 11.2.5

11.3 Section 11.2.5

12.1 2.9×10^{41} kg ($\simeq 1.4 \times 10^{11}$ M$_\odot$) (equation (11.5))

12.2 3.1×10^{12} s $\simeq 10^5$ yr

12.3 $P(\text{neutral}) \simeq 2.1 \times 10^9 k$ (k = Boltzmann's constant) (equation (10.15))
        $P(\text{WIM}) \simeq 2.0 \times 10^9 k \simeq P(\text{neutral})$
        $P(\text{HII}) \simeq 10^{12} k \simeq 500 \times P(\text{neutral})$
        $P(\text{coronal}) \simeq (5.0 \times 10^9 \text{ to } 10^{10}) k \simeq (2.5 \text{ to } 5) \times P(\text{neutral})$ (Section 12.3)

12.4   Intersection of lines gives $(B - V)_0 = -0.028, (U - B)_0 = -1.12$ (Section 12.2) (Note: observed values of $B - V, V$ not needed until exercise 12.5)

12.5   1.4 kpc (Section 12.2 and equation (8.4))

13.1   Integration gives $4\pi \cos 75° = 0.259 \times 4\pi$
        Cruder method gives $0.262 \times 4\pi$ (Section 13.1)

13.2   Several kpc (Section 13.1)

13.3   Linear diameter $\simeq 3$ Mpc; average separation $\simeq$ cluster diameter/ $(2500)^{1/3} \simeq 230$ kpc (Section 13.1)

14.1   694 sec $\simeq 11.6$ min (Section 14.2.1)

14.2   182 pc ($m - M = 5 \log d/10 + A_V$ – cf. equation (8.4))

14.3 $\bar{d} = \sum_i v_i / \sum_i \mu_i$ (equation (14.7))

14.4   —

14.5 0.084 arcsec yr^{-1} (equation (14.10))

14.6   2 Mpc (equation (8.4))

14.7   80.5 pc (equation (14.11))

14.8 $M_V(\text{max}) = -15.7$ (equations (1.7), (8.4))

15.1   Section 15.2

15.2 $\dot{R}^2 > 0$ at all times if $\Lambda > 0, k \leq 0$ (Section 15.2)

15.3   —

15.4 $v \propto \frac{1}{\lambda} \propto \frac{1}{R} \implies \frac{v}{T}$ independent of R and $B_v \propto \frac{1}{R^3}$ – reduces in magnitude but doesn't change frequency dependence (Section 15.3.2.3)

Appendix 3
General bibliography

The books and articles listed by chapters in Appendix 4 are intended as further reading for the particular material of those chapters. There are many other books that give a general introduction to astrophysics, or provide useful background material, and I list a personal selection here.

A3.1 Low-level introductory texts

There is a huge range of books designed for the American first-year college course for liberal arts majors. These are largely non-mathematical, and provide useful background for the present book as well as an excellent introduction to astronomy for the general reader. Examples are:

Exploration of the Universe, by G.O. Abell, D. Morrison and S.C. Wolff, Holt, Rinehart & Winston, 6th edition 1991 (earlier editions remain useful).

Realm of the Universe, by G.O. Abell, D. Morrison and S.C. Wolff, Saunders, 5th edition 1992 (earlier editions remain useful).

Astronomy: Fundamentals and Frontiers, by R. Jastrow and M.H. Thompson, John Wiley, 4th edition 1984; this book contains an appendix on distance measurements.

Universe, by W.J. Kaufmann, Freeman, 3rd edition 1991; this is just one of many books by this prolific and successful author.

Survey of the Universe, by D.H. Menzel, F.L. Whipple and G. de Vaucouleurs, Prentice-Hall Inc. 1970. Although this book is now rather out-of-date, Chapter 5 contains a useful discussion of time and the seasons.

Astronomy: from the Earth to the Universe, by J.M. Pasachoff, Saunders, 4th edition 1993; Pasachoff is another prolific author. Again, earlier editions of this and similar books remain useful.

Universe, by M. Rowan-Robinson, Longman 1990, is (unusually for a book at this level) by an English author, whose earlier popular book *Cosmic*

Landscape (OUP 1979) provided a readable description of the Universe as seen over the whole electromagnetic spectrum. Another UK publication at this level is a collection of essays to mark the centenary of the British Astronomical Association:

Images of the Universe, edited by C. Stott, Cambridge University Press 1991.

The last two books are aimed more at the general reader than at liberal arts students. Two more American college texts are:

The Dynamic Universe, by T.P. Snow, West Publ. Co., 4th edition 1991;

Introductory Astronomy and Astrophysics, by M. Zeilik, S.A. Gregory and E. Smith, Saunders 1992.

At a slightly higher level is:

The Physical Universe: An Introduction to Astronomy, by F.H. Shu, University Science Books/OUP 1982, which is a well-written, qualitative but strongly physics-based introduction to the whole of astronomy. It contains useful problems, and some mathematics.

The Cambridge Atlas of Astronomy, edited by J. Audouze and G. Israel, Cambridge University Press, 2nd edition 1988, is not an atlas at all but a pictorial encyclopædia with explanatory text written by a great variety of people. It is a well-illustrated, large-format 'coffee-table' book, which is packed with useful information.

A3.2 Textbooks

Astrophysics (2 volumes: I *Stars*, II *Interstellar Medium and Galaxies*), by R. Bowers and T. Deeming, Jones and Bartlett 1984. A comprehensive coverage of astronomy at a rather more advanced level than the present book; it is a valuable reference source and contains many problems.

Introduction to Astrophysics: the Stars, by J. Dufay, Newnes 1964. This small book, long out of print, is one of the few to treat absorption in the Earth's atmosphere at a suitably elementary level.

Fundamental Astronomy, by H. Kartunnen, P. Kröger, H. Oja, P. Poutanen and K.J. Donner, Springer-Verlag 1987. This Finnish volume is an unusual but successful example of a book written by a committee. It is a compendium of carefully related chapters by different people who between them cover the major topics in astronomy and astrophysics, at a level suitable for undergraduates. Graduate students new to astronomy would also find it useful.

Astrophysical Techniques, by C.R. Kitchin, Adam Hilger, 2nd edition 1991. This provides a fuller coverage of much of the material in Chapters 1–5 of the present book; chapter 1 gives a particularly good account of non-optical techniques.

Observational Astrophysics, by P. Lena, Springer-Verlag 1988. Despite its name, this is very different from the present book: it is an advanced text on observational techniques for graduate students, but it contains some useful material at a lower level.

High Energy Astrophysics, by M.S. Longair, Cambridge University Press (first published 1981; now being republished in a much expanded three-volume version: Vol.1, *Particles, photons and their detection*, appeared in 1992). This book is much more wide-ranging than its title suggests, and is an excellent introduction to the whole of astronomy, at an advanced under-graduate level.

Astronomy: Principles and Practice, by A.E. Roy and D. Clarke, Adam Hilger, 3rd edition 1988. This covers observational astronomy at an elemen-tary level, giving clear explanations of many topics. It includes much more on spherical astronomy than the present book, and would be particularly useful as supplementary reading for Chapter 7.

Astronomy: Structure of the Universe, by A.E. Roy and D. Clarke, Adam Hilger, 3rd edition 1989. This gives an elementary but wide-ranging account of the material in Chapters 8–15 of the present book; it also includes mate-rial on the solar system.

The New Cosmos, by A. Unsöld and B. Baschek, Springer-Verlag, 4th English edition 1991. This is a terse but clear account of the whole of astron-omy, from the solar system to cosmology and from instruments to biology and the origins of life. It is most suitable for advanced undergraduates or first-year graduate students, but it could be a useful reference for other users.

A unique book, which provides a broad introduction to many of the physical concepts used in astrophysics, and one that is particularly useful for advanced undergraduates or beginning graduate students, is

Astrophysical Concepts, by M. Harwit, 2nd edition, Springer-Verlag 1988.

A3.3 Reference books

Astrophysical Quantities, by C.W. Allen, Athlone Press, 3rd edition 1973. Ever since its first publication in 1955, 'Allen's A.Q.' has been recognized as the 'astronomer's bible'. It is a remarkable compendium of facts and figures, from the names of the constellations to the energy levels of hydrogen, from the Earth's atmosphere to clusters of galaxies and from the values of constants to the properties of variable stars. Unfortunately, the author's call for a successor to provide a 4th edition was unsuccessful, and the 3rd edition is now seriously out of date in certain areas, especially the planetary system and results from space observations. None the less, it remains a uniquely useful collection.

An attempt to provide a theoretician's equivalent to A.Q. is:

Astrophysical Formulae, by K.R. Lang, Springer-Verlag, 2nd edition 1980. This fat volume provides a summary of virtually all the simple formulae that one could ever need, covering many topics under the broad headings of radiation, gases, nuclear astrophysics and high-energy particles, astrometry and cosmology.

In the 1960s and 1970s an ambitious series of books was produced under the general title *Stars and Stellar Systems*; the series editors were G.P Kuiper and B.M. Middlehurst and the books were published by Chicago University Press. The intention was to provide a complete review of the current state of research in astronomy, at the graduate level, with each chapter written by an expert in the field. Nine volumes were planned but only eight appeared (the missing volume, Vol. IV, is *Clusters and Binaries*). The full list is:

Telescopes, edited by G.P. Kuiper and B.M. Middlehurst, 1960.

Astronomical Techniques, edited by W.A. Hiltner, 1962.

Basic Astronomical Data, edited by K.Aa. Strand, 1963.

Galactic Structure, edited by A. Blaauw and M. Schmidt, 1965.

Stellar Atmospheres, edited by J.L. Greenstein, 1960.

Nebulae and Interstellar Matter, edited by B.M. Middlehurst and L.H. Aller, 1968.

Stellar Structure, edited by L.H. Aller and D.B. McLaughlin, 1965.

Galaxies and the Universe, edited by A. Sandage, M. Sandage and J. Kristian, 1975.

A much briefer, modern equivalent is

The Astronomy and Astrophysics Encyclopædia, edited by S.P Maran, Van Nostrand Reinhold/Cambridge University Press 1992, which contains graduate-level entries on a vast range of topics.

A3.4 Practical books

There are many books which guide the amateur astronomer into observing. Three of the best are:

Guide to the Stars: Exploring the Sky with Binoculars, by L. Peltier, Cambridge University Press 1986. Love of the sky radiates from the pages of this book, which introduces the constellation patterns by unique pairs of pictures: on the left-hand page is a conventional 'atlas' picture with a white background, showing the stars as black dots, joined by the outlines of the constellations; on the right-hand page is a drawing of the sky as it really appears, with white stars on a black background, with a horizon of trees and the Milky Way stretching overhead.

Norton's Star Atlas, edited by I. Ridpath, Longman, 18th edition 1989. This is the best-known star atlas for the amateur, though it does not show stars much fainter than the naked-eye limit, and so is not really suitable for serious work. The sky is covered in eight double-spread atlas pages, which alternate with double-page lists of interesting objects on the maps. There is also an extensive text, which is in itself a useful general introduction to astronomy. Earlier editions had a sewn binding, which opened flat and was more convenient for use at the telescope.

The Monthly Sky Guide, by I. Ridpath and W. Tirion, 3rd edition, Cambridge University Press 1993. After a useful quick guide to finding constellations and how they change with the seasons, and to judging angular separations of stars, this is simply a set of maps of the visible sky, one for each month of the year, each with three pages of text on interesting objects for the month, sometimes with more detailed maps.

Perhaps the most useful single accessory for naked-eye observing is a *planisphere*: a circular map of the sky, covered by a rotating plastic sheet which is transparent only in an outline which represents the part of the sky that is visible. One part has dates around the edge, the other part has times, and the planisphere is set by rotating the plastic cover until the required date and time of observation coincide. Planispheres are made for a variety of latitudes, and published (for example) by George Philip Ltd; a useful pack, containing a planisphere, wall chart and booklet, is

Philip's Stargazer. The Complete Astronomy Map and Guide Pack, George Philip 1991.

The same publisher has also brought out a new aid for the beginner: a 60-minute audio tape, consisting of an introduction followed by four narratives describing the sky for each of the seasons, for use while actually observing the sky:

The Sky at Night – a Guided Tour of the Constellations, by P. Moore, George Philip 1989.

Serious observers will need an atlas that has a fainter limit than Norton's. An excellent start can be made with:

Sky Atlas 2000.0, by W. Tirion, Cambridge University Press 1981, which is spiral bound for ease of use. It contains 26 charts, showing 43 000 stars (black on a white background) down to visual magnitude 8.0, and 2500 deep-sky objects (colour-coded). A *Field Edition* (unbound) is available, with white stars on a black background. The atlas can also be obtained in a laminated version suitable for heavy use, each chart being sealed with a plastic film.

There are many guides to interesting objects, and how to observe them; two classics are:

Burnham's Celestial Handbook (three volumes), by R. Burnham Jr, Dover 1978;

Amateur Astronomers' Handbook, by J.B. Sidgwick, 4th edition prepared by J. Muirden, Pelham Books 1979.

The latter book also contains much useful information on building telescopes and on observing techniques. A pair of more recent books, which may also become classics, are:

Visual Astronomy of the Deep Sky, by R.N. Clark, Sky Publishing Corporation/Cambridge University Press 1991;

The Deep Sky Field Guide to Uranometria 2000.0, by M. Cragin, J. Lucyk and B. Rappaport, Willmann-Bell 1993 (*Uranometria 2000.0* is a two-volume sky atlas, published by the Sky Publishing Corporation, charting stars down to at least 9th magnitude, and including some 10 000 nebulae, star clusters and galaxies, on which the *Field Guide* concentrates).

A guide aimed more at undergraduates than at amateurs is

Observational Astronomy, by D.S. Birney, Cambridge University Press 1991. It starts with a useful introduction to coordinate systems, spherical trigonometry and time.

Practical work can also be done in the laboratory. Some of the best resources for this are films made from Schmidt plates, and their use is well-described in

Exercises in Practical Astronomy using Photographs: with Solutions, by M.T. Brück, Adam Hilger 1990, which is particularly useful for teachers planning practical classes for undergraduates.

Those considering making their own telescope will find a useful beginners' guide in

Building and Using an Astronomical Observatory, by P. Doherty, Patrick Stephens 1986.

A3.5 Problems

Astronomical Methods and Calculations, by A. Acker and C. Jaschek, Wiley 1986. This curious book does not fit easily into any category, but it is a valuable source of problems – try some!

A3.6 The solar system

As I explained in the preface, there was no space to include the solar system in this book. For completeness, however, I include here a short list of introductory books:

The New Solar System, edited by J.K. Beatty and A. Chaikin, Sky Publishing Corporation/Cambridge University Press, 3rd edition 1990;

Solar Astrophysics, by P.V. Foukal, Wiley 1990;

Wanderers in Space, by K.R. Lang and C.A. Whitney, Cambridge University Press 1990;

The Traveller's Guide to the Solar System, by R. Miller and W.K. Hartmann, Macmillan 1981;

Exploring Planetary Worlds, by D. Morrison, Freeman (Scientific American Library) 1993;

The Planetary System, by D. Morrison and T. Owen, Addison-Wesley, 1988;

Guide to the Sun, by K.J.H. Phillips, Cambridge University Press 1992;

Orbiting the Sun: Planets and Satellites of the Solar System, by F.L. Whipple, Harvard University Press 1981;

Astrophysics of the Sun, by H. Zirin, Cambridge University Press 1988.

A nice popular account of recent developments in solar physics can be found in

The Variable Sun, by P.V. Foukal, *Scientific American*, **262**, No.2, 26, February 1990.

A3.7 Magazines

Excellent popular accounts of the most recent developments in astronomy can often be found as news items or longer articles in magazines. A selection of the more important magazines is:

Astronomy

Astronomy Now

Mercury

New Scientist

Scientific American

Sky & Telescope.

A few specific articles, mostly from *Scientific American,* are cited under individual chapters.

Appendix 4
Further reading, arranged by chapter

This list gives suggestions of books and articles that supplement the material in particular chapters. More general further reading can be found in Appendix 3; references below to section numbers are to sections in that appendix.

Chapter 1
An elementary description of the sky, with useful charts, can be found in:

The Greenwich Guide to Stargazing, by C. Stott, George Philip/National Maritime Museum 1987.

The account of Bouguer's method is based on the one in chapter 1 of Dufay (1964) (section A3.2).

Chapter 2
See also the list for Chapter 3.

A useful small book is:

The Astronomical Telescope, by B.V. Barlow, Wykeham 1975 (out of print). Alternative, briefer accounts can be found in chapter 1 of Kitchin (1991) and chapter 8 of Longair (1992) (section A3.2). An interesting account of the historical development of the use of telescopes is given in:

The History of the Telescope, by H.C. King, Dover 1979 (originally published by Charles Griffin & Co. Ltd 1955).

For a good account of aberrations in lenses and mirrors, see

Fundamentals of Optics, by F.A. Jenkins and H.E. White, McGraw-Hill, 3rd edition 1957.

Interesting reviews of plans for new telescopes can be found in a lecture, published as:

The revolution in ground-based telescopes, by R. Angel, *Quarterly Journal of the Royal Astronomical Society*, **31**, 141, 1990 and in three magazine articles:

Mirroring the cosmos, by C.S. Powell, *Scientific American*, **265**, No.5, 80, November 1991

Spinning a giant success, by L.J. Robinson, *Sky & Telescope*, July 1992, p.26

Guide to the world's largest telescopes, by R.W. Sinnott and K. Nyren, *Sky & Telescope*, July 1993, p.27.

A more technical book, suitable for advanced undergraduates or graduate students, is

Astronomical Observations: an Optical Perspective, by G.A.H. Walker, Cambridge University Press 1987.

Recent technical papers on active and adaptive optics are:

Active optics: I. A system for optimizing the optical quality and reducing the costs of large telescopes, by R.N. Wilson, F. Franza and L. Noethe, *Journal of Modern Optics*, **34**, 485-509, 1987.

Adaptive optics for astronomy: principles, performance and applications, by J.M. Beckers, *Annual Review of Astronomy and Astrophysics*, **31**, 13, 1993.

Chapter 3

See also the list for Chapter 2.

An introduction to the ideas involved in detectors, especially in photography, can be found in a collection of *Readings from Scientific American*:

Lasers and Light, edited by A.L. Schawlow, Freeman 1969.

A useful text on solid-state physics, which will assist understanding of the discussion on energy bands in conductors and semi-conductors is:

Introduction to Solid State Physics, by C. Kittel, John Wiley Inc., 4th edition 1971.

The application of photography and solid-state electronic devices to astronomy is well described in:

Low Light Level Detectors in Astronomy, by M.J. Eccles, M.E. Sim and K.P. Tritton, Cambridge University Press 1983.

Chapter 4

Elementary accounts of radio astronomy include

The Radio Universe, by J.S. Hey, Pergamon 1971.

Radio Astronomy, by F.G. Smith, Penguin, 3rd edition 1966.

A more historical approach is taken by

The Invisible Universe Revealed: the Story of Radio Astronomy, by G.L. Verschuur, Springer-Verlag 1987.

Clear, but somewhat dated, accounts of the basic ideas, at an intermediate undergraduate level, are given in:

Radio Astronomy, by J.D. Kraus, McGraw-Hill 1966;

Radio Astronomy, by J.L. Pawsey and R.N. Bracewell, Oxford University Press 1954;

Radio Astronomy, by J.L. Steinberg and J. Lequeux, McGraw-Hill 1963. A more recent discussion at the same level can be found in chapter 8 of Longair (1992) (section A3.2).

Much more technical accounts are given in the graduate monographs:

Radio Telescopes, by W.N. Christiansen and J.A. Högbom, Cambridge University Press, 2nd edition 1985;

Astrophysics Part B: Radiotelescopes (*Methods of Experimental Physics, Vol.12B*), by M.L. Meeks, Academic Press 1976;

Tools of Radio Astronomy, by K. Rohlfs, Springer-Verlag 1986 (paperback 1990).

Chapter 5

A good general account of all non-optical techniques can be found in chapter 1 of Kitchin (1991) (section A3.2).

An early, relatively popular account of infrared astronomy is given in:

Infrared: the new astronomy, by D.A. Allen, Keith Reid Ltd 1975.

The high-energy techniques are well covered at an undergraduate level in Longair (1981, 1992) (section A3.2). Other relatively non-technical accounts can be found in:

Cosmic X-ray Astronomy, by D.J. Adams, Adam Hilger Ltd 1980;

X-ray Astronomy, by J.L. Culhane and P.W. Sanford, Faber & Faber 1981;

Gamma-ray Astronomy, by R. Hillier, Oxford University Press 1984. A more technical book is

Gamma-Ray Astronomy, by P.V. Ramana Murthy and A.W. Wolfendale, Cambridge University Press 1986.

Recent accounts of the Compton Gamma Ray Observatory can be found in magazine articles by K. Hurley (*Sky & Telescope*, December 1992, pp.631–636) and C.S. Powell (*Scientific American*, **268**, No.5, pp.69–76, May 1993), who also discusses ROSAT results.

Cosmic rays are described in

Cosmic Rays, by M.W. Friedlander, Harvard University Press 1989.

There have been many popular articles about the solar neutrino problem and about the neutrinos from SN 1987A. An example is:

The solar neutrino problem, by J.N. Bahcall, *Scientific American*, **262**, No.5, 26, May 1990.

A splendid popular but physics-based account of the supernova neutrinos is given in:

End in Fire, by P. Murdin, Cambridge University Press 1990.

At a technical level, the graduate monograph

Neutrino Astrophysics, by J.N. Bahcall, Cambridge University Press 1989, is the most comprehensive and authoritative account.

A popular account of gravitational waves can be found in:

Catching the wave, by R. Ruthen, *Scientific American*, **266**, No.3, 72, March 1992.

Chapter 6

There are now many examples of fine, colour and false-colour pictures of the Universe at many wavelengths. In addition to the illustrations in the introductory texts mentioned in section A3.1

The New Astronomy, by N. Henbest and M. Marten, Cambridge University Press 1983

gives a broad coverage of images taken in all electromagnetic bands, while

Colours of the Stars, by D. Malin and P. Murdin, Cambridge University Press 1984

concentrates on the techniques of colour photography; David Malin is the acknowledged world expert in the application of colour photography to astronomy. More recent books presenting a pictorial account of the Universe include:

Images of the Cosmos, by B.W. Jones, R.J.A. Lambourne and D.A. Rothery, Hodder & Stoughton 1994;

A View of the Universe, by D. Malin, Sky Publishing Corporation 1993.

A special issue of *Scientific American* in April 1990, on the theme 'Exploring Space', included IRAS views of the sky. Some of the first results from the Hubble Space Telescope are reported in

Early results from the Hubble Space Telescope, by E.J. Chaisson, *Scientific American*, **266**, No.6, 18, June 1992.

Chapter 7

The classic text on positional astronomy is:

Textbook on Spherical Astronomy, by W.M. Smart, 6th edition (revised by R.M. Green), Cambridge University Press 1977. It contains far more material than I have attempted to cover here; although it is now rather dated in its approach, chapters 1, 2 and 6 still present the basic ideas clearly. A simpler treatment, with many examples, can be found in chapters 6–8 of Roy and Clarke (1988) (section A3.2).

Other books in the classical tradition are:

Spherical Astronomy, by R.M. Green, Cambridge University Press 1985.

Positional Astronomy, by D. McNally, Muller 1974.

More modern texts use vector methods to derive the important formulae. An example is the graduate monograph:

Vectorial Astrometry, by C.A. Murray, Adam Hilger 1983.

Many useful tables of positions of the Sun, Moon and planets, and of other phenomena that change with time, can be found in the annual publication, *The Astronomical Almanac*, produced jointly by the US Naval Observatory in Washington, DC, and the Royal Greenwich Observatory's Nautical Almanac Office and published by HMSO about six months before the year to which it refers. The definitive account of how to use it, including careful definitions of all the relevant quantities, can be found in:

Explanatory Supplement to the Astronomical Almanac, edited by P.K. Seidelmann, University Science Books, Mill Valley, California, 1992. (The previous edition was published in 1961, so this new edition was much needed!)

A flavour of what it is like to observe with a large telescope can be found in the article:

A night on the Anglo-Australian Telescope, by J.A.J. Whelan, *Quarterly Journal of the Royal Astronomical Society*, **17**, 306, 1976.

Chapter 8

Atoms, Stars and Nebulae, by L.H. Aller, Cambridge University Press 3rd edition 1991
gives an excellent elementary account of the observational properties of stars, as does volume 1 of

Introduction to Stellar Astrophysics (three volumes: 1 *Basic Stellar Observations*, 2 *Stellar Atmospheres*, 3 *Stellar Structure and Evolution*), by E. Böhm-Vitense, Cambridge University Press 1989, 1989, 1992.

Stars, Nebulae and the Interstellar Medium, by C.R. Kitchin, Adam Hilger 1987, covers the properties of stars and their atmospheres, stellar evolution and the interstellar medium, including a chapter on the Sun, and an unusual emphasis on the interaction between stars and their surrounding gas.

Most books on stellar structure contain similar accounts (see references for Chapter 10). More technical accounts of magnitude systems and stellar classification systems can be found in Strand (1963) (section A3.3).

Chapter 9

See also the list for Chapter 10.

A fine popular account of stars is given in

Stars: their Birth, Life and Death, by I.S. Shklovskii, Freeman 1978.
It contains many suggestions for further reading. More recent accounts are given in:
  Stars, by J.B. Kaler, Freeman (Scientific American Library) 1992;
  100 Billion Suns, by R. Kippenhahn, Princeton University Press, paperback edition 1993.
A splendid introduction to star formation is given in
  In Darkness Born, by M. Cohen, Cambridge University Press 1988.
  Some relevant magazine articles are:
  Accretion disks in interacting binary stars, by J.K. Cannizzo and R.H. Kaitchuck, *Scientific American*, **266**, No.1, 42, January 1992
  Planetary nebulae, by N. Soker, *Scientific American*, **266**, No.5, 36, May 1992
  The early life of stars, by S.W. Stahler, *Scientific American*, **265**, No.1, 28, July 1991.

  Because stars are fairly well understood, the basics have not changed much since the 1960s and
  Astrophysics and Stellar Astronomy, by T.L. Swihart, Wiley 1968
can still be recommended for its clear exposition. Another classic text, first published by Wykeham in 1970 but recently revised and republished is
  The Stars: their Structure and Evolution, by R.J. Tayler, Cambridge University Press 1994.

  Most books on binary stars are fairly technical. Relatively straightforward introductions are
  Binary and Multiple Systems of Stars, by A.H. Batten, Pergamon 1973;
  Double Stars, by W.D. Heintz, D. Reidel 1978.
An excellent, but much more difficult, discussion of interactions in close binaries is:
  Interacting Binary Stars, edited by J.E. Pringle and R.A. Wade, Cambridge University Press 1985.
A more recent survey, at a similarly technical level, suitable mainly for graduate students, can be found in the lectures given at the 22nd Advanced Course of the Swiss Society for Astrophysics and Astronomy, Saas-Fee, 1992:
  Interacting Binaries, by S.N. Shore, M. Livio and E.P.J. van den Heuvel, Springer-Verlag 1994.
  A nice account of variable stars, which concentrates on the observations, is:
  Variable Stars, by C. Hoffmeister, G. Richter and W. Wenzel, Springer-

Verlag 1985. This covers pulsating and eruptive variables in particular, including cataclysmic binaries.

Chapter 10

See also the list for Chapter 9.

For the first-year undergraduate, one of the best books on stellar structure is still the one by Tayler (1970, 1994) (see references for Chapter 9). However, there are several other suitable books at a slightly more advanced level, for example, Volume 1 of Bowers and Deeming (1984) (section A3.2). Others include Volume 3 of Böhm-Vitense (1992) (see references for Chapter 8),

Introduction to Stellar Atmospheres and Interiors, by E. Novotny, Oxford University Press 1973, and

The Physics of Stars, by A.C. Phillips, Wiley 1994, which contains useful problems at the ends of chapters. An unusual and effective approach, emphasizing the physical principles and making extensive use of approximate arguments to allow quantitative treatment of complex topics, is adopted in:

Physics of Stellar Evolution and Cosmology, by H.S. Goldberg and M.D. Scadron, Gordon & Breach 1981.

At the advanced undergraduate/first-year postgraduate level, there are several classic texts:

The Internal Constitution of the Stars, by A.S. Eddington, Dover 1950; re-issued by Cambridge University Press in 1988 (with a foreword by S. Chandrasekhar). First published by CUP in 1926, this book was the first monograph on stellar structure. Although many details are now known to be wrong, the writing is a model of clarity and reveals great physical insight.

An Introduction to Stellar Structure, by S. Chandrasekhar, Dover 1957. This highly mathematical book, first published in 1939 by the University of Chicago Press, is a distillation of the author's original papers on the subject, including his discovery of the existence of a maximum mass for a white dwarf.

Stellar Structure and Evolution, by M. Schwarzschild, Dover 1965 (first published, Princeton University Press 1958). This was the first modern text on stellar structure, and the last before the advent of fast computers. It is still authoritative on the general physical principles, but lacks any detailed account of post-main-sequence evolution.

Stellar Structure and Evolution, by R. Kippenhahn and A. Weigert, Springer-Verlag 1990. This is the replacement for Schwarzschild's book as a modern monograph on stars. Written by two of the pioneers of numerical calculations of stellar evolution, it contains both details of these calculations and a clear physical explanation of them. It is definitely a graduate text.

By far the most detailed textbook on stars is still:

The Principles of Stellar Structure, by J.P. Cox and R.T. Giuli, Gordon & Breach 1968. In two volumes the authors cover all the basic physics, and then apply their results to stellar structure and evolution. This is an extremely comprehensive account of the state of the subject in the late 1960s, and remains a useful reference, especially for the underlying physics. Its main drawback is that every topic is treated in such generality that it is difficult to extract the simple cases that are treated in most textbooks. It is not very suitable for undergraduates.

Other graduate texts include:

Principles of Stellar Evolution and Nucleosynthesis, by D.D. Clayton, McGraw-Hill 1968. This clearly written book emphasizes the role of nuclear physics in stellar interiors.

Fundamentals of Stellar Astrophysics, by G.W. Collins II, Freeman 1989. This covers all the basic physics and applies it to stellar structure and evolution; it includes a long section on stellar atmospheres.

Chapter 11

The first description of Hubble's classification of galaxies is found in his classic book:

The Realm of the Nebulae, by E. Hubble, Dover 1958 (first published, Yale University Press 1936).

Later versions of the classification can be found in, for example,

Chapter 2 of *Evolution of Stars and Galaxies*, by W. Baade, Harvard University Press 1963;

Chapter 1, by A. Sandage, of Sandage *et al.* (1975) (section A3.3);

Section 27 of *The New Cosmos*, by A. Unsöld and B. Baschek, Springer-Verlag, 4th edition 1991.

A splendidly illustrated version is

The Hubble Atlas of Galaxies, by A. Sandage, Carnegie Institution of Washington 1961.

Many recent books give dramatic colour photographs of galaxies, for example

Galaxies, by T. Ferris, Stewart, Tabori & Chang 1982.

Superb black-and-white images of galaxies, and some colour ones, are included in

Exploring the Southern Sky, by S. Laustsen, C. Madsen and R.M. West, Springer-Verlag 1987, subtitled 'A Pictorial Atlas from the European Southern Observatory'.

The main properties of galaxies are described at an introductory level in

Galaxies, by P.W. Hodge, Harvard University Press 1986.

Galaxies: Structure and Evolution, by R.J. Tayler, revised edition, Cambridge University Press 1993 (originally published by Wykeham 1978), is one of the few books available at the undergraduate level.

At a much more advanced level, amongst the best graduate textbooks at the time of writing are:

Galactic Astronomy: Structure and Kinematics, by D. Mihalas and J. Binney, 2nd edition, Freeman 1981;

Galactic Dynamics, by J. Binney and S. Tremaine, Princeton University Press 1987.

A recent semi-popular account of dark matter (Section 11.2.5) is given by:

The Hidden Universe, by R.J. Tayler, Ellis Horwood 1991.

Interacting galaxies are described in

Colliding galaxies, by J. Barnes, L. Hernquist and F. Schweizer, *Scientific American*, **265**, No.2, 26, August 1991.

One of the best recent accounts of AGN (Section 11.5) is the set of systematic graduate lectures given in 1990 at the 20th Saas-Fee Advanced Course sponsored by the Swiss Society for Astrophysics and Astronomy:

Active Galactic Nuclei, by R.D. Blandford, H. Netzer and L. Woltjer, edited by T.J.-L. Courvoisier and M. Mayor, Springer-Verlag, 1990.

An earlier book, concentrating on quasars, is

Quasar Astronomy, by D. Weedman, Cambridge University Press 1986.

At a more elementary level, there are two good articles in *Scientific American*:

The quasar 3C273, by T.J.-L. Courvoisier and E.I. Robson, *Scientific American*, **264**, No.6, 24, June 1991;

Black holes in galactic centers, by M.J. Rees, *Scientific American*, **263**, No.5, 26, November 1990.

Chapter 12

Our Galaxy is well described at an introductory level in

The Milky Way, by B.J. Bok and P.F. Bok, Harvard University Press, 5th edition 1981

and in a beautifully produced recent book

The Guide to the Galaxy, by N. Henbest and H. Couper, Cambridge University Press 1994.

At a slightly more advanced level is

The Milky Way, by L. Kühn, John Wiley 1982.

A more technical account is contained in the lectures given at the 19th

Advanced Course of the Swiss Society of Astrophysics and Astronomy, Saas-Fee, 1989:

The Milky Way as a Galaxy, by G.F. Gilmore, I.R. King and P.C. van der Kruit, Geneva Observatory 1989; University Science Books 1990.

A non-mathematical account of the interstellar medium was given by one of the foremost authorities on the subject in the Silliman Memorial Lectures:

Searching between the Stars, by L. Spitzer Jr, Yale University Press 1982. A more recent description is given in

The Fullness of Space, by C.G. Wynn-Williams, Cambridge University Press 1992.

The physical and chemical processes in the interstellar gas are well described in the advanced undergraduate textbooks:

The Physics of the Interstellar Medium, by J.E. Dyson and D.A. Williams, Manchester University Press 1980;

Interstellar Chemistry, by W.W. Duley and D.A. Williams, Academic Press 1984.

An elementary decription of the centre of the Galaxy is given in:

What is happening at the center of our Galaxy?, by C.H. Townes and R. Genzel, *Scientific American*, **262**, No.4, 26, April 1990.

Chapter 13

Most introductory texts on general astronomy or on galaxies contain a section on clusters of galaxies (e.g. Hodge 1986 – see references for Chapter 11).

The reality of large-scale structure in the Universe is discussed in an interesting article on optical pattern recognition:

Filaments: what the astronomer's eye tells the astronomer's brain, by J.D. Barrow and S.P. Bhavsar, *Quarterly Journal of the Royal Astronomical Society*, **28**, 109, 1987.

At a technical level, most results are still available only in the original papers or in conference reports. The best single review on clusters, although it is now somewhat out of date, is:

Clusters of galaxies, by N.A. Bahcall, *Annual Review of Astronomy and Astrophysics*, **15**, 505, 1977.

The large-scale structure of the universe is discussed in:

The Large-Scale Structure of the Universe, by P.J.E. Peebles, Princeton University Press 1980.

This is a graduate monograph, which is an excellent introduction to many of the techniques, such as the use of correlation functions, but many of the results have been superseded by later work. Some of the more recent work is described in the popular article

Textures and cosmic strings, by D.N. Spergel and N. Turok, *Scientific American*, **266**, No.3, 36, March 1992.

Chapter 14

Most elementary books on astronomy contain some account of distance measurements, usually scattered through the book. Jastrow and Thompson (1984) (section A3.1) include a complete appendix on the distance scale. At the more advanced level, there is one book and a large number of detailed reviews. The book, which critically reviews the various methods and tries to resolve the discrepancy between two conflicting distance scales, is:

The Cosmological Distance Ladder, by M. Rowan-Robinson, Freeman, New York, 1985.

One of the most recent reviews, which concentrates on seven of the most reliable distance indicators and finds a distance to the Virgo cluster of between 15 and 20 Mpc, is:

A critical review of selected techniques for measuring extragalactic distances, by G.H. Jacoby and 8 others, *Publications of the Astronomical Society of the Pacific*, **104**, 599, 1992.

Useful articles at a more elementary level are:

The expansion rate and size of the Universe, by W.L. Freedman, *Scientific American*, **267**, No.5, 30, November 1992;

The extragalactic distance scale: agreement at last?, by P. Hodge, *Sky & Telescope*, October 1993, p.16.

Chapter 15

The classic popular text on the physics of the Big Bang remains:

The First Three Minutes, by S. Weinberg, André Deutsch 1977; with a new afterword: Flamingo 1993.

Other books at a similar level are:

The Left Hand of Creation, by J.D. Barrow and J. Silk, Basic Books 1983; Unwin Paperbacks 1985;

Gravity's Lens: Views of the New Cosmology, by N.Cohen, Wiley 1989;

Bubbles, Voids and Bumps in Time: The New Cosmology, edited by J. Cornell, Cambridge University Press 1989 (paperback 1991) (a collection of lectures given in 1987 at the Hayden Planetarium in Boston).

Rather more detail can be found in

The Big Bang, by J. Silk, Freeman, 2nd edition 1989.

Two magazine articles are

Quantum cosmology and the creation of the Universe, by J.J. Halliwell, *Scientific American*, **265**, No.6, 28, December 1991.

Universal truths, by J. Horgan, *Scientific American*, **263**, No.4, 74, October 1990.

Textbooks on cosmology include:

Introduction to Cosmology, by J.V. Narlikar, 2nd edition, Cambridge University Press 1993;

Principles of Physical Cosmology, by P.J.E. Peebles, 2nd edition, Princeton University Press 1993;

Cosmology, by M. Rowan-Robinson, 2nd edition, Oxford University Press 1981;

Cosmology, by D.W. Sciama, Cambridge University Press 1971;

Gravitation and Cosmology, by S. Weinberg, Wiley 1972; this book is somewhat more advanced than the others.

Appendix 5
Acknowledgements

Grateful acknowledgement is made for permission to reproduce illustrations obtained from the following sources:

The Royal Greenwich Observatory: Figs. 2.8, 2.11, 2.12 (a) and (b), 2.15, 7.2(a)

The Royal Observatory Edinburgh: Fig. 6.2

The Anglo-Australian Observatory: Figs. 2.9, 3.11, 6.9, 7.2(b)

The Royal Astronomical Society: Figs. 3.15, 3.16, 3.24, 9.11, 9.12, 9.21, 9.23(a), 9.23(b), 10.13, 11.2(a), 11.2(b), 11.3 to 11.7, 13.1, 13.2

Fig. 2.16: Fig. 2 of the article by Roger Angel, *Q.J.R.astr.Soc.*, **31**, 141, 1990

Figs. 2.17, 2.18: Figs. 3 and 2 of J.M. Beckers *et al*, *Telescopes for the 1980s*, eds. G. Burbidge and A. Hewitt, p.63, Annual Reviews Inc., 1981

Fig. 3.11: Cover picture, by Richard Robinson, on *AAO Newsletter*, No.46, July 1988

Fig. 3.13: Courtesy of David J. Stickland

Fig. 4.5: G. Hutschenreiter, Max Planck Institut für Radioastronomie, Bonn

Fig. 4.6: Courtesy of Daniel R. Altschuler, Director, Arecibo Observatory

Fig. 4.12: Adapted from Fig. 10.1 of chapter 10, by E.B. Fomalont and M.C.H. Wright, in *Galactic and Extra-Galactic Radio Astronomy*, edited by G.L. Verschuur and K.I. Kellermann, Springer-Verlag 1974

Fig. 5.3: Fig. 1.3.8 of Chris R. Kitchin, *Astrophysical Techniques*, Adam Hilger Ltd, 1984

Fig. 6.1: © Lund Observatory

Fig. 6.3: Fig. 2 of the article by Ken A. Pounds *et al*, *Mon.Not.R.astr.Soc*, **260**, 77, 1993

Fig. 6.4: Courtesy of NASA/Goddard Space Flight Center

Fig. 6.5(a): Mullard Radio Astronomy Observatory

Fig. 6.5(b): Harvard-Smithsonian Center for Astrophysics

Fig. 6.6: D. Fabricant and P. Gorenstein, *Astrophys.J.*, **267**, 535, 1983

Fig. 6.7: Courtesy of NASA/Goddard Space Flight Center

Fig. 6.8: Courtesy of NASA/Goddard Space Flight Center

Fig. 6.10: Fig. 2 of the article by Ladd *et al*, in *Astrophys.J.*, **410**, 168, 1993

Fig. 6.11: Courtesy of Patrick Thaddeus and Thomas M. Dame

Fig. 6.12: Fig. 7 of the article by J.M. Dickey and F.J. Lockmann, *Ann. Rev. Astr. Astrophys.*, **28**, 215, 1990

Fig. 6.13: G. Haslam, Max Planck Institut für Radioastronomie, Bonn

Fig. 8.4: Fig. 7 of Chapter 13, by W. Becker, on p.254 of *Basic Astronomical Data*, ed. K.Aa. Strand, University of Chicago Press, 1963

Fig. 9.2: Fig. 1 of the article by R.C. Smith, *Observatory*, **103**, 29, 1983

Fig. 9.6: Fig. 1 of the article by D.J. Stickland, *Observatory*, **113**, 204, 1993

Figs. 9.13, 9.14: Figs. 16, 34 of the article by H.C. Arp, *Handbuch der Physik*, **51**, 75, Springer-Verlag 1958

Fig. 9.16: L. Goldberg and L.H. Aller, *Atoms, Stars and Nebulae*, Blakiston 1943

Fig. 9.20: Fig. 44 of M. Petit, *Variable Stars*, John Wiley & Sons Ltd, 1987

Fig. 9.22: M.T. Friend, DPhil thesis, University of Sussex, 1988

Fig. 9.24: Fig. 1 of the article by R. Barbon, F. Ciatti and L. Rosino, *Astr. Astrophys.*, **25**, 241, 1973

Fig. 9.25: Figs. 1 and 2 of the article by R. Barbon, F. Ciatti and L. Rosino, *Astr. Astrophys.*, **72**, 287, 1979

Fig. 10.2: Fig. 12 of the article by H.C. Arp, *Handbuch der Physik*, **51**, 75, Springer-Verlag 1958

Figs. 10.6, 10.7, 10.8: Figs. 22.7, 30.1, 30.2 and 30.3c of R. Kippenhahn and A. Weigert, *Stellar Structure and Evolution*, Springer-Verlag 1990

Fig. 10.9: Fig. 7 of the article by C. Hayashi, *Ann. Rev. Astr. Astrophys.* **4**, 171, 1966

Figs. 10.10, 10.11, 10.12, 10.15: Figs. 31.4, 31.6, 32.10, 36.2 of R. Kippenhahn and A. Weigert, *Stellar Structure and Evolution*, Springer-Verlag 1990

Fig. 11.8: From *The Hubble Atlas of Galaxies*, A. Sandage, Carnegie Institution of Washington, 1961

Figs. 11.10 and 11.11: Figs. 47 and 84 of R.J. Tayler, *Galaxies: Structure and Evolution*, CUP 1993

Fig. 11.12: Courtesy of Halton C. Arp

Fig. 11.13: Fig. 1b of the article by R.A. Perley *et al*, in *Astrophys.J.*, **285**, L35, 1984

Fig. 12.3: Fig. 1 of the article by D. Crampton and Y.M. Georgelin, *Astr. Astrophys.* **40**, 319, 1975

Fig. 12.6: Fig. 2 of the article by R. Genzel and C.H. Townes, *Ann. Rev. Astr. Astrophys.* **25**, 377, 1987

Fig. 12.7(b): Fig. 2a (bottom panel) of the article by D.C. Morton in *Astrophys.J.*, **197**, 85, 1975

Fig. 12.8: Fig. 3 of the article by T.M. Dame, E. Koper, F.P. Israel and P. Thaddeus in *Astrophys.J.*, **418**, 730, 1993

Fig. 13.3: Plate I of the article by M. Seldner *et al*, in *Astr. J.*, **82**, 249, 1977

Fig. 13.4: Fig. 1a of the article by V. de Lapperent *et al*, in *Astrophys.J.*, **369**, 273, 1991

Fig. 13.5: Figs. 2 and 4 of the article by M. Rowan-Robinson *et al*, in *Mon.Not.R.astr.Soc.*, **253**, 485, 1991

Fig. 14.7: Fig. 9 of Chapter 20, by A. Blaauw, on p.407 of *Basic Astronomical Data*, ed. K.Aa. Strand, University of Chicago Press, 1963

Fig. 15.5: Courtesy of David G. Wands

Fig. 15.6: Fig. 2 of the article by A. Sandage in *Q.J.R.astr.Soc.*, **13**, 282, 1972

Fig. 15.7: Fig. 7.4 of M. Rowan-Robinson, *Cosmology*, 2nd edition, Oxford University Press 1981

Fig. 15.9: NASA/Goddard Space Flight Center; supplied by J.C. Mather on behalf of the COBE Science Team

Index

431